LOCOMOTIVE
TO AEROMOTIVE

O. Chanute

LOCOMOTIVE TO AEROMOTIVE

Octave Chanute

and the Transportation Revolution

SIMINE SHORT

Foreword by Tom D. Crouch

UNIVERSITY OF ILLINOIS PRESS
Urbana, Chicago, and Springfield

Frontispiece: Octave Chanute. Chanute Papers,
Manuscript Division, Library of Congress,
Washington, D.C.

Library of Congress Cataloging-in-Publication Data
Short, Simine
Locomotive to aeromotive : Octave Chanute and
the transportation revolution / Simine Short ;
foreword by Tom D. Crouch.
p. cm.
Includes bibliographical references and index.
ISBN 978-0-252-03631-6
1. Chanute, Octave, 1832–1910.
2. Civil engineers—United States—Biography.
3. Aeronautics—United States—Biography.
I. Title.
II. Title: Octave Chanute and the
transportation revolution.
TA140.C425S56   2011
629.04092—dc22   2010051966
[B]

This book is dedicated to Jim,
my favorite glider pilot, who is just as interested in
the history of the sport as in flying.

"From the locomotive to the aeromotive," shouted the noisiest of all, who had turned on the trumpet of publicity to awaken the Old and New Worlds. . . .

A flying machine must therefore be constructed to take advantage of these natural laws, . . . But as has been said, it is not necessary to copy Nature servilely. Locomotives are not copied from the hare, nor are ships copied from the fish. To the first we have put wheels, which are not legs; to the second we have put screws, which are not fins. . . . Besides, what is this mechanical movement in the flight of birds, whose action is so complex?

—Jules Verne, *The Clipper of the Clouds* (1887)

CONTENTS

FOREWORD

Octave Chanute and
"The Course of Human Progress"

OCTAVE CHANUTE DIED in his home at 1138 Dearborn Avenue on the morning of Wednesday, November 23, 1910. The family immediately wired the news to Wilbur Wright, who boarded a train in Dayton, Ohio, in order to reach Chicago in time for the funeral, scheduled to take place at 4 P.M. on Friday, November 25. It is safe to assume that he spent much of that trip considering his complex relationship with Chanute, which had begun with a single letter more than ten years before.

"For some years I have been afflicted with the belief that flight is possible to man," Wright had written to introduce himself to Chanute on May 13, 1900. "My disease has increased in severity and I feel that it will soon cost me an increased amount of money if not my life." Chanute understood completely, having himself been infected by the flying machine bug a quarter-century before. He responded on May 17, remarking that he was "quite in sympathy" with Wilbur Wright's desire to begin aeronautical experiments, and offering advice on some publications that might prove useful.

Sixty-eight years old in the spring of 1900, Octave Chanute was one of the most successful and distinguished civil engineers in the nation. He began his career in 1849 as a chainman on a surveying crew, the lowest rung on the professional ladder. Over the decades that followed, he worked his way up to a position as chief engineer with a series of railroads that were opening the West and binding the nation together. He bridged great rivers, built the stockyards in Kansas City and Chicago, pioneered wood preservation, helped plan an elevated railway system for New York City, and supervised engineering projects that would serve the needs of urban dwellers in the new western cities.

Perhaps because he had learned his skills on the job, Chanute was dedicated to fostering professional standards and a spirit of cooperation and information sharing among engineers. During the course of his long career, he served as both vice president and president of the American Society of Civil Engineers and as president of the Western Society of Engineers. A fellow of the American Association for the Advancement of Science, Chanute received honors from such international organizations as the Institute of Civil Engineers of Great Britain and the Canadian Society of Civil Engineers.

As Wilbur Wright knew, Chanute had for some years been taking time from his professional duties and obligations to pursue the great interest of his life — heavier-than-air flight. Originally attracted to the field by his observations of the impact of high winds on bridges and roofed structures, he quickly shifted his attention to the possibility of achieving winged flight. True to form, Chanute set out to read what had been written on the subject, and to make contact with a handful of experimenters scattered around the globe who were working toward the development of a practical airplane.

Within a few years, Chanute stood at the center of an informal international community of flying machine experimenters and enthusiasts. He shared the information that he gathered with other workers in the field, and offered encouragement, advice, and occasionally even financial aid. Emerging as an international authority in the field, Chanute lectured on the subject, wrote articles for professional journals, and published *Progress in Flying Machines* (1894), a book that quickly became the "bible" for aeronautical experimenters everywhere. His efforts not only illuminated the field, but established aeronautics as an acceptable area of study for professional engineers.

Not content with his mastery of flight theory, Chanute funded the design and construction of a number of gliders, the best of which were test flown by his younger associates during trials that he supported in the sand dunes along the southern shore of Lake Michigan, east of Chicago, in the summer of 1896. When Wilbur Wright wrote to Chanute for the first time four years later, he had no doubt that he was addressing the world's authority in flying machine studies.

From the outset, Octave Chanute served an essential role as a source of information and a sounding board for the Wright brothers' ideas. His encouragement in the fall of 1901 convinced the brothers to continue their work at a time when they might otherwise have abandoned the field. And it was Chanute who spread word of their successes in 1902 and 1903 to the rest of the world, igniting new interest in heavier-than-air flight in France and elsewhere in Europe.

Tensions began to grow between Octave Chanute and the Wright brothers after 1903. For their part, the Wrights believed that their old friend had portrayed them as his "pupils" in his talks with the Europeans, and that he failed to understand or appreciate the originality and importance of their ideas and achievements. Chanute, disturbed when the Wrights brought suit against their rivals for patent infringement, suggested that "the desire for great wealth" had warped their "usual sound judgment," and he remarked that his old friends had failed to appreciate "such aid as I may have furnished." The hard feelings continued to fester into the spring of 1910, when Wilbur wrote a warm and conciliatory note to Chanute: "My brother and I do not form many intimate friendships," he began, "and do not lightly give them up. I believe that unless we could understand exactly how you felt, and you could understand how we felt, our friendship would tend to grow weaker instead of stronger. Through ignorance or thoughtlessness, each would be touching the other's sore spots and causing unnecessary pain. We prize too highly the friendship which meant so much to us in the years of our early struggles to see it worn away by uncorrected misunderstandings, which might be corrected by a frank discussion."

Chanute took a step back, as well, explaining that he planned a trip to Europe, after which, "I hope . . . that we will be able to resume our former relations." But his hopes did not materialize. Chanute died before the men resolved their differences, and now Wilbur was on his way to his old friend's funeral, where the family invited him to give the eulogy. "By the death of Mr. O. Chanute," he began, "the world has lost one whose labors had to an unusual degree influenced the course of human progress. If he had not lived the entire history of progress in flying would have been other than it has been. . . ."

Wilbur Wright offered praise for his friend's technical contributions, noting that the biplane glider that Chanute had designed in 1896 would ". . . influence flying machine design so long as flying machines are made." He focused, however, on the human qualities that had enabled the engineer to forge a community of experimenters that laid the foundation for the invention of the airplane:

> His writings were so lucid as to provide intelligent understanding of the nature of the problems of flight to a vast number of persons who would probably never have given the matter study otherwise, and not only by published articles, but by personal correspondence and visitation, he inspired and encouraged to the limits of his ability all who were devoted to the work. His private correspondence with experimenters in all parts of the world was of great volume. No one was too humble to receive a share of his time. In patience and goodness of heart he has rarely been surpassed. Few men were more universally respected and loved.

We have waited a long time for a solid biography of Octave Chanute. Simine Short has given us a book worth waiting for. She succeeds in situating the details of Chanute's long life and extraordinary career squarely in the context of his time.

As the author demonstrates, there was far more to the man than his contributions to aeronautics and his involvement in the Wright brothers' work. The nineteenth century was the age of the engineer. Technical professionals revolutionized transportation and industry, reshaping the economy, society, and everyday life in the process. There are far too few biographies of the engineers who managed that revolution. Chanute was one of those men. He came of age in the era of the canal builders and died having helped give birth to the air age. Chanute spent most of his working life as one of the technical professionals who managed the creation of the American railroad network, and on the design and construction of the bridges, stockyards, and other facilities that supported the railroads.

Beyond that, few men were more committed to the establishment of engineering as a profession on par with law or medicine. Chanute devoted much time and energy to the newly established professional societies created to set standards and serve the needs of civil engineers. He firmly believed in the importance of sharing information with other professionals. Chanute wrote countless letters to colleagues, published articles on a wide range of issues in professional journals, lectured, arranged programs for the annual meetings of professional organizations, and represented the engineering profession on the committees planning Chicago's 1893 World's Columbian Exposition and the 1904 Louisiana Purchase Exposition in St. Louis.

Simine Short has done a great service in providing this well-rounded and detailed account of one of the outstanding engineering careers of the nineteenth century. I feel certain that Wilbur and Orville Wright would approve, as well, for the author also underscores the personal qualities that led the inventors of the airplane to describe their old friend as "universally respected and loved." In short, this is a biography that not only fills a gap in the history of technology, but is a book worthy of its subject.

Tom D. Crouch
Senior Curator, Aeronautics
National Air and Space Museum
Smithsonian Institution

PREFACE

GAZING UP THE STEPS of the United States Capitol, thousands of visitors to Washington, D.C., every day are struck by the grandeur of this symbol of the American people. Nowhere is the pageant of America's early history so well showcased as inside the Capitol's great ceremonial rotunda. Sixty feet above its floor, the eight-foot-high frescoed *Frieze of American History* encircles the rotunda, giving the illusion of a sculptured relief and depicting nineteen scenes, from the "Discovery of America" to the "Birth of Aviation."

Above the rotunda's west door is the final scene portraying Orville and Wilbur Wright's first powered airplane flight in 1903 and three of their acknowledged precursors: the Italian Leonardo Da Vinci and two Americans, Samuel P. Langley, secretary of the Smithsonian Institution, and Octave Chanute, one of the renowned civil engineers of the nineteenth century, developer of a successful series of gliders and mentor to the Wright Brothers. In this scene Chanute, formally attired with long suit, starched collar, and trademark goatee, holds his *Katydid* multiplane glider. He stands closest to the Wrights, possibly because of the dates and depth of his aviation involvement but also possibly because of his close and continuing friendship with the brothers.

How did Chanute, a self-educated French immigrant, earn a place in the *Frieze of American History* among others whose names are better known? How did Octave Chanute rise from being a penniless immigrant to rank among the elite American engineers of the nineteenth century? How did he learn from others and then give back by mentoring his juniors? What inspired him to spend a lifetime in transportation and engineering, building world-class structures against a backdrop of dynamic engineering and social change? How did he learn to work with others to achieve what one man alone could not accomplish? Why did he work with the new engineering societies of the nineteenth century to elevate the professionalism of civil engineers? How did he rise to the highest executive ranks of one of America's largest railroads? How did he capitalize on his early perception of the need to preserve natural resources? Finally, why, when others would have retired to a life of leisure, did he pursue his ultimate transportation passion—aviation—until his final days? How did Octave Chanute become the pivotal person in the Wright Brothers' quest for powered, controlled flight? Why, when the Wrights flew their powered flight of December 17, 1903,

"Birth of Aviation." *Frieze of American History,* United States Capitol, Washington, D.C. Courtesy Architect of the Capital Collection.

did their sister Katharine immediately wire the news to Chanute, making him the first, and for some time the only, person outside the Wright family to know of the epic event? This book aims to answer these questions and to chronicle the amazingly productive eight decades that earned Octave Chanute a place in the *Frieze of American History.*

ACKNOWLEDGMENTS

BECAUSE OCTAVE CHANUTE had so many intertwining and diverse interests, I knew that I would need help compiling this biography. My husband Jim provided not only help but also constant support throughout this project. Especially in the last three years, Jim's patience in the various stages of writing, editing, trimming contents, and proofreading was extremely helpful and encouraging. My personal "Advisory Committee" consisted of Albion Bowers (NASA Dryden Flight Research Center); Dr. Gary Bradshaw (professor, Department of Psychology, Mississippi State University); Dr. Leonard C. Bruno (science manuscript historian in the Manuscript Division, Library of Congress); Dr. Tom D. Crouch (senior curator, Aeronautics, Smithsonian Institution, National Air and Space Museum, Washington, D.C.); Dr. Kevin Kochersberger (research associate professor, Virginia Tech, JOUSTER project, Institute for Advanced Learning and Research, Danville, Va.); Dr. Jeffrey Oaks (professor, Department of Mathematics and Computer Science, University of Indianapolis); and David Young (retired transportation editor, *Chicago Tribune*). Without their input, the writing of this biography would have been much more difficult, if not impossible. Especially, I would like to recognize Len Bruno, Tom Crouch, and Jeff Oaks as friends who patiently answered questions and suggested improvements after reading and rereading the text, from start to finish. They were true and valuable mentors in my learning process as a historian. In the final stages of this project, I appreciated the knowledge and expertise of several staff members at the University of Illinois Press, who contributed so much. It all started with my first inquiring telephone call to Laurie Matheson, senior acquisitions editor, and then submitting the first draft of the manuscript. Receiving thoughtful suggestions for improvement made me tackle several rounds of revisions, but eventually the contract was signed. Tad Ringo, senior editor, became the project manager. Other staff members contributed to the design, production, and last, but not least, the promotion of this book. A special thanks goes to each and every one, and I am proud of the final product. To make the book more complete and human, Joseph Hodges and his wife Jean, of Denver, graciously allowed me to look through the Chanute family papers, collected and assembled by Joe Hodges's mother, Mrs. Elaine Chanute Hodges, and by Octave A. "Ox" Chanute, the two great-grandchildren of Octave Chanute. Many photos in this book were reproduced from Elaine's files. My special thanks go to them and to their descendants, who have graciously shared their memories and files.

People Who Helped with This Book

Because my research stretched over more than a decade, please accept my apologies if I have missed listing the name of someone who was kind in the past. To the best of my knowledge, I have asked all copyright holders for permission, but I would be glad to hear from anyone I may have inadvertently missed.

Bart Ryckbosch and John Zukowsky, Art Institute of Chicago; Maxim Avdeev, Sidney, Australia, always available to help with translating articles written in the Russian language; Gary Bradshaw, arguably the first to establish a Web site in 1994 when no one else thought of doing this: a "Virtual Museum covering the invention of the airplane," http://invention.psychology. msstate.edu/; Larry O'Neal, Baxter Heritage Center, Baxter Springs, Kans.; George Rogge and Gregory Reising, Chanute Aquatorium Society, Gary, Ind.; Jean Conklin, St. Joseph, Mich., who researched Augustus M. Herring for her upcoming book; Center For Research, Chicago, Ill.; Ruth Ports, Chanute, Kans., Office of Tourism; Joanna Welch, Chanute, Kans. Public Library; Cornell University, Olin Library, Ithaca, N.Y.; Traff Doherty, Glenn Curtiss Museum, Hammondsport, N.Y.; members of the "Date Nail Group": Cheryl and Charles Johnson, Kila, Mont.; Rolland Meyers, Oakland, Calif., and Charles Sebasta, Caldwell, Tex.; Bob and Lynne Davis, Seymour, Tenn.; Paul Dees, Seattle, Wash., an aeronautical engineer and hang glider pilot who shared his experiences of first building and then flying his Chanute-type reproduction; Judy Brown, Marilyn Chang, John Irwin, Jim Kroll, Lori Swingle, and Brent Wagner, Denver Public Library, with the Pearl I. Young Papers, donated by Elaine Hodges in the early 1970s; Downers Grove, Ill. Public Library, and the staff of the Interlibrary Loan Division; Franklin County Historical Society, Ottawa, Kans.; Steve Repp, Galena, Ill. Public Library; Jay Dickerson, *Galena* (Ill.) *Gazette*; Gene Glendinning, Barrington, Ill., who freely shared his research on the Chicago & Alton Railroad; Patricia Goitein, Peoria, Ill., a longtime friend who shared her knowledge on the Civil War, slavery, and life in Peoria in the nineteenth century; Francis E. Griggs, professor emeritus, Merrimack College, N.Y.; Robert J. Havlik, retired librarian at the University of Notre Dame; Bonnie Lewandowski, Homer Township, Ill. Public Library; Barbara Heflin, Web master, Illinois State Archives, who patiently explained the platting system and where lands were located; Debra Francis, Institute of Civil Engineers Archive and Library, London, England; Jo Daviess County Historical Society, Galena, Ill.; Daniel Coleman and Rebecca Powers, Kansas City, Mo. Public Library; Lin Frederickson, Kansas State Historical Society, Topeka, Kans.; Kathy Lafferty, University of Kansas, Spencer Research Library, Lawrence, Kans.; Reinhard Keimel, Technisches Museum

Wien and Aviatika Museum, Vienna, Austria, always available for technical and historical information. For this book, Reinhard Keimel drew the Chanute glider designs as they developed from the ladder kite to the biplane; La Crosse, Wisc. Public Library; Mark Hall, Kristi Finefield, and Daniel De Simone, Library of Congress; Ola May Earnest, Linn County Historical Society, Pleasanton, Kans.; Bernd Lukasch, Otto-Lilienthal-Museum, Anklam, Germany; Barbara Anderson and Arnold Natzke, Livingston County, Courthouse, Pontiac, Ill.; Abigail Dixon, Louisiana State University; Cam Martin, NASA Dryden Flight Research Center, Edwards, Calif.; Wade Myers, National Park Service, Harpers Ferry Center, Harpers Ferry Center, W.Va.; William Baxter, Greg Bryant, Kate Igoe, Allan Janus, Russell Lee, Paul McCutcheon, and Patricia Williams, National Air and Space Museum, Washington, D.C.; Bill Gallagher, Shirley Sliwa (deceased), and Peter Smith, National Soaring Museum, Elmira, N.Y.; Benjamin Neff, Lincoln, Neb.; Newberry Library, Chicago; Bill Nicks, Lenexa, Kans.; Stephan Nitsch (deceased), Langenhagen, Germany, who built and flew his Chanute-type glider in the early 1990s and again for television crews in 2003. Nitsch also built, flew, and then donated his Lilienthal replicas to the Lilienthal-Museum in Anklam, Germany; Betty Roberson, Elaine Sokolowski, and Linda Aylwort, Peoria, Ill. Public Library; Prescott, Wisc. Public Library; Rensselaer Polytechnic Institute, Folsom Library, Troy, N.Y.; Paul Schweizer (deceased) and William Schweizer and their families, Elmira, N.Y.; at Proquest Company, Susan Brooks arranged for me to be a member of the beta-testing crew in 2000, and later arranged access to the digitized newspapers that are now available at many public libraries; journals and magazines, digitized by Proquest, are available at the University of Chicago Library; Dennis Parks (formerly from the EAA Boeing Library/Archives), Janice Baker (librarian), and Ben Sarao, Seattle Museum of Flight; Seattle Public Library; Martin Simons, Adelaide, South Australia; Peter Selinger, Stuttgart, Germany; Ellen Alers and Mary Markey, Smithsonian Institution Archives, Reference Services; Steve Spicer, Gary, Ind., who created the Chanute Web site to help publicize the one-hundredth anniversary of gliding in his hometown in 1995 (http://www.spicerweb.org/ Chanute/Cha_index. aspx); Streatorland Historical Society, Streator Ill.; Urszula Kerkhoven, Kathleen Zar (deceased), and Barbara Kern, University of Chicago Library System, John Crerar Library. We collaborated in 2001 in arranging "Octave Chanute, Chicago. Aeronautical Pioneer, Engineer, Teacher" (http://www.lib.uchicago. edu/e/crerar/ exhibits/chanute-main.html), and Barb Kern maintains the digital display on their Web site; John Kimborough and Sean Dempsey; Debra Levine and Jay Satterfield, Special Collections, Regenstein Library (Univ. of Chicago); Joe Gerdeman in the Interlibrary Loan Division (Univ. of Chicago); Raymond Gadke in microfilm collection (Univ. of Chicago); David Boutros, Western Historical Manuscript Division, University of Missouri, Kansas City, Mo.; Mark Stauter, University of Missouri, Rolla, Mo.; Ann and Jim Kepler,

George Wagner, and Maggie White, Western Society of Engineers, Chicago; Susan McNeil-Marshall and Ruta Jancys, Woodridge, Ill. Public Library; John Armstrong, Wright State University, Special Collections, Dayton, Ohio; Darrell Collins, Wright Brothers National Memorial, Manteo, N.C.

If this book contributes something worthwhile to a better understanding of Octave Chanute and the early civil engineers, the opening of the West, the growth of American railroads, and the development of mechanical flight, then it is these people who deserve much credit for making this book become reality.

CHAPTER 1

The Formative Years

MISSISSIPPI DELTA. December 1838—After more than two months of monotony at sea, the *Havre Paquet* approached the wide Mississippi River delta. At the river entrance, Captain Robert H. McKown took on a local pilot to guide the sailing vessel, towed by a steamboat tug, over the final 120 miles to New Orleans.

The New York and Havre packets made regularly scheduled North Atlantic crossings three times each month; the average travel time between the French port Le Havre and New York was forty-four days, with fourteen additional days to New Orleans.[1] Even though screw-driven steamers were already cruising between the harbors along the Atlantic coast of Europe, sailing vessels such as the *Havre* were still commonly used to cross the ocean.

America may have appeared raw and uncivilized to Europe's aristocracy, but intellectuals looked for new and fascinating experiences and hoped to write about what they had seen. Common people considered the New World the "promised land"; they anxiously crossed the Atlantic to escape political oppression, to seek religious freedom, or to seek a better life in a country where "the air is more free";[2] others came for a combination of these reasons. One of the passengers aboard this Havre packet was the cultivated, forty-two-year-old Joseph Chanut, who had received an unsolicited offer to teach in one of the three major colleges in antebellum Louisiana and was now immigrating to his new future with his oldest son Octave. As his marriage in France was not as he had envisioned, Joseph felt attracted by the prospect of a better life; he accepted the challenge to teach in a country with which he was not familiar and to take responsibility as a single parent. Understandably, Joseph must have wondered if he had made the right decision. Would he like being vice president of a southern college and would his colleagues accept him as their peer?

His six-year-old Octave Alexandre had his own thoughts about his life and future. Octave's mother had told him that he would go with his father, whom he barely knew, and leave the security of his home, then living with his mother, his grandmother, and two younger brothers. Many family members had come to see him and his father leave Paris,[3] and he undoubtedly wondered if he would ever see them again and why he had to go with his father.

Crossing the Atlantic Ocean was a long and, at times, hazardous journey, but Captain McKown had made everyone feel comfortable. The passenger list of this Havre packet shows twelve passengers in the cabin section;[4] eight male, three female, and one child, with six of them French and the other half American residents. About fifty French, Swiss, and German passengers traveled in the steerage section, probably dreaming of a golden future in America. Most likely Octave became acquainted with fellow passengers and members of the crew, who were known to give youngsters an education in seafaring while crossing the Atlantic.

After a brief stop in New York, the packet continued its trip south past the fabled and stormy Outer Banks, around Florida, and finally to New Orleans. On arrival, the passengers recorded their names, ages, and professions at the Customs House. Then, father and son Chanut located the steamboat to take them to their final destination, Jefferson College in the St. James Parish, sixty miles upriver.

A new life, so different and not Parisian at all, began for Joseph and Octave. During the next twelve years, their paths would slowly diverge. Joseph Chanut could not reinvent himself and eventually returned to France. Young Octave Alexandre Chanut sought a different direction. As he matured, he became a successful civil engineer and accumulated wealth; he earned United States citizenship and anglicized his name to Octave Chanute, married, and raised five children with his wife. By every definition, he became a true American.

Antebellum Louisiana and Jefferson College

In 1803, Napoleon, emperor of France, sold "New France" to the United States, and part of this purchase became the state of Louisiana in 1812. During the next fifty years, more than a half-million immigrants from the East Coast, Europe, Haiti, and Cuba settled in this new possession, looking for freedom and fortune. Joseph Chanut was one of these immigrants hoping to utilize his native tongue for his and his son's personal betterment.

Most newcomers to Louisiana were fascinated by the medley of culture, manners, language, and complexion. The Creoles, or the *ancien population*, descended from French and Spanish colonials and had arrived first. The "foreign French" were natives of France, having more lately arrived in Louisiana; they considered the Creoles "provincial," and in turn, the Creoles called them "foreigners." The Anglo-Americans, mostly enterprising merchants and lawyers, had recently migrated south from the East Coast, anxious to make a name for themselves. The last and largest group in this melting pot of humanity consisted of African American slaves. The 1840 census reported a total of 352,000 inhabitants living in Louisiana, with more than 168,000 of them slaves. Members of each faction tried to live with the other; however, they struggled constantly for power, wealth, and prestige, with education providing one of the sources of power.

The Louisiana parishes along the Mississippi River between New Orleans and Baton Rouge were characterized by the growing wealth and prestige of the antebellum plantation aristocracy of the 1830s. In the St. James Parish, plantation owners made their money by growing sugar cane and tobacco, and they wanted a nearby college for the proper education of their sons. Governor André Bienvenu Roman, who had grown up on his father's sugar plantation in the St. James Parish, solicited funding from private sources and in 1831 chartered an institution of higher education in his home parish, established on the east bank of the Mississippi and named in honor of Thomas Jefferson.

In 1835, the Louisiana state legislature granted a total of $363,775 to support three colleges, with local plantation owners set to administer Jefferson College. This private college prospered from the beginning, receiving $48,775 to defray the expenses of its buildings, and another $15,000 annually for the next ten years to pay the salaries of its professors. Claudius Crozet, the first president of Jefferson College, hired professors from the United States Military Academy at West Point and the University of Paris, France, to teach literature, the sciences, and mathematics.[5] Thomas R. Ingalls succeeded Crozet as president two years later, and Jefferson College soon became an establishment of the first rank, where boys from the surrounding plantations could acquire "every branch of useful knowledge." The goal was to give the students of various origins a common education, with a discipline similar to France. On one day, everyone spoke French, and the next day, English. The Creole boys were allowed to speak their mother tongue only in Spanish classes and during recreation times.

The newly hired vice president, Joseph Chanut, arrived with his son late in December 1838 and took residence in the recently erected three-story faculty living quarters. At that time, the staff consisted of ten assistant professors, three prefects, a treasurer, and a librarian. About 150 students were enrolled, including fifteen nonpaying pupils.

Without input on how to bring up his son, Joseph used the same strict methods his parents did in the early nineteenth century in France. When he needed to leave his son alone, he normally locked the door of their living quarters. This prevented seven-year-old Octave from disappearing, when he, like other inquisitive lads, was motivated to discover his new world amid the surrounding sugar plantations or to mingle with other children his age or any of the college's sixty slaves.

Joseph home-schooled his son, and his French-speaking colleagues supplied a teaching curriculum according to their expertise, usually communicating in their mother tongue. They not only taught the youngster to read and write, but also to tell the truth and observe the general rules of etiquette. Manners of early-nineteenth-century French society formed an integral part of the boy's

daily life. Thus rigidly protected from the outside world, Octave Alexandre
Chanut grew up a young man who did not swear and was generally ignorant of
the more common American slang words. In later life, his daughters made mild
fun of him, but he had no intention to learn these words, which he believed a
well-bred person ought not use.

In 1841, the school was at its zenith, with 210 students and twenty nonpaying
boys,[6] but the situation changed abruptly, as reported by a Catholic teacher
from the University of New Orleans:

> For ten years Jefferson shone in the heaven of literature and science. Surely, if
> in those days of literary prosperity the faculty was a precious joy to Louisiana,
> the pupils had no enjoyment either of the church or its pastors. But one day
> God abandoned it, and whilst the holy office was summoning the faithful to
> church, while the students were devoting themselves to laughter and sport, on
> March 6, 1842, Sunday, about 10 o'clock in the morning, the cry "Fire, Fire"
> resounded suddenly in the establishment. Everyone was eager to render aid.
> Vain efforts. The conflagration, lit perhaps by an imprudent hand, perhaps by
> a vindictive one, embraced the superb athenaeum, and a few hours sufficed
> the pitiless element to achieve its work. At nightfall, heaps of ruins attested
> that the devastation had been as rapid as complete.[7]

No deaths occurred, but several staff members lost all their personal effects
except for what they were wearing, as the fire also destroyed the faculty living
quarters. Officials sent students and teachers to New Orleans or Baton Rouge
to continue classes for the remaining academic year. Recovery from this disaster
was slow, as the state cut its funding and the fortunes of the plantation economy
declined because of the weakening sugar industry. Receiving only intermittent
paychecks, Joseph resigned in early spring of 1844 and moved with his son to
the Crescent City, or *La Nouvelle Orléans*.

New Orleans: The Crescent City

The former capital of French Louisiana, New Orleans, was one of the wealthi-
est cities in the Union. The great staples of the western states moved down the
Ohio and Missouri Rivers into the Mississippi, flowing into New Orleans for
further distribution and export. In the early 1840s, the city extended about five
miles parallel to and "crescent" with the river and recorded a colorful population of about 100,000 inhabitants.

Most likely, Joseph joined three other former professors from Jefferson College
at the newly organized Orleans High School on Esplanade Avenue,[8] a public
school offering a liberal education, including the English language. In exchange
for lodging and food, Joseph also worked as a tutor on the nearby Pandely plantation. Many years later, Lizzie Chanute, Octave's middle daughter, reminisced

about taking a boat ride down the river, from which her father pointed out the plantation where he had lived with his father.[9] Being close to the river, Octave could watch flatboats, keelboats, sailing vessels, and steamboats passing by. As before, the doors to their living quarters remained locked and the books in the library provided his chief source of enjoyment. A side benefit of confining Octave to the library was to avoid the yellow fever, cholera, and smallpox that were prevalent and frequently fatal during the "sickly season."

New Orleans was a city with a great deal of social life and elegance. Having been paid well by Jefferson College up to the time of the fire, Joseph had saved some money, which he thought would be sufficient to support his whole family. Hoping that his wife Elise might have forgotten some of their personal difficulties and would adjust to married life with him, he sent her money to come with their two younger boys to New Orleans. She arrived late in October 1844,[10] but did not enjoy the life of a housewife. Joseph later reported that she made a hell of their home and behaved like a mad woman.[11] She stayed only a short time and returned to France without telling anyone, taking along Joseph's papers, diplomas, money, other valuables, and the two younger boys, leaving Octave with his father. The fifty-year-old Joseph Chanut was mentally discouraged and financially ruined, going through a midlife crisis.

Octave's Father, Joseph Chanut

Joseph Chanut was born in August 1796 in Clermont-Ferrand in central France.[12] His father was a baker, so he, the first-born, learned his father's trade to eventually take over the business. But Joseph disliked this work and wanted to become an educator and writer. In the early 1820s, he accepted a teaching position at *L'Académie de Poitiers* in Paris, which allowed him time to research the history of his mother country. Chanut's first book, *Histoire de France*, discussed recent history, starting with the execution of Louis XVI in 1793 and closing with the "July Revolution" of 1830; it was published later in 1830.[13]

With both of Joseph's parents deceased, his maternal aunt arranged a marriage between the thirty-five-year-old bachelor and the nineteen-year-old Elise Sophie, the only daughter of the widowed Catherine Louise Madeleine Gérin de Bonnaire. They married on April 19, 1831, in the Catholic parish church Saint Sulpice, where Joseph's uncle, Antoine Chanut, had been an abbot and director of the parochial school until his death two years earlier. Elise and Joseph's first son, Octave Alexandre, was born on February 18, 1832, and christened in Saint Sulpice. The following year, the young couple moved with Elise's mother to Sceaux, a southern suburb of Paris. A second boy, Emile Frédéric, was born there in May 1833.

Married life became more and more difficult for both partners. Joseph enjoyed his work as a professor of history at the *Lycée Henri-IV*, but his salary barely supported his growing family. Elise became frustrated, as she needed to care for two small boys and could not be part of Paris society anymore. A third child, Leon, was born in May 1835, and two weeks later Elise requested a legal separation. Joseph moved back into the city,[14] while his estranged wife and the three children continued living in Sceaux. Accepting a flattering offer from Jefferson College allowed Joseph the opportunity to distance himself from France and his wife.

Following the unfortunate demise of Jefferson College and his unhappy attempt to revive his relationship with Elise, Joseph considered yet another move. The beginnings of the war with Mexico in the spring of 1846, the fear of a possible occupation of New Orleans, and his growing distaste for the southern plantation way of life also factored into his worries about his future. Rather unexpectedly, he was elected an honorary member of the Historical Society of Pennsylvania, so he decided to move to New York with his fourteen-year-old son in tow.

From New Orleans to New York

In the mid 1840s, wealthy plantation owners considered the eight-day voyage along the Atlantic Coast from New Orleans to New York City more pleasurable than the one-month-long inland journey. However, Chanut's financial situation did not permit the expensive mode of travel by sea, so he selected the inland way via the Mississippi and Ohio Rivers at a cost of about $23 a person. Not only was this route significantly lower in cost, but Joseph also hoped that venturing into the heartland of the continent and meeting people of different cultures would prove a learning experience for both of them. Little did he know that this trip would be far more formative to the development of his son than of himself, leaving Octave with unforgettable impressions that shaped his future.

Joseph and Octave packed their accumulated books and other belongings early in August 1846 and boarded the Mississippi packet for a slow trip upstream against the current of the mighty river. They passed Convent and Jefferson College on the eastern shore, perhaps recalling bittersweet memories. Villas of wealthy plantation owners, residences of common people, and slave quarters were part of the changing scenery as father and son traveled north. Joseph had educated his son about Thomas Jefferson and the Franco-American relations; looking over the western bank of the Mississippi, one could only imagine the vast extent of land acquired from France in the Louisiana Purchase.

Gradually the river narrowed and reddish bluffs rose above the muddy waters; forests on either side of the river were interspersed with plantations and farms. Octave saw his first Indians, gliding swiftly and noiselessly with their "dugouts" in and out among the numerous islands dotting the river. Some of the islands were covered with small timber, while others seemed newly formed, with nothing but sand and a few bushes.

The teenaged Octave absorbed this new adventure with its opportunities to meet a wide variety of passengers; some headed for Ohio or Illinois, attracted by cheap land, while others had traveled this route before and shared stories about the vast land and its people. A correspondent to the *Chicago Tribune* reported on a similar trip upriver: "Passengers seemed to be variously employed; some in canvassing the columns of the daily newspapers, others eloquently descanting on the greatness of the West, or some favorite project of railroad or canal building, while the remainder were either playing cards or were wrapped in the arms of Morpheus or some other 'extinguisher' of dull care."[15]

The average speed of the packet was about eight miles an hour, requiring a week to travel the 1,046 miles upriver between New Orleans and the mouth of the Ohio River. The Chanuts disembarked at Cairo, Illinois, while the packet continued to St. Louis. Now aiming for Pittsburgh, another one thousand miles upriver to the northeast, the Chanuts boarded their next steamer. Unlike the Mississippi, the water of the Ohio was clear, but clusters of sandbars in the river channel made navigation difficult at times. Most likely Octave listened to the crew members' colorful stories of river pirates living in caves along the river, who attacked boats, killed passengers, and stole their belongings.

The Chanuts reached the smoky industrial center of Pittsburgh. Situated at the confluence of the Allegheny and Monongahela Rivers, Pittsburgh and its neighboring Allegheny City were important transportation hubs, with thousands of travelers and immigrants passing through all year.[16] As their steamer landed at the wharf for the passengers to disembark, the teenager and his father could see Pittsburgh's latest engineering marvel, opened a few months earlier, the John A. Roebling–designed wire suspension bridge across the Monongahela River. Its graceful structure stretched 1,500 feet across the river, supported in an ingenious manner by eighteen towers and joined by cables.[17]

Walking along the Monongahela wharf, they located the basin with the canal packet lines at 2nd Avenue. Joseph had read about the four hundred–mile-long Pennsylvania Main Line across the Allegheny Mountains, built a decade earlier at a cost of more than seventeen million dollars. At the agent's office, an eye-catching sign nailed above the entrance read: "Through to Philadelphia in three days and a half." Octave's glimpse into new aspects of transportation continued.

The commercial success of the Erie Canal, built between 1817 and 1825, threatened the trade from Philadelphia into the interior, and Pittsburgh was in danger of losing its status as a western depot. To regain business, the Common-wealth of Pennsylvania appropriated two million dollars in 1828 to examine the most practical route for connecting the East Coast with the Ohio River valley.[18] Six years later the Pennsylvania Main Line opened for business,[19] showing the ingenuity of early-nineteenth-century civil engineers. For a quarter century this composite system of canal boats and railroads passing through lift locks and tunnels, and crossing streams on small bridges and aqueducts, was the most efficient method of transporting passengers and freight across the mountain barrier of the Alleghenies.[20] Young Octave experienced firsthand some of its novel feats of transportation and technology.

Early in the morning, the Chanuts stowed their belongings on the *Pioneer* canal packet. Shortly after their departure, a stationary engine pulled the boat through a tunnel. The same ropes were then attached to two horses, which pulled the boat via a one-mile-long aqueduct across the Allegheny River, and finally on a canal dug parallel to the river.[21] Two plodding horses then pulled the boat over the 104-mile stretch to Johnstown, Pennsylvania. Octave joined other passengers, often walking briskly on the towpath instead of sitting on the boat.

The packet passed Blairsville during the night. Because the quarters on the twelve-foot-wide canal boat were tight, crews put down rudimentary "beds" so the passengers could sleep. These beds were narrow shelves, furnished with minimal bedding and a thin curtain for privacy. With everyone settled, one could hear the trample of the horses, the water splashing against the hull of the boat, and the other passengers snoring.

The following morning, the packet reached the end of the canal at Johnstown. Now the boats were separated and each unit was hoisted out of the water and transferred onto flatcars of the Allegheny Portage Railroad, to cross the densely forested Allegheny Mountains. At that time, the locomotive was a crude machine that could not exceed fifteen miles an hour, nor climb a steep grade, so engineers introduced a rope and pulley system to wind over the next thirty-six miles. The cars were unhitched from the locomotive at the foot of incline No. 1 and fastened to a rope attached to a stationary steam-driven winch at the head of the incline that pulled the cars majestically up the steep slope. Between inclines No. 1 and 2, the tracks went through the nine-hundred-foot-long Staple Bend Mountain, the first railroad tunnel built in the United States. Travelers traversed the fourteen-mile-long second level in one hour, crossing the Conemaugh River via the Horseshoe Bend. This viaduct went in a perfect eighty-foot diameter semicircle, carrying the tracks seventy feet above the water surface and ascending over its whole course.

The train travelled over the final incline and reached Blair's Gap Summit, 2,326 feet above sea level, during the early afternoon. A stone tavern, the Lemon House, invited passengers to relax during the one-hour wait while the crew reconnected the cars for the downhill trip. Curious travelers visited the engine house, with its thirty-horsepower steam engine, and inspected the thick wire ropes that pulled the cars uphill. Octave was probably not the first to wonder how the descending and ascending trains could move at the same time in opposite directions on the double-track roadbed.

The tracks on the eastern side of the Alleghenies were much steeper; another passenger had described this downhill trip earlier: "The descent of 1,398 feet on the eastern side of the mountain is much more fearful than the ascent on the western side, for the inclines are much longer and steeper, of which you are made aware by the increased thickness of the ropes and you look down instead of up."[22]

In late afternoon, they reached the canal basin in Hollidaysburg on the Juniata River, about seven miles south of modern-day Altoona. The boat sections were now hoisted from the railroad cars and reassembled to glide eastward on the water for the next 172 miles to Columbia, Pennsylvania. British writer Charles Dickens described this portion of the trip: "During the night the boat was gliding almost noiselessly past hills; the shining of the bright stars, undisturbed by noise of wheels or steam or any other sound than the rippling of the water as the boat went on, all these were pure delights."[23]

The final eighty-two-mile trip utilized the Columbia Railroad, one of the earliest railroad lines in the United States. The railroad still used cars resembling stagecoaches pulled by horses that were exchanged every twelve miles. The narrow-gauge, double-tracked line paralleled the river and descended slightly into Philadelphia. New coaches, called "flying machines," had recently been introduced between Philadelphia and New York, but Chanut selected the Philadelphia & Trenton Railroad, rolling along the Delaware River via Bristol and over a substantial bridge into Trenton. Arriving at the depot in Jersey City, New Jersey, crews transferred passengers and freight to a steam ferry to cross the Hudson River to New York.

The engineering achievements of the Pennsylvania Main Line were an inspiration to the fourteen-year-old Octave, who never forgot the difficulties crossing the Allegheny Mountains. Years later, fellow engineer Theodore Condron[24] recalled Chanute frequently talking of this journey with its wonders, delights, hardships, and inconveniences, proudly noting that he was part of many steps in the evolution of improved rail transportation during his long career.

Life in New York City

New York was the most commercial city in the Union, and Joseph knew several French nationals who had emigrated from Paris for a variety of reasons. One acquaintance, Charles Coudert, had been an officer under Napoleon Bonaparte;[25] after Napoleon's abdication, Coudert participated in a conspiracy to put Napoleon's son on the throne and was sentenced to death; he managed to escape and emigrated with his wife to America in 1824. Soon after arriving in New York, the Couderts opened a boarding school for about thirty pupils that provided income to the family and a classical French education to the children of other French immigrants. State law required each private school to accept a small percentage of nonpaying students, or "indigent children," which meant that Octave could be accepted in the Coudert Lyceum. This was fortunate because the events in Joseph's life and the inland trip had left him without means to pay tuition.

Early in September 1846, Octave entered Coudert's school for an education different than what he had received from his father and other professors in Louisiana. An advertisement stated, "Boys will be received at any stage of their education and be thoroughly prepared for college or business. The instruction is practical and covers every branch of French Education. In the English Department a solid English education will be pursued with a complete classical education. The discipline is parental, but firm. While every effort will be made to secure the esteem of the pupils, they will be required to observe habits of industry and self-control, and to yield a ready submission and cheerful obedience to the regulations which may be imposed for their common advantage."[26]

Living at the lyceum of Papa Charles, Octave became part of the Coudert family, quickly learning the English language and making friends with their oldest son Frédéric and other children. Joseph stayed initially with the Le Barbier family, but then moved to a small apartment at 39 Walker Street. To support his simple lifestyle, he earned money by tutoring, translating, and writing for the local French-language newspapers.

The challenges of moving freight and passengers across the Alleghenies were vividly fixed in Octave's mind. He wondered how he could improve travel to open the country. His father made it clear that he had no money to send him to college, so the sixteen-year-old teenager studied newspapers and magazines to determine how best to achieve his career goals.

After the United States won the war with Mexico in the spring of 1848, present-day Arizona, California, Nevada, Utah, and parts of Colorado and New Mexico were incorporated into the nation, but there remained little means of transportation or communication between the East and the West. The frontier town of Chicago sat at the edge of civilization, with more than 2,100 miles of land, unpopulated by settlers, separating it from burgeoning California. To reach the West Coast, one could travel by sea down the East Coast, cross the malaria-infested Isthmus of Panama on a mule, and continue by sea to California. Or one could sail around South America, which took about three months. Alternatively, one could travel for an undetermined time by covered wagon through mostly uncharted territory across the continent. Octave was convinced that there had to be faster modes of travel.

An editorial in the *Merchant's Magazine* described the benefits of the railroad in early 1848: "The railroads are the chief means by which the whole commerce of the earth, its movement and its population, are to be connected together and the ends of the world literally united. No man can over-estimate their value."[27] Reading reports about the chief engineer of the Hudson River Railroad, John B. Jervis, who had already accomplished several marvelous engineering feats, Octave was impressed. Knowing that he had to learn from the experts, he selected the Hudson River Railroad as his path into the future after graduating in August 1848 with a degree similar to a high school diploma.

CHAPTER 2

The University of Experience

APART FROM THE EAST COAST, North America was thinly populated in
the early nineteenth century, with few roads and little communication existing
between the populace in the East and settlers living west of the Allegheny Mountains. To open up the inaccessible, or what some easterners called "worthless,"
wilderness required an effective transportation system. The railroad promised
to provide that, but first, miles of iron needed to be laid cheaply and rapidly
over a rough and seemingly impenetrable landscape.

The sixteen-year-old Octave Chanut had read that engineers working for
the expanding railroads needed to possess universal knowledge. To become a
civil engineer, not a military engineer and not just a surveyor, he had much to
learn, but he felt sure that intelligent and earnest work would provide his key to
success.[1] In due time, Octave hoped to make a name for himself, earn income,
and find reward in wealth and status.

His first order of business was to find a job. To succeed when calling on a
future employer, Octave needed to follow standard etiquette. He gave himself a
haircut and dressed as well as his meager means allowed. Before leaving home,
his father reminded him to wear his black gloves; however, there was a gaping
hole in one of the fingers. Knowing how to sew buttons and mend holes in
his socks, the teenager took along needle and thread; he would have plenty of
time to fix the glove while he traveled on the steamer upriver to Sing Sing, or
modern-day Ossining, New York, to meet with his hoped-for future employer.

Apprenticing on the Hudson River Railroad

The construction of a railroad to connect New York City with the state capital,
Albany, first came under consideration in 1833, but was dropped owing to lack
of funding. The citizens of Poughkeepsie, located about halfway between New
York and Albany, revived the project in 1846 and hired John B. Jervis as chief
engineer. He began construction a year later.

Forty years earlier, Robert Fulton's steamboat had moved up the Hudson by paddle wheel at a speed of four miles per hour. In September 1848, the steamboat with Octave Chanut onboard traveled a little faster, arriving at Sing Sing in six hours. He headed at once to the headquarters of the Hudson River Railroad and introduced himself to the chief engineer, Jervis. Octave explained that he wanted to become a civil engineer, and asked for employment. Looking at the slender teenager, wearing his best clothes, Jervis did not think that he should hire him, so his reply was simple and discouraging; there were no vacancies. Determined to begin his professional career, Octave then offered to work without pay to prove himself an industrious worker.

Jervis knew that he needed good people to reach Poughkeepsie within the next six months, as mandated by the shareholders. Looking at Octave again, he thought he recognized him as the fellow passenger, traveling on the same boat that morning, who had busily repaired his glove. When asked, Octave shyly admitted that he did mend his glove. This provided enough justification for Jervis, who assigned him to Henry Gardner's survey party with the comment, "Well, I think you are careful and industrious and deserve a trial."

The career of Octave's first supervisor, Henry A. Gardner,[2] was typical for civil engineers learning the trade in the early nineteenth century. Beginning in 1836 as a survey rodman, Gardner gained experience in locating railroads through a country of varying obstacles and moved to New York a decade later to become assistant chief engineer for the Hudson River Railroad under Jervis. When Octave began his "employment," Gardner's survey party was working near Poughkeepsie. The party consisted of a corps of axemen who cut away trees and bushes, a transitman who, with his chainmen and flagmen, recorded the distances and angles of the line, and the leveler, who recorded the grade. The chief engineer then finalized the location of the new line.

Gardner took a liking to the teenager, who reminded him of his own career beginnings, and he assigned Octave to work with each member in his group. This on-the-job apprenticing provided opportunities to observe and ask questions, but also to participate in hands-on applications and calculations. The mystery of the level, the taking of sights, and the computations were all new experiences to Octave; as a good mentor, Gardner made himself available to answer questions.

Late in December 1848, Gardner recommended hiring Octave as assistant chainman for $1.12½ a day. Even though this was the lowest-paid job in the full surveying party, Octave felt like the richest man in the world. Soon, perhaps, he would be a real civil engineer and could add the title "C.E." to his name, just like Mr. Jervis and Mr. Gardner!

Early in 1849 Jervis reported to the Board of Directors that the workers had solved most of the engineering difficulties in the railroad's construction. Crews had graded all culverts, bridges, tunnels, and the roadbed for double track,[3] but the construction of the 836-foot New Hamburg tunnel south of Poughkeepsie was not yet finished. It had to go through a rocky outcrop, then across the Wappingi Creek on a causeway, and finally over a bridge.[4] Octave learned firsthand of the obstacles nature regularly presented to railroad builders.

An episode highlighting the hazards of an engineer working on the railroad took place at Stuyvesant Point, north of Poughkeepsie: "A party of engineers were engaged in the survey, and in that party was a slight young gentleman, who was carrying the hind end of the chain; at that point, it was extremely difficult along the river. The man who held the forward end of the chain was a tall lank Yankee, and the position was so difficult that it was impossible for him to turn around." The young man holding the hind end of the chain slipped and fell into the icy water. Fellow engineers pulled the injured worker out and took him to Poughkeepsie. The next morning, work began in the same locale. James C. Spencer again ran the party, and the injured worker was back on the job. In an imperative manner, Spencer told the man to hurry up drawing the chain. The young man very quietly put the handle of the chain on a projecting stone, turned around and said: "Mr. Spencer, if you think I am going to break my neck for nine shillings a day, you are sucked in."[5] This young man was Octave Chanut, as Richard Morgan recalled years later. From this inauspicious beginning, the friendships and fortunes of Morgan, Spencer, Gardner, and Chanute were to grow through a long association in railroading.

The locating of the line progressed steadily toward Albany, with most members of the survey party now living in Hyde Park, New York, about ninety miles north of New York City. The 1850 census listed the eighteen-year-old Octave living in the Marshal Tavern with two of his coworkers, twenty-seven-year-old Thomas Meyer and twenty-three-year-old James Spencer.

While Octave learned the duties of a civil engineer, his fifty-five-year-old father decided to return to Paris twelve years after coming to the New World. Working far upriver, Octave could not take time off to see his father to the boat, so he wrote an emotional letter. "Because I am young, I will not feel my isolation with as much bitterness as you do. I am going to work very hard so that in a few years I will be able to join you with a small fortune."[6] Joseph left New York in late December 1850, and father and son never saw each other again.

The Hudson River Railroad opened over its entire length on Wednesday, October 8, 1851. Two trains with invited guests left New York's Chambers Street Depot and headed to the Greenbush terminus across the Hudson from Albany. An enthusiastic reporter described what was the first train ride for many invitees: "The iron horse progressed, winding around the rocky base of some rugged hill, the breaking of the stillness of the quiet vale, now following the shore of the noble stream, then dashing through the tunneled mountain with an exultant roar of strength and triumph, and perforating St. Anthony's nose with confident impunity. . . . The track seems to have been laid in the most substantial manner, and even when running at the rate of forty miles an hour, the motion was so perfectly easy that reading was almost as easy as riding."[7]

To connect New York with Albany was an impressive achievement, present-ing a new era in transportation. Travel time by steamer took thirty-six hours, but the railroad covered the 144-mile distance in three hours, fifty-five minutes. The budding civil engineer Octave Chanut was part of this momentous event that many people did not believe possible. Still, during Chanute's lifetime, but fifty-nine years later, Glenn Curtiss flew an airplane over the same route in two hours, thirty-two minutes.

An editorial in the *Brooklyn Eagle* highlighted the benefits of the railroad, especially during the winter season: "Before the opening of railroads to the in-terior from New York, and after the Hudson River had closed, poultry, venison, butter, eggs, milk and such like commodities were seldom brought to this city from any remote place. Now let the reader walk through Fulton or Washington markets, and his eyes will convince him that besides travel, railroads contribute essentially to the comforts of city life."[8]

The Board of Directors was anxious to see whether the railroad could meet the competition from the Hudson steamers. As hoped, traffic exceeded expecta-tions and brought in much-needed revenue, even though the charge per ton of freight was higher than on the steamers. A news release announced that more than 600,000 passengers traveled on the Hudson River Railroad between New York and Albany during its first nine months of operation.

Working and learning in his railroad apprenticeship, Octave passed through an orderly progression of railroad duties, including chainman, axeman, target-man, rodman, surveyor, and then resident engineer of a subdivision. At each stage, he moved into a larger sphere of action and acquired a thorough educa-tion, with a different superior reviewing his work. He proved himself a good worker, eager to learn new aspects in his job, and quickly realizing that one saved time, money, and anguish if he did a job right the first time.

Generally, the building of a railroad progressed quickly, as new roads had to open for traffic at minimum cost and in the shortest possible time. Almost four years after being hired at first without pay, supervisors promoted the twenty-year-old to division engineer, in charge of the completion of the terminal facilities and maintenance of way between Hudson and Albany, a thirty-five-mile stretch.[9] Chanut moved to Albany in late 1852. As soon as he could, Octave Chanut filed his intent to apply for citizenship. After all, he received a paycheck, was Americanized, and would be a real civil engineer in the not-too-distant future.

Land Grants and Grabs in Illinois

When Illinois was admitted to the Union in 1818, settlers moved west but were not interested in buying public land.[10] They usually squatted, built a cabin, and cultivated a patch of land, establishing a preemption right—or the right to purchase, if and when the land came on the market at the government price.

With grandiose tales of the railroads' potential, leaders envisioned many transportation projects between the Alleghenies and the undeveloped West. Speculative fever rose in the 1840s, and civil engineers, including many of those working with Octave, saw career opportunities in opening the country for settlement. Stephen A. Douglas, Illinois congressman and later Democratic senator, dreamed of a great westward expansion and introduced a bill for a central railroad between Dubuque, Iowa, in the North and Memphis, Tennessee, in the South. President Millard Fillmore signed the Land Grant Bill in January 1850, allotting 2,600,000 acres to Illinois, Mississippi, and Alabama, with railroads receiving thousands of miles of rights of way out of the federal public lands. Selling the even-numbered sections along the proposed railroad line would help pay for its construction, while the odd-numbered sections, still owned by the government, were sold at double the government minimum price.

Enterprising politicians chartered the Illinois Central Railroad a few months later. To sell the granted land, the railroad had to attract buyers, but the charter did not authorize the railroad to plat towns along its line. To circumvent the law, four directors formed a separate company, "The Associates," who then platted the land around the depots and sold town lots at a premium to settlers by advertising them as "communities with an unbounded future." Actual conditions varied from the glowing pictures used to attract settlers, but the scheme worked. According to historian Paul Gates, the town lot business and town site promotion was one of the most popular and profitable forms of speculation in Illinois history.[11] Chanut would soon participate in the trading of town lots, acquiring some of the "wealth" he sought and profiting from the insider knowledge of the strategic location of railroads, depots, and stations.

Thomas Clarke described the usual process of locating a railroad line: "It is not only to place and build a railway, but it needs to get the greatest amount of business out of the country through which it passes and at the same time be able to do that business at the least cost on the capital invested. In a new country, like Illinois, the shortest, cheapest and straightest line possible, consistent with the easiest gradients that the topography of the land will allow, is the best."[12]

The railroad offered many advantages, so immigration increased, and settlers purchased land and placed farms under cultivation. The 1850 Illinois census recorded about 851,000 residents; by 1860 the population had doubled to 1,712,000 inhabitants.[13] With this increase in people, land prices rose and public land, previously offered in vain at $1.25 per acre, averaged $5 in 1852. Four years later, land sold for $20 and increased steadily.

Chicago & Mississippi Railroad

The Alton & Sangamon Railroad received a charter in 1847 to link the Mississippi River port of Alton with Springfield, Illinois, at the Sangamon River. In

June 1852, the Illinois legislature authorized the extension of the railroad from Springfield via Bloomington toward Chicago.[14] Eastern capitalists recognized the possibilities for profits. Henry Dwight, a thirty-two-year-old financier from New York, bought the road, changed its name to Chicago & Mississippi Railroad, and hired four engineers with a store of railroad experience from the Hudson River Railroad: Oliver Lee as the chief engineer plus Henry Gardner, Richard Morgan, and James Spencer. With this group, it was apparent that their friend and coworker Octave Chanut would soon follow.

The town of Joliet, Illinois, about forty miles southwest of Chicago, became the northern terminus. To circumvent the high switching charges of Chicago, the Chicago & Mississippi combined its business with two other railroads to establish a connection with the western terminus of the Michigan Central at Lake Station (today Gary), Indiana. Gardner, in addition to his construction duties on the Chicago & Mississippi, became chief engineer of the forty-five-mile Joliet & Northern Indiana Railroad Company and offered Octave Chanut the position of assistant engineer for construction. The twenty-one-year-old Octave, with no family to hold him back, left Albany in late March 1853 and moved to Joliet. After the "Joliet Cut-Off" was finished, Gardner assigned Octave to other projects on the Chicago & Mississippi.

Financial problems soon flared, as the construction cost ran much higher than Dwight had anticipated.[15] In December 1853 Gardner received promotion from superintendent of construction to chief engineer, and his construction crew worked from dawn to dusk to complete the Chicago & Mississippi as rapidly as funding allowed. In his new assignments, Octave Chanut continued to learn from engineers who had developed their skills in the East.

At a time when real estate dealings consisted of little more than a purchase and sale, buying land based on known railroad development, and then quickly reselling it, presented an appealing get-rich scheme. Gardner, Morgan, and Spencer acquired many acres in the area where they located the railroad line[16] and then sold the just-acquired land to their employer.[17] As Octave still harbored the dream of retiring with independent means by the time he reached his 30s, his three coworkers easily enticed him to become involved in real estate. Taking a conservative approach, Octave invested his hard-earned money in land warrants, ranging from $.50 to $1.15 an acre. He bought his first property, 160 acres, in July 1853 just south of Dwight, Illinois, but not near the railroad line, and six additional quarter sections the next day. The Livingston County Deed Records show that Octave made his first sale on October 1, 1853, selling seven quarter sections for $1.00 to Alphonse Le Barbier, the son of his father's friend who had helped them when they came to New York in 1846.[18] Octave acquired a taste for real estate deals and became skilled in the art of sales and speculation. By paying strict attention to the business, by quick and at times instant reinvestment, he soon increased his little fortune.

Three years almost to the date of filing and after one year of state residency, Octave Alexandre Chanut became a United States Citizen. On April 17, 1854, the twenty-two-year-old French immigrant was sworn in at the McLean County Courthouse in Bloomington, Illinois, with Richard Morgan as his witness. He anglicized his name to Octave Chanute, dropping his middle name, Alexandre, and adding the letter "e" to his family name.[19] Chanute's oldest daughter, Alice, later recalled that a schoolmaster at the Coudert School had teased her father by calling him "Naked Cat," as his family name could be pronounced *chat nut* in the French language.[20] She thought that this might have been her father's reason for changing his name. Another reason could have been to adopt a name similar to his Scandinavian ancestors, whose name was reportedly Canute.

Coming from the Hudson River valley, Chanute was impressed with the Illinois countryside, where the mostly flat plains rolled on as far as the eye could see, with occasional groves of trees and flowers everywhere. More than once he saw the seemingly endless covered wagon trains with immigrants, of which he had heard so much. Watching and hearing the heavily loaded wagons struggle over the slightly elevated eight-foot-wide plank road, traveling twenty-five miles in one day, was an unforgettable sight. Would travel over this "National Road," coming from Cumberland, Maryland, and ending at the Mississippi River, soon be replaced by something more efficient?

Octave's goal was not only financial security, but he also sought a part in the transportation evolution. In his personal learning process, he gravitated to men of studious and constructive brainpower, individuals who contributed as civil engineers. One of these men was his former supervisor at the Hudson River Railroad, Richard Morgan. A well-known civil engineer, highly experienced in canal and railroad work, Morgan was now the superintendent of the Chicago & Mississippi and part owner of the Peoria & Oquawka Railroad. Knowing of Octave's ambitions, Morgan recommended that the directors of the Peoria & Oquawka hire his protégé as chief engineer for their just-authorized Eastern Extension.

This would provide a great opportunity for the nascent engineer to hold responsibility for every part of the new railroad's construction. Locating the line involved detailed field examination; he needed to prepare plans for roadbeds, track, bridges, culverts, buildings, and other structures; he had to prepare plats accurately, showing the dimensions and areas of all land required for right-of-way, stations, or terminals. Then he needed to hire people to build the road and work under his direction, and once the road opened, he was responsible for all matters connected with the maintenance of the roadway, bridges, and buildings. Chanute felt anxious to tackle such a challenge.

A New Life in the West

In the summer of 1854, Chanute resigned from the Chicago & Mississippi to join the Peoria & Oquawka as chief engineer and moved to Peoria. The town, with its 17,000 citizens, had much to offer a young engineer, including a social and cultural life, so Chanute selected the Peoria House as his new home.[21] This hotel, located at the corner of Adams and Hamilton, had opened in the late 1830s as the "Planters' House" and soon became the largest and best hotel in the state. In the early 1850s the owners introduced substantial improvements, including a dinner bill of fare, something new to Peoria.[22] Lawyers and politicians boarded here regularly; this was where society met.

Having grown up in antebellum Louisiana, Chanute disliked slavery, but in central Illinois, especially around Peoria, the Whig and Democratic Parties were generally divided on this question. In the upcoming congressional elections, voters had to choose between the proslavery Stephen Douglas and the mostly antislavery Abraham Lincoln. The young lawyer Lincoln objected to the recently signed Kansas-Nebraska Act, as it allowed settlers to decide whether they would accept slavery. Lincoln's Peoria speech[23] in October 1854 set the stage for his political future. Chanute most likely stood with other railroaders in front of the courthouse, a few steps from his room at the Peoria House, and listened to the lawyer of his former employer, the Chicago & Mississippi Railroad.

Standing five feet, six inches tall, the confident twenty-one-year-old engineer decided to grow a half beard with a tiny goatee, which was just as curly as his main hair. He thought this style of beard complemented his face and made him look more sophisticated and grown-up. Chanute wore a half beard throughout his life. As a young bachelor, Octave was reportedly "a dandy in dress, wore wonderful waistcoats with jeweled buttons and was courtly in manners."[24] No wonder he aroused the interest of Annie James when her brother Charles took her to a Saturday dance in one of Peoria's town halls. Annie was a charming little person with animated eyes, and Octave soon made her acquaintance.

The James family was well known in Fredericksburg, Virginia, at the beginning of the nineteenth century. Annie's grandfather, William James, was a Baptist minister, and her father, Charles Pearson James, owned a merchant tailoring business. The family moved to Peoria in 1851, where her father opened a shop, advertising a "variety of goods, all of superior quality, and all would be made up in the newest and most fashionable manner at the shortest notice."[25] A year later, Annie's father died and her older brother, Charles James Jr., took over the business. The family worked hard to make ends meet, sewing clothes for shops in the bigger cities, like Marshall Field in Chicago, and uniforms for the military.

Octave and Annie received their marriage license from the clerk of Peoria County on March 11, 1857, made out to Octave Chanut (without his middle name and without the "e" at the end of his family name) and Annie R. James.[26]

Octave Chanute
(circa 1856). Courtesy
of the Chanute Family.

Annie Riddell James
Chanute (early 1860s).
Courtesy of the
Chanute Family.

Robert Farris, the pastor of the Second Presbyterian Church, married them the next day.

In the early 1850s, the Presbyterian Church had split into the "Old School" and the "New School." Although there were theological differences, the main difference concerned their attitude toward slavery. The New School was decidedly antislavery, while the Old School Presbyterians were more tolerant. The Second Presbyterian Church, an Old School church, was home to the conservative, moneyed element in Peoria, and its members were active in politics.[27] Annie may have felt more comfortable in this church, because of her family's past ownership of slaves back in Virginia as well as the social status of its members, while Octave might have felt less comfortable because of his personal opposition to slavery.

The newly married couple bought a house for $2,200 at 121 Hamilton Boulevard, near the James tailor shop and home at the edge of town. The proud husband wrote a long letter to his father, reporting on his happiness being married to Annie and about his newly purchased home. Joseph replied to his daughter-in-law, just hinting about his own personal difficulties: "But in this gay, brilliant and charming Paris for the wealthy, the battle of life is sometimes hard for other people and for a long time, my mind, my time have been engrossed by cares and literary labors."[28] He continued in perfect English how proud he was to have her as his daughter-in-law and that he looked forward to meeting her. "Be happy!" On February 6, 1858, Annie and Octave's first child arrived, and they baptized him Joseph Arthur.[29] Wanting to share the joyful news, Octave considered sending a telegram to his father, but this communication link had collapsed soon after the cable was laid across the Atlantic Ocean.

Several former "Hudson River Boys," including Chanute, had enjoyed the sport of rowing in New York and wanted to introduce it in Peoria where the Illinois River expanded to the four-mile-long Lake Peoria. The *Peoria Daily Transcript* of June 8, 1858, reported on the group's latest project, the *Foam* rowboat: "This beautiful row-boat, built by Harvey Spencer, after a drawing by O. Chanut, Esq., was launched yesterday afternoon. She is calculated for four sweeps, is elegantly furnished and reflects much credit upon the builder as well as the designer."[30]

Living in Peoria, Chanute became involved in various civic pursuits. Among others, he was a charter member of the Illinois Lodge No. 263, formed in Peoria as a chapter of the Grand Lodge of the Free Masons in October 1858. Annie's family enjoyed prominence in Peoria society and regularly attended social gatherings during the winter. Because she enjoyed dancing, her husband agreed to join the organizing committee for the first "Grand Fancy Dress-Ball" of the season, held on January 7, 1859. A newspaper advertisement gave the rules: "No one but subscribers will be admitted. Every person must be dressed in mask until 12 o'clock, when the sound of the Bugle will give the signal to unmask. No Lady or Gentleman will be admitted without tickets. Stringent regulations will be enforced to guard against the admission of any improper persons. Tickets for the Gentlemen are $1.50 each. For Ladies, accompanying them, admission is free."[31]

Octave and Annie's second child, Alice Elise, was born on December 24, 1859.[32] Both parents felt strongly that their children needed to be educated the way they were brought up in Virginia and Louisiana, always observing the one hundred "Unclassified Laws of Etiquette."

Peoria & Oquawka Railroad: The Eastern Extension

The Peoria & Oquawka Railroad was chartered in 1849,[33] but promoters soon ran out of money. In June 1852 the Illinois legislature amended the charters of several Illinois railroads to attract eastern capital; the Peoria & Oquawka received authorization to extend from Peoria east to the Indiana state line and to take possession of state-owned land for the right-of-way.[34] The always vocal Chicago press voiced strong concerns in an editorial about this new "Cut-Off," as the Peoria & Oquawka would connect the Mississippi River with eastern cities, bypassing Chicago and threatening its commerce.[35]

At about this time, Boston investors, with John M. Forbes at the helm, sought railroad properties with strategic advantages for their envisioned coast-to-coast railroad system, and the Peoria & Oquaqua or Oquawka (Forbes reportedly exclaimed, "Phoebus, what a name!")[36] would fit well into their scheme. In 1854 Forbes signed a contract with the railroad's president for a joint venture[37] and funded the road's construction. Little did Chanute realize that working with Forbes and the "Boston Party" would prove a major part of his developing career.

The Eastern Extension began on the east bank of the Illinois River opposite Peoria. Even though few people lived along the planned road, directors looked for heavy passenger and freight business with good future earnings. With Morgan's help, Chanute located and then supervised the construction of the first fifteen-mile section, reaching the intersection with the Chicago & Mississippi in the fall of 1854. Chanute named the terminus after William Cruger, the superintendent, who owned much land in the surrounding Cruger Township.

With the financial help of John Forbes, the road purchased a Rogers locomotive and tender that delivered timber, ties, and rails when construction started again in the spring. The Eastern Extension now ran as an "airline," meaning that the route followed a straight line nearly exempt of curves and grades over a continuous prairie. The next station sat about twenty-five miles east of Peoria, and Chanute named it after Charles Secor, president of the railroad.

The road reached the junction point with the Illinois Central late in 1856, and Chanute erected a passenger and freight depot, recorded as Peoria Junction. The Illinois Central had platted the surrounding area a year earlier and recorded it as El Paso. Their standard plat map, with the exact size of blocks, lots, and street width, had all the streets laid out parallel or perpendicular to the railroad route. The "parallel streets," going from east to west, were named after trees and the "perpendicular streets," going from north to south, were numbered. The street along the railroad tracks with its depots was named First Street or Center Street and was located between Chestnut and Oak Streets.[38] Being new to laying out towns, Chanute quickly saw the benefit of their platting scheme that sped up and reduced surveying costs, so he adopted their standard plat for the towns along his roads in Illinois and later in eastern Kansas and Nebraska.

The area around El Paso was at first uninhabited prairie not worth $0.50 an acre,[39] but this changed with the arrival of the two intersecting railroads. Daily trains with freight and passengers began running and new settlers moved in. Ten years later, the prospering town of El Paso became a county seat and recorded about three thousand citizens. "The State of Illinois never stood in a more advantageous position to invite settlers than at this moment,"[40] reported the *Chicago Tribune*.

To provide efficient commerce, the Peoria & Oquawka, west of Peoria, needed to connect with its Eastern Extension via a bridge across the Illinois River. It was too time-consuming and labor-intensive to transfer passengers and freight from the railroad cars onto a ferry to cross the river and then unload everything to another railcar. However, building a bridge across a navigable river was a hot political issue, as steamboat owners strongly opposed the bridging of "their" rivers.

This political opposition became more apparent when a bridge disaster happened ninety miles to the northwest. The Chicago & Rock Island Railroad was the first to reach the Mississippi River from the east, and its president, John Jervis (who had hired Chanute in 1848), had erected a bridge at Rock Island, Illinois, across the Mississippi to connect with railroads in Iowa. Two weeks after the bridge opened in May 1856, the steamboat *Effie Afton* hit a pier, burning the boat and part of the bridge.[41] Furious steamboat owners filed suit against the bridge owners, who claimed the "accident" intentional. Defending the bridge owners, Abraham Lincoln gathered technical expertise from knowledgeable civil engineers like Jervis[42] and Chanute, and then argued in court that a person had as much right to build a bridge to cross a river as another person had to travel on the river.[43] In his closing statement, Judge McLean declared that the steamboat was not managed by competent river pilots, and dismissed the case.

The Rock Island court case was essentially a contest between the commercial rail interests of Chicago and the maritime interests of St. Louis. "The jury in the Rock Island bridge case failed to agree and were discharged. The case will be tried over again. Meantime the bridge is left standing, despite the Chamber of Commerce at St. Louis and all the prejudiced sore-head pilots on the river,"[44] was the editorial comment in the *Chicago Tribune*.

While the Rock Island case was fought in court, Chanute began his first major bridge construction project. He designed a drawbridge with a straight bottom and curved top chord with two 150-foot-long spans; the 290-foot pivot draw rested on a round, 31-foot-diameter pier, leaving a clear waterway of 127 feet on either side. A 3,100-foot-long trestlework carried the railroad tracks over the flat swampland on the eastern bank of the Illinois.

Built for just over $500,000, the Peoria Bridge opened with much fanfare on April 6, 1857, with Superintendent Cruger personally driving the first train across. Everyone, including chief engineer Chanute, watched with some anxiety as the locomotive pulled its tender, three heavily laden cars weighing eighty tons and a passenger car with officials, across the bridge. A report in the local newspaper the next day noted: "The bridge stood without a bend, break or squeak, and scarcely seemed to feel the tremendous weight which it upheld. After unloading the timber, the train went back and the feat was done."[45] Three years later, Chanute wrote to the *American Railway Review* that a previously published article was incorrect. The 280-foot-long Nashville Bridge across the Cumberland River was not the longest drawbridge; the draw of his Peoria Bridge was ten feet longer and exceeded in length any other drawbridge in the United States and perhaps the world.[46]

In late August 1857 unscrupulous brokers tried to pocket fraudulent fortunes on Wall Street, leading to a general economic panic that spread quickly across the nation.[47] In mid-October the Illinois Central announced that its difficulties

Peoria Railroad Bridge, built by Chanute in 1857, and wagon bridge across the Illinois River. From Perspective map of Peoria, Illinois (1867), Map Division, Library of Congress, Washington, D.C. No. G4104.P4A3 1867 .R8 Rug 29.

were not intrinsic to the company, "but from extraordinary conditions on the money market." Workers received their wages and were discharged. Chanute's Eastern Extension was also affected by the panic, but construction continued and the line reached Livingston County, about seventy miles east of Peoria, in late October.

The 1850 Livingston County census recorded a population of 1,552, with the first settlers blazing their claims and "squatting."[48] With the path of the Eastern Extension known, new settlers arrived and land values started to rise. The Livingston County Deed Record Book[49] shows a $6,000 bond between a John L. Miller and Caleb Patton for a quarter-section of land, finalized in mid-October 1857. Patton wanted the Eastern Extension tracks and a station to be built on his property, so he made an attractive purchase deal with Chanute. The county recorded the deed on Saturday, October 31, 1857 and Chanute's men laid the tracks through their jointly owned land on Sunday.

Using the standardized plat, Chanute laid out the town, recorded as Fairbury. He subdivided the 169.5 acres into city blocks, 400 feet by 316 feet deep; each block was then divided into fourteen town lots, 50 feet by 158 feet. Summing up the land deal, Chanute paid about $65 for one acre of his half interest, thus his cost of one town lot was about $13. The first two town lots sold in March 1858 for $155, with proceeds going to Chanute and wife, Patton and wife, and Andrew Cropsey and wife, the newly hired agent for the railroad. During the next decade, a town lot usually sold for $50.

Naturally, Chanute expected every member in the community to appreciate the benefits of the railroad, but hostility soon developed between the railroad company and the growing town, with each trying to annoy the other and not always stopping at annoyance. The author of the *History of Livingston County* recalled a story: "A train passed through the town one very dry, windy day at full speed, with fires and steam at a high stage, and emitting great blazing cinders. When the train arrived at Forrest, the next station, the engineer looked back and saw the dense smoke, then just remarked that he set the d——-d town of Fairbury on fire as he came through."[50]

The aftereffects of the 1857 panic continued to plague many railroads, and wages on Chanute's railroad also fell behind schedule, but the crew reached the intersection with the Chicago branch of the Illinois Central before the arrival of winter. Landowners gave the railroad company a half interest in their various lands in consideration of running the road to this point and establishing a station. Chanute platted the town of Gilman and received a trust-deed that included 340 town lots and twenty-five acres of adjoining land.[51] These real estate deals provided income to the Chanute family when income from the railroad business was not particularly stable.

The concerns that the Peoria Railroad Bridge structure interfered with river traffic slowly vanished as railroad commerce increased, but the inevitable happed in late spring 1859. The steamboat *Sam Gaty*, bound for St. Louis with a large passenger list and much freight, ran against one of its piers.[52] To recover damages, the Illinois River Packet Company sued the Peoria Bridge Association. Chanute, defending his engineering capabilities, testified six years later: "I was the engineer who superintended the erection of the defendant's bridge. The bridge is built on the most approved plan. I believe the dangers and risks of navigation are increased by the bridge, but boats, managed with ordinary care, in ordinary times, can pass the bridge without danger."[53] In designing the bridge, he had made sure that there was sufficient space between the piers for the passage of boats. Sidney Breese, justice of the Supreme Court of the State of Illinois, concluded: "The right of pontage is as important to the public, as the free navigation of the river. Pilots and captains must use a little more care and prudence than they have been accustomed to use."[54] The Peoria County jury found for the defendant.

The Eastern Extension struggled to meet its financial obligations, and at times there was not enough money to buy wood for the locomotives. The owners sold the road in June 1859 to two local investors from Peoria, including track, rolling stock, shop, tools, among others.[55] Chanute may have wondered about his and his growing family's future, because an additional twenty miles of track had to be laid through Iroquois County to reach the Indiana state line.

The last station in Iroquois County would be Middleport. The story of this village is an example of railroad economy and community involvement during turbulent times. Villagers had initially offered land for a depot, but their attitude changed.[56] Accordingly, a group of landowners that included Chanute and Morgan suggested running the railroad through undrained swampland as an airline south of Middleport. Again, Morgan proved a good teacher, as Chanute learned how to drain swamps, while fighting wild onions and having a good many bloody struggles with mosquitoes. Next, he erected a timber trestle bridge across Sugar Creek and platted South Middleport. The first train arrived early in November 1859, and newly arriving settlers renamed their town Watseka, in honor of the Indian heroine *Watch-e-kee*. Undoubtedly, Chanute and Morgan prospered just like the citizens of Watseka, while Middleport disappeared from the maps.

Just after Christmas 1859 the Eastern Extension reached the Indiana border, 112 miles east of Peoria, an economic milestone for the struggling railroad, but also an engineering achievement for Chanute. At that time, the total railroad mileage in Illinois was 2,867 miles, a significant increase from 110 miles in 1850.

On December 30, subscribers from Logansport, Indiana, accepted the invitation to travel gratis over the newly opened line as excursionists.[57] The financing bonds, issued by neighboring White County in Indiana, were conditional on the road operating by December 31. Some bond subscribers wanted to escape paying their subscriptions and blockaded the tracks with a boarding car hous-

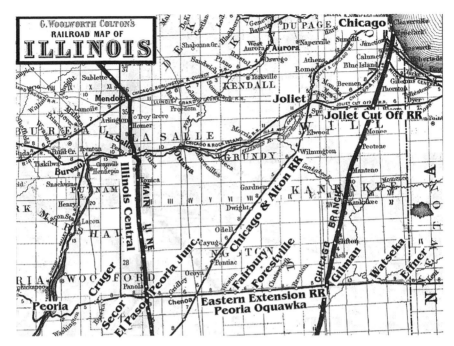

Railroads in Northern Illinois (1861): the Joliet Cut-Off runs between Joliet, Illinois, and Lake Station, Indiana; the Chicago & Alton (formerly Chicago & Mississippi Railroad); the Eastern Extension of the Peoria & Oquawka Railroad between Peoria and Effner on the Indiana state line. From G. Woolworth Colton's Railroad Map of Illinois, Map Division, Library of Congress, Washington, D.C. No. G4101.P3 1861 .C6 RR 204.

ing seventy-five Irish workers. After a long parley, the excursionists returned to Monticello, Indiana, took on the sheriff, and returned to the scene of the "blockade." A fight ensued, with one Irishman shot and slightly wounded. Finally, the bond subscribers relented and allowed the excursionists to proceed. Traveling through heavy snow as far as Gilman, they arrived in good spirits in Peoria, where the elated officers of the road met them. The railroad's income from the county bonds was safe.

The 1860 census[58] recorded Octave, his wife Annie, and their two children living in their house, valued at $15,000; Octave's personal estate was given at $2,000. Two female Irish servants, a nineteen-year-old and a fourteen-year-old, lived with the family. According to the census, the fathers of both girls worked for the railroad, possibly for Chanute.

Because handwritten letters were the main form of day-to-day communication, Chanute acquired a letter copying press from the financially ailing Eastern Extension in lieu of pay.[59] His first recorded letter, written on September 5, 1860, to Samuel Gilman, a director of the Eastern Extension, explains why he could not accept additional assignments:

> (1) The account as heretofore kept is so incomplete and so much mixed up that I can neither obtain a starting point, nor be responsible for the correctness of the months that have been nominally closed. (2) I can not attend to it without neglecting entirely the other matters now in my charge, which I conceive to be of more importance to the running of your road and your interests. (3) And this has greater weight with me than any of the objections, I find that my want of familiarity with the manner in which freight business is conducted, will form a serious obstacle to my doing it quickly and well, giving others a good deal of additional trouble. I am willing to devote my evenings to assist in straightening it up, but for the reasons given above I can not accept charge of it.[60]

As chief engineer, the twenty-eight-year-old Chanute felt responsible for making the railroad profitable. In another long letter to the president of the Eastern Extension, he suggested erecting a hotel and an "eating house" with waiting rooms for passengers in El Paso, Illinois, and Reynolds, Indiana,[61] assuming that the connecting railroads would pay half the cost of construction and maintenance. But money remained tight and nothing happened.

Later in September, Chanute drove over the line in a handcar to pay the workers, and he noticed the poor workmanship of the roadbed. "The condition of the track between Chenoa and Gridley is disgraceful, even making all allowances for the light and crooked iron in the track. The surface of the earth shows that less work has been done on that section than on any other this summer, although the force has been larger. Discharge the foreman for that reason as soon as you can obtain another, a Yankee if possible."[62] Writing to the contractor, Chanute also stressed that all corrections had to be made before the arrival of winter.

Chanute joined his old friend James Spencer to help entertain the Prince of Wales, later King Edward VII of England, especially because discussions of engineering projects provided part of the evening entertainment. After laying the cornerstone of Parliament Hill in Ottawa, Canada, the royal party arrived in Chicago on September 28, 1860. The prince wanted no honors and came simply as Lord Renfrew to see the city, especially the Water Works in Lake Michigan. The following day, the royal party left for Dwight, where Spencer and Morgan had opened their homesteads for a three-day hunting excursion for prairie chicken and quail.[63]

With financial problems continuing, New York investors bought the Eastern Extension in February 1861, combined it with their road coming from Toledo, Ohio, and renamed it Logansport, Peoria & Burlington Railroad. But regular paychecks were still not forthcoming; in frustration, Chanute wrote to Secor, closing his letter with, "I have written to you because I had and still have every confidence in your intentions and whatever fight I may make will be against your firm, not against you." To recover some pay, he sued the railroad, claiming a balance of $218.56 for surveying between February and April 1860. According to the Peoria County Court files, the plaintiff took a judgment and authorities seized the defendant's office furniture.[64] Even though his employer promised back pay, Chanute resigned in August 1861. "I do not think that I have been treated with justice. Self preservation is the first law of nature,"[65] he wrote to Secor after not receiving a paycheck for five months.

Aeronautics in Illinois

After installing the telegraph cable across the Atlantic in 1858, the public wondered what wonderful invention would come next. An unnamed writer in the *Scientific American* theorized: "Since the whales and porpoises have been astonished with the ocean cable, we are now bound to astound the gulls and eagles."[66]

The Montgolfier brothers had discovered the art of floating in the air with a balloon, but to soar in the sky, changing direction, speed, and altitude at will, required a mechanical flying machine. Chanute wrote in 1900 that he became interested in manned flight in the 1850s. For him, it presented the attraction of an unsolved engineering problem, "which did not seem as visionary as that of perpetual motion. Birds gave daily proof that flying could be done."[67] Several events could have triggered his initial curiosity in aeronautics. Periodicals piqued readers' imaginations with articles like "Flying, the Subject of Great Importance,"[68] and the sensation-hungry press colorfully described the increasing number of balloon ascensions, with their all-too-frequent tragedies.

Most likely, Joseph Chanut sent his son a thirty-two-page French language pamphlet on the history of aerial locomotion.[69] This attractively illustrated pamphlet, now part of the Chanute Collection at the University of Chicago Library, contains many of Chanute's penciled notations. Some markings are from the late 1880s, when he worked on his book *Progress in Flying Machines*, but others are in a much earlier handwriting.

In the mid-1850s, Silas M. Brooks popularized ballooning in Illinois and Iowa.[70] On July 25, 1856, the *Peoria Weekly Republican* carried the following advertisement: "The citizens of Peoria and surrounding country are respectfully informed that Mr. S. M. Brooks, the great American Aeronaut, who has made more successful voyages throughout the heavens than any other man living, will have the honor of ascending in his Mammoth Balloon, the Hercules, from this city on Thursday, 31 July 1856. Admission 25 cents. Raised seats 50 cents."[71] The *Illinois Gazette* from Lacon, Illinois, reported that Brooks set up a large circus tent, filled with about four thousand paying spectators, and began with a lecture on aeronautics, followed by the balloon being inflated. After releasing the ropes, "away heavenward the great Hercules went. As he rose from the ground, the Professor bade his audience 'good night' and majestically ascended."[72] The balloon landed about four miles to the east, and Brooks brought it back to Peoria. Brilliant fireworks finished the entertainment, with Chanute and other rowing club members most likely part of the enthralled crowd. With balloons the only craft capable of flight in those days, any ballooning event could have triggered Chanute's interest in aeronautics.

A Wartime Passage to New Orleans

Since its beginnings, the United States had developed into two distinct regions with opposing interests. The North was chiefly a manufacturing and commercial region with small farms, large urban areas, and an influx of inspired European immigrants. The South had huge plantations, worked with slave labor, and operated little industry. The philosophical gap widened more with the election of a northern president, Chanute's acquaintance, Abraham Lincoln.[73] Tension increased, and the bloody war began in April.

The beginnings of the Civil War provided an unanticipated adventure for the unemployed twenty-nine-year-old engineer. Chanute's maternal grandmother, Catherine de Bonnaire, had died in France and bequeathed her estate to Octave's mother Elise, who had moved to New Orleans in 1854. Because the Union had blockaded the harbor, the letter with the sad news was mailed to the French embassy in Chicago and forwarded to the dead letter office. There, Chanute picked it up.

After some soul searching, Chanute decided to hand-deliver his uncle's letter to his mother and bring her north, so that she could return to France and settle her mother's estate. People told Chanute that no one could pass the line between the northern and southern armies, but he tried nevertheless. Taking the Illinois Central to Cairo in early November 1861, he showed his uncle's letter to the commanding officer and stated his reasons for traveling. After being searched, he gained permission to cross the lines and arrived in New Orleans three days after leaving Peoria.

Elise appreciated her oldest son's efforts, but hesitated to cross the enemy lines; instead, she arranged for a power of attorney from the French embassy in New Orleans, "signed" by the emperor of France with an ornate red seal, as her uncle had suggested. Spending almost four weeks in New Orleans, Chanute experienced firsthand the bitterness of southerners against northerners. "We'll kill the Yankees like wild beasts whenever we meet them," was a typical statement.

Chanute's return trip proved adventurous, even though he pretended to be a southerner and traveled only in daylight hours. Leaving New Orleans on the Mississippi Central, he changed trains in Memphis and continued on the just-opened Louisville & Nashville Railroad. At Bowling Green, Kentucky, all passengers were ordered off the train, including Chanute. The officer in charge had just received intelligence reports that the Union army planned an attack at Woodsonville, and that no one should travel in that direction.[74] Trying to reach home, Chanute told his story to the commanding officer, but was not permitted to pass. In desperation, Chanute asked, "General, what would you do were you placed in my position?" The general answered, "I would run the blockade, Sir." Chanute replied: "Then you advise me to do so?" The general answered, "No, Sir, I forbid you doing it."[75]

With this questionable advice, a little daring, and a good deal of coolness, Chanute went back to the hotel to fetch his travel bag. If he were picked up running the blockade, he would simply ask for the same Confederate general. With no trains running, he took the next stage to Evansville, Indiana, filled with "Jewish drummers," or traveling salesmen. Pickets stopped them twice to see passes, so Chanute showed his impressive French language power of attorney with the big red seal and was allowed to proceed. A few miles south of Evansville, the mules ran away, upsetting the stage. No one was hurt, so Chanute let the other passengers help the driver, shouldered his bag, and started walking. He reached Evansville on the other side of the lines by dusk; the soldiers of the North did not take notice, as they were busy preparing their meals.

Only a few days after Chanute passed through northern Kentucky, the indecisive battle at Woodsonville took place on December 17, when Confederate troops burned the Green River Bridge to prevent the Northern army from crossing the river.[76] Officials reported approximately 130 casualties, but thanks to the Confederate general's advice, Chanute was not among them.

The *Peoria Transcript* reported on Chanute's romantic ten-day trip over a zigzag route.[77] He then wrote a long letter to his family in Paris, giving his view on the Civil War and life in the South, which he deemed on the brink of an abyss. Though his father usually submitted his comments to *La Presse* and *Revue Contemporaine*, Chanute's letter appeared on two consecutive days on the front page of *L'Opinion Nationale*, a widely read and well-informed newspaper on American affairs:

> The growth of the cultivation of cotton in the ten States, the enormous profits which assured them the monopoly of it, inspired them with the idea to spread it—and from thence comes the covetousness many times manifested to seize hold of the Island of Cuba and Mexico whose lands are admirably adopted to this culture. Blacks were to be secretly exported to cultivate the new lands. The South knew that the North would never consent to advancement in this direction under the rule of the Union. . . . The revolution in South Carolina began after Abraham Lincoln was elected to be the next President of the United States. The war that followed has been carried on if not for the restoration of the Union at least for the defense of the constitutional idea, to know that one state can not withdraw from the Union. This Constitution has been formed to imprint a contract solemnly accepted and to give a salutary lesson to the Western States who might be tempted to follow a revolutionary march.

Chanute ventured to predict that the North would lose all the first battles:

> Accustomed to danger and to holding arms, the Southern men hold every chance to put to rout those of the North, who are slower and softer, given up to industry and commerce. In the long run, they will make better soldiers than the Southerners, as they are more submissive to discipline, calmer in danger and more patient in defeat. . . . I excuse the violence and roughness of Southern manners. Living in the midst of an inferior and submissive population, that can be ill treated with impunity, one can easily take on the semblance of a tyrant. Moreover there is always present the terrible fear of a black insurrection, to be suppressed at any price and immediately. Add to this the heated blood owing to a hot climate, and you will grasp the small price attached to human life.
>
> I am convinced that the North has only to take one of two courses: either to recognize the independence of the Confederate States of the South or make war for the emancipation of the slaves, turn upside down the social order of the South and not flinch before fire and devastation which would be the inevitable consequence. It is towards this terrible extremity I greatly fear that passions and events are now tending.
>
> You have asked for a consciousness and complete exposition of the situation. I have tried to forget that I am a Northerner, so as to only remember the duty I owe to the glorious in the past and up to now respected name of an American citizen, and to pay my adopted country the feeble tribute of my affection and devotion.[78]

Letters went back and forth between Chanute and his mother's family in Paris to postpone the settlement of his grandmother's estate. In the fall of 1862 Elise received permission to cross the lines and travel north to Peoria; Annie was pleased to meet her mother-in-law and the two children were excited seeing their paternal grandmother. Chanute then took his mother and her belongings, including her dog Roquette, to New York for the trip back to France.

Pittsburgh, Fort Wayne & Chicago Railroad

The Pittsburgh, Fort Wayne & Chicago Railroad opened its 468-mile line from Philadelphia to Chicago on Christmas Day 1858. The road sold in the fall of 1861, and the new owners hired John Jervis as the general superintendent. Throughout his career, Jervis believed that the most important qualification in a manager was the ability to discriminate character "and place in all positions men suitable for the duty, both in regard to capacity and fidelity,"[79] so he hired engineers he knew and trusted. Knowing that Henry Gardner had been without professional engagement since the 1857 panic, Jervis hired him as the assistant general superintendent. Soon after Chanute returned from New Orleans, Jervis offered him the position of division engineer for operations on the line between Chicago and Fort Wayne, Indiana. This presented Chanute a good opportunity to earn a paycheck again and to advance his career, because Jervis and Gardner were valuable teachers, always willing to share their knowledge. Jervis, who had just published his treatise on "Railway Property,"[80] advised Chanute to use this book as a guide whenever he might encounter problems.

The Civil War brought resources that Jervis used wisely to improve the railroad's operating efficiency, but the war effort also put a heavy burden on the equipment and rolling stock. During the winter, shortly after Chanute took over, an incident happened that at first appeared to be sabotage, as reported by other railroads:

> Some 3,400 English rails, weighing 58 pounds per yard broke under the trains on a stretch of about 56 miles. They were laid in 1854 and had stood perfectly until that time. About 15 miles were relaid with rerolled iron, many new ties were put in and great attention was paid to the surfacing, draining and ballasting. During the ensuing summer and autumn, few breakages occurred. As soon as cold weather set in, the rails started to break again under the trains in an alarming rate. Passenger trains were slowed to a speed of 20 miles/hour, and freight trains to 12 miles/hour, over this portion of the line. Watchmen were patrolling day and night and section gangs were held in readiness at all times to repair or replace any breakage.[81]

Fortunately, only one passenger train jumped the track and no passenger was hurt. Chanute learned the hard way that rails broke under the heavier loads in cold weather, when the iron was less flexible and the roadbed became uneven owing to freezing. Having an office in Fort Wayne, Chanute learned something else during the winter of 1862: ice skating on the Maumee River, a sport he had not tried before.

Money was tight and the James family's merchant tailoring business did not fare well. In January 1862 they decided to sell the business and placed classified ads in the *Chicago Tribune* and in New York papers. The business finally sold a year later, but the family did not receive enough money to pay all its outstanding bills, putting an extra burden on the already tight budget.

Octave and Annie's third child, Gertrude Debonnaire, or Gertie, was born in November 1862. She lived only a few months, and the family buried her in Springdale Cemetery in Peoria. Ten months later, on July 22, 1864, their fourth child, a daughter named Elizabeth Chadwick (known as Lizzie) came into the world. Elizabeth C. Chadwick, Annie's sister, helped with the birth. Because Octave could not come home, Elizabeth wrote a short note: "At Ann's request I write to inform you of the safe arrival of a little daughter who was born this morning. Ann is as well as could be expected and we feel very grateful that she had so little trouble."[82]

Working so far away from home, Chanute found it difficult to balance family life and career, as he could not be with his growing family on every weekend. His oldest daughter Alice recalled one weekend when her father and her uncle

Charles made an "elephant" with the help of some blankets and white wrapping paper. It was so realistic that she and her brother Artie were filled with fear; the two men had to reveal themselves to calm down the excitement, but then the children enjoyed riding on the back of the "elephant."[83]

Chanute resigned from the Pittsburgh & Fort Wayne after thirteen months employment, leaving Gardner to take on the chief engineer position for the Western Division in addition to his role as assistant general superintendent. For additional income, Chanute continued buying and selling town lots and land in the range of $40 to $120 per acre. He also accepted consulting projects, one of which was with the City of Peoria to develop facts about water works. One city official, however, objected, claiming that Chanute's "head was too crammed with science, there was no room for commoners or common knowledge."[84] Fortunately, opinions differed and Chanute submitted his plans with probable cost at the next regular town meeting.[85] The matter rested until February 1868, when Chanute suggested hiring William Civer as the chief engineer.

Ohio & Mississippi Railroad

The Ohio & Mississippi was a six-foot-wide, broad-gauge railroad, running between Cincinnati and the Mississippi River at Illinoistown, later renamed East St. Louis, Illinois. Chanute became chief engineer of maintenance of way with the Western Division in September 1863; he held responsibility for the 148-mile section between St. Louis and Vincennes, Indiana, on the Illinois-Indiana state line. Chanute established an office in East St. Louis, with good rail connection to Peoria, and collaborated with Mendes Cohen, the general superintendent, and George Blanchard, the general freight agent, playing an important part in the difficult and perilous problem of transporting troops and materiel for the war effort. Changes came in upper management of the Ohio & Mississippi Railroad early in 1864, followed by another reorganization, so Chanute resigned after only half a year, accepting a call from the Chicago & Alton.

Chicago & Alton Railroad

Subsequent to Chanute's leaving the Chicago & Mississippi Railroad a decade earlier, the road had been sold twice. After the last sale in October 1862, the new board, consisting of Illinois residents,[86] changed the road's name to Chicago & Alton and rehired some of the former engineers. Chanute accepted the chief engineer position in the spring of 1864, initially keeping his office in St. Louis.

General superintendent Robert Hale reported to the stockholders: "A perfect track is the basis of economy and is of the greatest importance," thus the twenty-three-mile section between East St. Louis and Alton became Chanute's first assignment. The line needed to be resurveyed and construction began in

May 1864, in the midst of the Civil War. Even though he encountered difficulties in procuring laborers, work advanced rapidly to Chanute's high standards.
Just before year's end, freight trains started to run and passenger trains began
service on New Year's Day 1865.[87] Two new freight houses and a passenger
depot with a covered platform were built in East St. Louis for the comfort of
the traveling public. Chanute's depot design was far ahead of that of other
railroads, and the company erected similar depots at principal stations along
the line in following years.[88]

   With the Alton to St. Louis section completed, Chanute moved his office to
Chicago and boarded at the Tremont House, the most popular hotel among
railroad men and politicians. This six-story hotel, located at the corner of Lake
and Dearborn Streets, offered luxuriously furnished rooms, with hot and cold
running water and seemingly perfect lighting and ventilation.[89] Coincidentally,
the manager and part-owner of the hotel, John B. Drake, also served as a director of the Chicago & Alton. This was the place to meet out-of-town directors,
prominent and influential businessmen, and railroad executives like James Joy,
now president of the Chicago, Burlington & Quincy Railroad.

   For Illinois and Chicago, the Civil War, with its movements of soldiers and
war materiel, provided profitable railroad business. In the words of historian
Elias Colbert: "Chicago became the paradise at once of workers and specula-
tors, and grew mightily. She prospered apace, while the red hand of war was
sweeping her (wayward) sister cities as with the besom of destruction."[90]

   Reliability of railroad traffic was crucial, but shortages of wood for fuel, ties,
and bridges caused serious problems for the Chicago & Alton. To ensure a
sufficient supply of hardwood, Chanute initiated the purchase of a tract of
timberland along the Illinois River, south of Wilmington. Frequently, breaking iron rails posed another concern. In 1855 Englishman Sir Henry Bessemer
had patented the "decarbonizing" of crude iron using a blast of air, and this
relatively inexpensive process for the mass-production of steel gained rapid
popularity in Europe. The first American-made Bessemer steel rails were rolled
in May 1865 at the North Chicago Rolling Mill along the North Branch of the
Chicago River. Unfortunately for Chanute and the Chicago & Alton, the process did not always produce a uniform product. The deterioration of rails could
be attributed to heavy traffic, but Chanute also thought that inferior quality
might present another reason for breakage.[91] Railroads at that time possessed
the financial resources to research the underlying causes, and Chanute took the
lead to improve the rail situation. Analytical investigations went hand in hand
with experimentation, as variations of minor components drastically altered the
quality of the rail.

   To conserve funding, Chanute considered reusing the worn rails by welding
a new piece to the head and, if needed, a new piece to the foot. Not finding
pertinent information in the available literature, and having heard of Alexander

Holley, who promoted Bessemer steel production in the United States, Chanute wrote him in November 1866. "In common with the rest of the profession, I found out some years ago that rails wore out, and in seeking a remedy, I am almost ashamed to say that I turned inventor. I do not want to make myself ridiculous by offering my patent before being well assured of their value; nor to make it known before they are perfected. I have determined to consult you on the subject, strictly confidential and ask for your professional opinion."[92] Holley replied by return mail, starting a lasting friendship.

The Union Rolling Mill Company erected its plant in 1863 at Ashland and 31st Street in Bridgeport, Illinois, fronting the Chicago & Alton tracks. The owners allowed Chanute to heat and reroll three old British rails, and the East St. Louis Rolling Mill rerolled a second batch. Even though the rails from these two initial experiments wore rapidly,[93] Chanute applied for his first patent, "Repairing and Rerolling Railway Rails."[94] The patent office granted the patent on January 22, 1867, but Chanute requested to place the document in their secret archives to delay publication;[95] neither he nor the Chicago & Alton wanted to publicize this invention.

Addressing rail breakage, Chanute submitted a short paper on "American Iron and Steel Railways" to the editor of the British *Railway Times* in April 1866. Three years later, he submitted revised papers to the *American Railway Times* and to *The Manufacturer and Builder*,[96] which were published early in 1870.

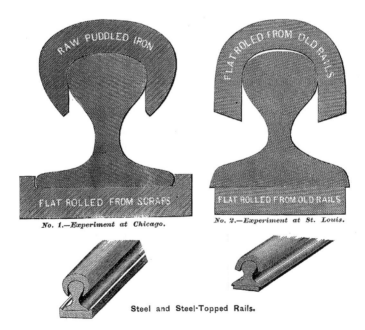

Steel and steel-topped rails, patented by Octave Chanute in 1867. From *The Manufacturer & Builder*, February 1870.

With the wartime growth of the railroad network, Chicago became the industrial center of the West. Iron ore came by rail from Michigan's Upper Peninsula or on boats from Lake Superior, and coal came from mines in central and southern Illinois. Many railroads brought other raw materials to town, fostering the rapid growth of manufacturing.

In January 1865, the 280-mile Chicago & Alton reported a 54 percent increase in net earnings over the previous year, paying a dividend to its stockholders.[97] Management also increased Chanute's annual salary to $2,700, effective January 1865. He knew that he helped make the railroad more profitable, and wrote to his friend Thomas Meyer in New York, "It would have been a little more satisfactory had it been done more promptly and for a little larger amount."[98]

The future looked safe and Chanute now wanted his family to live with him in Chicago. At first, Annie felt uncomfortable about leaving Peoria, but then agreed as long as their new home was close to a park where the children could play. Coming to Chicago to attend the Grand Ball, arranged by the employees of the Chicago & Alton late in January 1865, Annie liked what she saw, and the family moved to 454 West Adams Street, across from Jefferson Park. A few blocks to the north lay Lake Street, with the city's shopping district, easily reachable with an omnibus or horse-car. She could look into the windows of the different shops and stroll on sidewalks along the downtown streets, paved with wooden blocks; this offered an upgrade to the mostly unpaved roads of Peoria. Six years later, the Great Chicago Fire began a few blocks to the east, burning and smoldering for a week,[99] but by then the Chanutes lived in Kansas City.

To increase his income, Chanute became interested in the stock market. Educating himself, he tabulated the income and expenses of the railroads he knew and then invested $1,000 in Chicago & Alton stock. Considering further investment opportunities, Chanute heard of the recently located Vermillion Coal Basin near Pontiac, Illinois, adjacent to the Chicago & Alton tracks, and considered buying the mine. "But neither Mr. Blackstone, the President of the Road, nor Mr. Hale, the Superintendent, would take any interest in the mines, they were determined to keep aloof from any connection with an enterprise depending upon the operations of the Road for its success, nor did they permit any employee to be interested."[100] Morgan and Gardner then purchased the mine because Chanute did not want to jeopardize his job.

Always looking for improvements, Chanute noticed John Burnham's wooden water tank with an airtight, frost-proof liner that allowed water to be fed by gravity into the locomotive's tender without freezing during the winter months. This design provided a significant improvement over the old system, which required an attendant to keep a boiler heated to prevent the water in the wooden reservoirs from freezing during the cold Illinois winters. Because Burnham experienced difficulties patenting his design, he let Chanute use it at the Chicago & Alton, free of charge.

On March 4, 1865, President Lincoln took the oath of office for a second term, and crucial events followed quickly. The Confederates surrendered on April 9, and on April 14, John Wilkes Booth assassinated the president. Lincoln's funeral train traveled for thirteen days on a meandering route through 180 communities, retracing his political route to Washington. The train arrived in Chicago on May 1, and Chicagoans paid homage to the slain president at the courthouse, where his body lay in state.[101] The next evening, a torchlight procession escorted the remains to the Chicago & Alton depot, and the funeral train left for Springfield.

Security was high. The day prior, a special work train ran over the line and the crew inspected all the crossings and bridges carefully. That evening, workers manned telegraph stations, and flagmen assumed posts along the tracks; no other train was allowed on the line during the thirty minutes before or after the funeral train passed. The pilot train, with chief engineer Chanute onboard, went over the road ten minutes prior to the funeral train to verify that everything was safe. Officers of the Chicago & Alton traveled to Lincoln's final resting place in Pullman sleeping cars, attached to the funeral train.[102] As the train moved south, Chanute probably thought back to the times when he and Lincoln discussed river and railroad commerce, wondering how best to open the country. At the age of thirty-three, Octave Chanute undoubtedly felt heartened but humbled to be part of this sobering event in American history.

Later the next February, Mother Nature required Chanute's fullest attention. A heavy snowstorm with strong winds and frigid weather blocked all tracks south of Chicago for almost a week. Then the snow melted rapidly and the Kankakee River aroused anxiety in the citizens of Wilmington, Illinois, but also in Chanute. The ice backed the water to the chords of the railroad bridge, cutting several feet off the face of the pier, with one support chord hanging in midair. "Being continuous, the bridge did not fall, but railroad traffic was suspended. The current going under the bridge was about ten feet deep and running nine miles an hour."[103] Looking for a solution, Chanute inquired and received rubble stone from the prison quarry near Joliet. Prisoners loaded the cars during the day and unloaded them in the night, throwing the rocks into the fast-moving river and creating an island to support the bridge pier. Chanute wrote to his friend Meyer: "I had the satisfaction of succeeding and passing a train over the bridge in just twenty-four hours from the time I got started there."[104]

Union Stock Yards in Chicago

To keep up with the needs of the Civil War and the growing population, cities erected major slaughterhouses in Chicago, Cincinnati, and Louisville. In Chicago, the livestock business occurred in several yards in the Bridgeport area, and the South Branch of the Chicago River became a convenient dumping ground. Because citizens needed the water for consumption, the city council established a Board of Sewer Commission, then a Board of Health, and finally a Society for Cruelty to Animals to control the chaotic situation.

If Chicago was to become the distribution hub for moving meat to the East, consolidated stockyards were needed. Businessmen organized the Union Stock Yard and Transit Company in September 1864 to establish a single mart for the transaction of stock business and to stop drovers from driving cattle and hogs over crowded city streets from one yard to the other.[105] Timothy Blackstone, president of the Chicago & Alton, was elected president, and James Joy, president of the Chicago, Burlington & Quincy, became a director. Almost one million dollars were needed as start-up capital,[106] but the project received backing from nine railroads and members of the Chicago Pork Packers' Association. Boston investors then purchased 320 acres of swampland west of Halsted Street for $100,000, hoping that the housing boom would not reach the area for years. Because no one had designed consolidated yards in the past, the chief engineers of the railroads entering Chicago were asked to submit design proposals, with the winner then being appointed chief engineer of the project.

This project appeared intriguing. Knowing how to build railroads and drain swampland, Chanute was eager to enter the competition. He contacted the Van Nostrand publishing house and asked for literature on building stockyards. Familiar with the Chicago transportation system, he put the new information to good use and submitted his first competitive engineering proposal.

When the entries arrived, judges regarded the best solution as that of the Chicago & Alton's chief engineer, and appointed Chanute chief engineer of the yards in early May. He commented dryly: "This, in addition to my salary from the Chicago & Alton Road, will assist my income, but keep me dreadfully busy."[107] Chanute offered William H. Civer the resident engineer position and Russell Hough became the superintendent of construction.

Officials broke ground on June 1. The most important and possibly most challenging part was the draining of the marshy prairie, which skeptics claimed could not be done. Chanute ordered about thirty miles of drainpipes laid to drain the land into two discharge sewers, leading to the South Branch of the Chicago River. Once dry, he had the site laid out like a typical town.

Next, Chanute built a stone dam in the South Fork of the Chicago River and piped water from this reservoir for one mile into the yards. Because the site faced exposure to the vigorous Chicago climate, workers piped the water into six frost-proof water tanks for distribution to the animals. The patent office finally granted Burnham's patent after Chanute successfully clarified the value of this invention to the examiners; in gratitude, Burnham gave Chanute the design to Chanute as a gift to use at any future railroad and at the yards.[108]

Chanute's scheme then piped sewage back into the South Branch, but downstream of the reservoir, which the locals soon called "Bubbly Creek," owing to the decomposing animal remnants from the yards. Author Upton Sinclair later described this unflattering situation in his book *The Jungle*.

To increase efficiency in the yards, Chanute guided the railroads on a fifteen-mile-long elevated track around its perimeter. The yards were divided into four separate sections, with each railroad owning its own unloading area; a six-story hotel, a two-story exchange and an office building provided conveniences to the public doing business.[109]

The Union Stock Yards opened with great fanfare on December 26, 1865, just six months after construction began and seven months after the end of the Civil War. Railroads brought sightseers from Chicago, charging $ 0.20 for the thirty-minute ride. The first commercial livestock train, owned by the Chicago, Burlington & Quincy, arrived the next day. During the first week, nearly twenty thousand animals arrived to be sold and shipped east. President Blackstone reported to stockholders, "The successful operation of the Union Stock Depot, and the profits direct and indirect during the first two months after its completion, fully justified the most sanguine anticipations of its projectors."[110]

The cattle market and the meatpacking industry grew rapidly. In its first full year of operation, the yard handled 1,564,293 head of livestock, and within three years, the Chicago operations became the largest stockyards in the United States.[111] The packing industry claims that it implemented the first conveyor system, moving the meat to be processed past the worker, a principle of modern mass production.

Chanute had combined his talents in designing railroads and now stockyards to the benefit of farmers shipping and selling their cattle. The Chicago Union Stock Yards became a pride of the city and operated successfully for 106 years,

Chicago Union Stock Yards, as depicted in the Rufus Blanchard *Guide Map of Chicago* (1868).

closing in July 1971. The rugged limestone gate at the entrance to the yards at Exchange Avenue and Peoria Street is a visual reminder of Chicago's past supremacy in the livestock and meatpacking industries.

Making a Name for Himself and Building a Career

In the wake of the Civil War, Chanute grew in his career and broadened his knowledge base by working with engineers from related sciences, responding to their fresh input and establishing himself as one of the technological developers of the westward movement. The competition and subsequent construction of the Union Stock Yards, in addition to his regular job at the Chicago & Alton, proved stimulating, but to become better known in engineering circles, he needed to tackle more-challenging projects. A major bridge design competition in early 1865 provided the key to perhaps his most famous bridge project, two years later.

Octave Chanute
(circa 1865). Courtesy
of the Chanute Family.

Chicago-based railroads prospered during the Civil War, while the railroads in neighboring Missouri suffered destruction and little expansion. The North Missouri Railroad between St. Louis and the Iowa state line needed improvements, specifically a bridge at St. Charles across the Missouri River, about twenty-five miles west-northwest of St. Louis.[112] Directors advertised a design competition for a "First Class Railroad Bridge" in March 1865, offering $1,000 as prize.

To design such a bridge posed a challenge, because this would constitute the first railroad bridge across what many considered an unbridgeable river. Studying reports of previous bridge designs and combining this with his own experiences, Chanute submitted his proposal to the directors of the North Missouri Road in August 1865. It was an impressive-looking document, consisting of nine large drawings, a nineteen-page description, and a twenty-three-page cost estimate, bound in a book with the title embossed in gold.[113]

Chanute won third place for this, his first, competitive bridge design. Many years later, he discussed this competition with the editor of the *Railroad Gazette*:

> There is a reminiscence connected with the fact that Mr. Albert Fink received the second prize which illustrates how little was known of the art of bridge building by very well informed men, even as late as 1865. The North Missouri Railroad appointed a commission to award the premium. This consisted of two Engineer officers: General W. T. Sherman, General John Pope and a civilian, Thos. C. Fletcher, Governor of Missouri. These gentlemen were greatly puzzled how to decide on the merits of the fifteen or more designs submitted, and consumed a week in discussion. The award was finally put to rest by Gen. Sherman, who stated that when disabling Confederate Railroads, he had several times occasion to destroy bridges. This he did with artillery, and he remembered distinctly that upon coming to a Fink bridge, the first cannon ball struck the center post (cast iron) and the bridge tumbled down into ruin . . . thus making for that occasion, cannon balls the measure of merits of bridges.[114]

Charles Shaler Smith became the chief engineer for the St. Charles Bridge in April 1868, which opened for traffic in 1871 as the second bridge across the Missouri River; the bridge at Kansas City, erected by Chanute and finished in 1869, was the first (see "The Bridge at Kansas City" in chapter 3).

To open the western country to civilization, railroads needed to cross the mighty Mississippi River. Naturally, those interested in river navigation advocated high bridges, while those concerned in bridging the river with railroad tracks opposed them as impractical because of cost. To assure fairness with each interest group, Congress combined all requests for bridges across the Mississippi in the Omnibus Bridge Act, authorizing and regulating the construction of seven bridges (Quincy, Illinois; Burlington, Iowa; Prairie du Chien, Wisconsin; Keokuk, Iowa; Winona, Minnesota; Dubuque, Iowa; and St. Louis, Missouri). The general provisions stated that if built as high bridges, they should stand fifty feet above the extreme high-water mark (to clear steamboat chimneys), with spans not less than 250 feet in length, and one channel span not less than three hundred feet in length; if built as drawbridges, they should have two draw-openings of not less than 160 feet in the clear, with adjoining spans of not less than 250 feet, and should sit ten feet above high and thirty feet above low water. [115]

The story of building the bridge across the Mississippi at St. Louis provides some flavor of contemporary politics. The "St. Louis & Illinois Bridge Company" was organized in 1865 under the laws of Missouri, recognized by the Illinois legislature a few months later. In early 1867 James Eads became the chief engineer, and proposed a radical arch bridge, longer and higher than any in existence.

Interested Chicago businessmen, including Timothy Blackstone from the Chicago & Alton and Lucius Boomer of the American Bridge Company, chartered a competing "Illinois & St. Louis Bridge Company." The Chicago group then invited the country's most respected bridge, railroad, and hydraulic engineers to attend a "Bridge Convention" in August 1867, because several engineers considered the proposed design by James Eads impractical and unsafe. William McAlpine clarified the overall situation in his opening address: "We must give due attention to the relative importance of the navigation of the river and of the traffic which the bridge is to accommodate. The former is one of the most important inland water-channels of the world, and the latter is to form a connecting link in one of the most important lines of land transport across the continent." [116]

Chanute was a member of the Committee on Superstructure and Approaches; after lengthy discussion, its five members recommended using spans of less than 350 feet in the clear, if the federal law permitted a reduction in the overall length. Even though the twenty-seven prominent civil engineers met on four days, Chanute thought that the general question of bridging the great rivers could not be settled in such short sessions. Writing to Gouverneur K. Warren two years later, he agreed that the requirements of the river commerce should be fully guarded and preserved, but warned "it seems wasteful that bridge companies should be required to incur what may be needless expense, and sink capital in providing greater accommodation than is really needed. Long spans are expensive and hazardous, to require them without the clearest necessity might make the building of bridges onerous." [117]

After the convention, the two competing bridge companies consolidated under the name of the Illinois group. [118] But the health of Eads failed soon thereafter, and he asked Chanute to take over the St. Louis Bridge. This flattered him, but being in the middle of his own bridge project at Kansas City for a different set of investors and interest groups, Chanute replied cautiously: "While I should be happy to assist in carrying out your plans, I can make no fresh engagements before consulting my people who have sent me here all alone and have placed the entire management in my hands. Our bridge is not done, and although we have pretty well solved the difficulties of foundations, it will not be completed until next fall or winter. I shall consider your proposal as withdrawn until you make it again. I hope moreover that your health will improve, to enable you to retain the active charge of the work."[119] After recuperating for several months in Europe, Eads returned to St. Louis and finished the bridge project;[120] his two-level bridge opened with a gala celebration in 1874.

The bridge across the Mississippi at Quincy, Illinois, 130 miles to the northwest of St. Louis, was the first built according to the Omnibus Bridge Act.[121] James Joy had bought the thirteen-year-old Quincy Railroad Bridge Company charter to connect the Chicago, Burlington & Quincy, terminating on the eastern bank of the Mississippi, with the Hannibal & St. Joseph Railroad on the western bank; the same Boston investor group controlled both roads. Joy then advertised a design competition for a railway bridge between Quincy and Hannibal, Missouri.[122] Chanute entered his design, but Thomas C. Clarke submitted the winning plan for an iron bridge and Joy offered Clarke the chief engineer position in November 1866.[123] Over the next years, Chanute and Clarke collaborated to overcome the numerous difficulties on either man's bridge projects, with Clarke becoming the prime contractor for many of Chanute's later bridge commissions.

The Quincy Bridge opened for business in November 1868, with Chanute serving as a member of the engineering team.[124] Joy reported to the directors that this bridge had been built in the most substantial manner and "is a model of neat and beautiful works." At this point, Bostonians enjoyed an unbroken railroad line from the East, across the bridge at Quincy to either St. Joseph or Kansas City.

Westward Expansion

The post–Civil War years were critical for the growth of many communities in the West, and local politicians from Leavenworth, Kansas, and Kansas City competed for the first bridge across the Missouri River. Each group contacted James Joy, who then selected the latter city as his choice for the Bostonian's expansion scheme. In December 1866, Joy offered the project to build this bridge to the chief engineer of the Chicago & Alton, Octave Chanute. After discussing its pros and cons with his mentor, Henry Gardner, and the superintendent of the Chicago & Alton, Robert Hale, Chanute wrote to Joy[125]: "I have arranged so that I am free to accept your offer. As the season of less water should by no means be over, if it be desired to erect the bridge next year, I should propose to go at once to Kansas City, start the necessary surveys and accumulation of material and come back here to close up my work on this line, make my yearly report,[126] and leave all my business in good shape. Please advise me where and when I can meet you." A few days before his thirty-fifth birthday, Chanute resigned as chief engineer of the Chicago & Alton Railroad to build the first bridge across the Missouri River, and to open the route to the far West for the Boston Party and their expanding railroad empire.

CHAPTER 3

Opening the West

THE ENGLISH PHILOSOPHER Lord Francis Bacon wrote, "There be three things which make a nation great and prosperous; a fertile soil, busy workshops and easy conveyance of men and things from one place to another."[1] After successfully completing several hundred miles of railroads in Illinois, Octave Chanute began to realize his role as one to provide conveyance of men and things.

The settlement of the West made the need for railroads apparent, but building tracks required substantial funding. Congress granted public land for the aid of railroads, but railroad companies also looked for financial help from local communities, financiers, and overseas investors. John Forbes from Boston was one eastern capitalist who foresaw railroads as essential for the economic development and growth of the West. James Joy, a Detroit lawyer and a member of the investor group from Boston which was soon called the "Boston Party,"[2] managed and greatly influenced railroad competition west of Chicago. Many consider him one of the earliest professional railroad builders.

Building tracks required skilled railroading, because the route past the Mississippi River, the previously drawn boundary of the western frontier, was not an easy path. Nature supplied a variety of obstacles that required broad expertise. Even though bridges across rivers and valleys were primarily utilitarian, Chanute had educated himself on the aesthetic aspects of architecture, seeing a future for himself in building bridges, so he wanted them to look attractive.

To expand the Bostonians' railroad system into the as-yet-undeveloped trans-Mississippi area, Joy selected Chanute to design and build a lasting bridge across the unbridged Missouri River at Kansas City, his latest point of interest. Providing fresh talent in a challenging field, the confident civil engineer was slowly drawn deeper into the web of the "Boston Party," and some of the momentous episodes in his professional life resulted from his relationship with the Bostonians.

The Bridge at Kansas City

The offer to bridge the Missouri, the most difficult of all navigable streams, was a compliment for Chanute, but also a formidable challenge to his ambition as a civil engineer. Experts had described it to him as an unpredictable river, with bottomless quicksand,[3] but he accepted the challenge and told several friends that he would "take charge of an important bridge at a salary of $5,000 a year."

His wife Annie was at first concerned about a move to this unsophisticated steamboat-stopping place, because she had heard so much about the "Jayhawkers" and "Border Ruffians," quasi-military groups that looked for excuses to murder and plunder, as her father-in-law Joseph Chanut had described a decade earlier.[4] But her husband convinced her that authorities had restored law and order, and she reluctantly agreed. On his next trip to Kansas City, Chanute bought a colonial brick house with a large garden (in lieu of a park for the children to play) on a half city block for $5,000, overlooking the future bridge site, and the family moved west in late spring 1867. Annie gave birth to their fifth child, Charles Debonnaire Chanute, on October 13.

Even though many people considered civil engineers as pioneers in the frontier, they seldom identified themselves with frontier life. For them, the wilderness was not desirable until they had built the means by which the populace could enjoy articles of necessity and comfort. Perhaps illustrating this point and not wanting to miss their familiar necessities in Kansas City, Chanute ordered olive oil and stomach bitter from Chicago, delivered by the railroad. Having acquired a fondness for dark cigars, Octave ordered three thousand in bulk for $84, which reportedly lasted about a year, along with a white Riesling wine from northern Missouri that complemented evening meals.

In July 1866 Congress amended the Omnibus Bridge Act and added a clause to allow a bridge across the Missouri.[5] Two years earlier, the Missouri State Legislature had also authorized the Kansas City & Cameron Railroad to bridge the Missouri, so Joy acquired the unfinished, fifty-five-mile road with its bridge charter. Chanute then hired one hundred tracklayers, and the line opened for business in November 1867.

Needing good people for the bridge project, Chanute hired R. H. Temple from the Richmond & York River Railroad in Virginia as the principal assistant engineer, and W. C. Cranmer and C. H. Knickerbocker as assistant engineers. More important for the long term, he hired the ten-year-younger George Morison as his third assistant engineer. Morison had graduated from Harvard Law School in 1866, but quickly developed a dislike for practicing law; he explained several years later, "Not wishing to delay longer my entrance into active life, I determined to become a civil engineer, and to take such a place as I could find at once." He went west to work for Chanute. The two men soon became friends

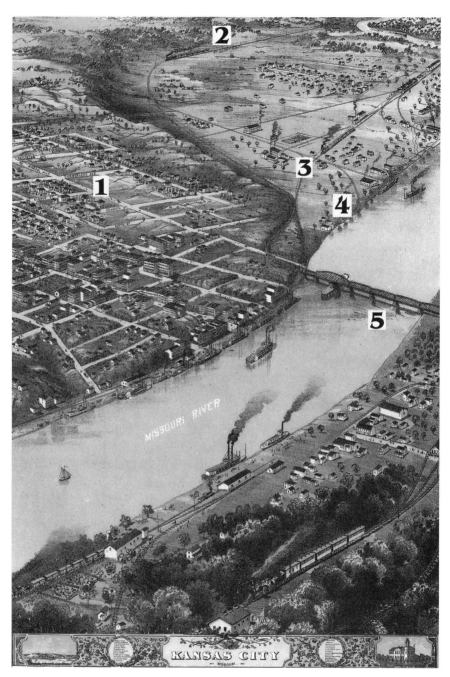

Bird's-eye view of Kansas City, Missouri (1869). (1) Chanute family home; (2) Missouri River, Fort Scott & Gulf Railroad; (3) Union Railroad depot; (4) machine shop of the bridge crew; (5) the Kansas City Bridge, designed by Chanute and opened on July 3, 1869. Perspective map (not to scale) drawn by A. Ruger, photographed by A. C. Christy. Map Division, Library of Congress, Washington, D.C. No. G4164.K2A3 1869 .R8 Rug 132.

and collaborated on many engineering projects in the next decade,[6] becoming known as builders of great bridges.

Joseph Tomlinson had left England in 1840 at age twenty-four and had built several bridges in Canada and the United States.[7] Hearing of the proposed bridge at Kansas City, he contacted Chanute and they corresponded on how best to bridge the Missouri. Chanute hired him in October 1867 as the superintendent of superstructure.

As work progressed, Chanute hired superintendents, foremen, laborers, a blacksmith, an auditor, and an accountant.[8] Because there was just one small foundry and machine shop in Kansas City, he bought a vacant building near the bridge site on the river and converted it to a shop for the crew. To streamline the work process Chanute designed special tools that required inventive thought.

His next-door neighbor, Howard Holden, living in the only other house on the same block, was cashier of the First National Bank of Kansas City, located at the corner of Delaware and Fourth Streets. He offered Chanute rental space on the second floor of his elegant bank so that the engineering staff could work without disturbance on foundations, piers, and superstructure.[9] The Holden and Chanute families quickly became friends and Annie appreciated seeing Mr. Holden in the evening with a lantern to check on her and the children when her husband was out of town.

River pilots, settlers, and members of the Wyandotte and Delaware Indian tribes still talked about the flood of 1844, when great volumes of water flooded the junction of the Kansas (or Kaw) River with the Missouri.[10] Chanute asked them for help locating high-water marks and systematically approached all difficulties. Quarrying stone for the bridge foundation began soon after his arrival, but there were no boats to move the stone to the construction site, so Chanute purchased the steamboat *Gypsey*, and the bridge crew built flatboats. To identify the currents in the river bend, he initially used flagmen, with transits on either shore; however this did not give reproducible results,[11] so he purchased a small boat, *Alice* (named after his oldest daughter), to make soundings in the river.

The Omnibus Bridge Act of 1866 dictated the bridge to stand ten feet above high and thirty feet above low water; the topography of the river with its banks confined the crossing. Because all railroads centered in the bottomland at the confluence of the two rivers, a high bridge crossing would have made it difficult to reach the grounds without steep grades, so Chanute adopted a drawbridge design for Kansas City.

Having studied the characteristics of wood and iron, he sent his bridge specifications to five prominent bridge companies, and reported to Joy: "I have put it into the shape of a circular to bridge builders; while specifying for the class of structure we should like to put up, I have endeavored to leave it loose enough to get the benefit of the builder's brain in saving money. We can always cheapen the work by using wood instead of iron."[12] Receiving nine sets of proposals, Cha-

nute then contracted in early November with the Keystone Bridge Company of
Pittsburgh. The Howe truss spans, with a curved upper chord, were to be built
for a fixed sum according to Chanute's plans, while the contractor would build
the draw as he proposed. The bridge company paid for iron by the pound and
timber by the cubic foot.

Combining style with engineering requirements, Chanute employed a double
triangular truss with the top chord, posts, and braces made of wood and wrought
iron; the details and connections were made from cast iron. He placed the
pivot pier in the center of the river channel, and numbered the piers from the
southern end of the bridge. The lengths of the spans were: a fixed span of 132
feet between the south shore and Pier No. 1; a pivot draw of 363 feet, with each
arm having a clear span of more than 160 feet; a fixed span of 250 feet from Pier
No. 3 to Pier No. 4. The remaining distance of 577 feet was at first divided into
three spans of equal length, but later changed to two 200-foot spans with straight
parallel chords, and one of 177 feet. The south shore span of 68 feet extended
over a street and the Pacific Railroad tracks, bringing the overall length of the
bridge to 1,400 feet.[13]

On August 21, 1867, the crew laid the cornerstone, with Masonic ceremonies,
before an estimated five thousand spectators. Chanute may have been a bit
optimistic when he assured the crowd that his crew would complete the bridge
construction within one year.[14] Immediately after the ceremony, Chanute rushed
to St. Louis to meet with other prominent civil engineers to discuss the merits
of the competing proposals regarding a bridge across the Mississippi at St. Louis
(see "Making a Name for Himself" in chapter 2).

Chanute had read of European engineers using wooden caissons to build their
piers and water-jets to remove loose material from confined areas; he used both
techniques on the Kansas City Bridge. Reviewing how others had laid bridge
foundations, he submitted his first scholarly paper to the widely read *Journal of
the Franklin Institute*.[15]

Pile driving began on the north bank early in 1868 but stopped soon thereafter because of high water levels. The unpredictable forces of wind, water,
and flowing river sand were new experiences for Chanute, who could almost
watch the rising spring waters wash away the south bank of the river bend. To
protect the channel approach to the draw span, he built an eight-thousand-foot
reinforcement of broken stone, or rip-rap, which, however, caused erosion of
the unprotected shoreline above and below.[16] Landowners then sued the Bridge
Company, requesting rip-rap along the complete riverfront.

Not wanting his crew to sit idle during the period of high water, Chanute
had them build the caissons and cribs. Because each pier site was unique, each
pier had to be designed differently. The three southerly piers rested directly on
rock, while the four northerly ones had pile foundations to overcome the effects of the flowing river sand. To position the timber caisson on solid bedrock,

passing through what seemed an undefined thickness of sand, gravel, slimy silt, and fast-flowing water, Chanute and Morison designed a "Dredging-Machine" consisting of four large dredges of the endless chain pattern. They later submitted the design for patent.[17]

The caisson work progressed steadily, with eight divers working in two shifts, day and night, removing boulders and debris from the riverbed as others lowered the caisson. Chanute harbored great concern about the safety of his men, because no one knew how much air pressure the workers needed inside their diving suits, and no one knew how much time the men needed to adjust when reentering normal atmosphere. At first, Chanute thought it necessary to have a physician in attendance "in order to relieve cases of asphyxia and treat the numerous instance of illness which may arise."[18] Fortunately, Kansas City had a hospital, but no accidents were reported from the divers or the men working with the caissons, passing through twelve feet of water and thirty-five feet of silt.

When James Eads, chief engineer of the bridge at St. Louis, returned from Europe in April 1869, he met with Chanute to discuss newly acquired knowledge. Eads then introduced pneumatic caissons as a novel feature in his pier construction, and Chanute's paper was undoubtedly a major source of information for Eads. The men working at his bridge in St. Louis encountered difficult working conditions, but Chanute felt distressed upon hearing in May 1870 that four of the fifty workers had died in one week of what became known as "caisson disease."

As expected, some river pilots navigating the Missouri were convinced that the Kansas City Bridge endangered their navigation. In October 1868, Major C. E. Suter of the Army Corps of Engineers and A. Bryson, in charge of river improvements, visited Kansas City to inspect the construction site for the secretary of war. They attached a letter, signed by eighty-two river pilots, to their report that "the Kansas City Bridge is not being built at an accessible point, as provided by the law, and greatly endangers navigation at all times, but especially during high water, when it is impassable with any degree of safety."[19] No follow-up was reported; ship owners may have realized the hopelessness of their opposition, as river traffic passed the construction site at all water levels.

Early in 1869, the bridge crew reached solid rock in the riverbed, and Chanute wrote with relief to Thomas Clarke, chief engineer of the Quincy Bridge: "I think we have conquered all difficulties of the bridge. All our piers are done, we are erecting superstructure this week."[20] But then the river fought back. Pier No. 4 was down sixteen feet through the silt when a sudden flood scoured one side and then toppled the pier.[21] To avoid future problems, Chanute decided to move the pier site by fifty feet.

Chanute had decided early on to build the substructure of ample dimension to accommodate increased weight of future traffic. To give the massive bridge an aesthetic look, he had the icebreaker piers above the waterline covered with a dense blue limestone, found in narrow veins nearby, but the remainder of the pier consisted of coarsely grained white or gray-colored stone.

While Chanute neared completion of his bridge across the Missouri, the first transcontinental railroad between San Francisco and Omaha opened in May 1869. Enterprising travelers could now journey from New York to San Francisco on various trains in six days, eight hours, assuming there were no unforeseen difficulties, and advertisements also claimed that one could now travel around the world in eighty-one days. Envisioning increased competition, Joy wanted his Kansas City Bridge finished more quickly. To reduce cost and speed construction, Chanute combined railroad and street traffic by laying wooden planks on the doublewide bridge; guards, stationed on either side of the swing span, could regulate river and railroad traffic and also collect tolls from street traffic.

On June 15, 1869, workers finished the southern approach to the bridge, and four men swung the draw for the first time, needing only two minutes ten seconds;[22] the northern approach was completed ten days later. To celebrate the upcoming opening of the first bridge across the Missouri River, the Kansas City Mayor mailed invitations to politicians, railroad officials, and other interested parties.[23]

The three engineers of the Kansas City Bridge: Joseph Tomlinson, Octave Chanute, and George Morison (July 1869). Note the ornamental details above the fixed spans. The guard in the background regulated river and road traffic and collected tolls from people crossing the bridge. Western Historical Manuscript Collection—Kansas City.

Mother Nature may not have felt at ease, however; after a few days of heavy rains, an earthquake rattled through the area on Friday, July 2, shaking doors and windows. But the sun was shining the next day for the engineering test, when two engines pulled two cars with a full load of one hundred tons of iron rails across the bridge. The *Kansas City Times* reported that "locomotives and trains passed backward and forward over the bridge and not a jar or a vibration was perceptible." The *Journal of the Franklin Institute* published Chanute's "Engineer's Test and Report,"[24] and other publications carried abstracts.

On Sunday, Independence Day, 1869, about 40,000 people came to Kansas City to celebrate. A procession formed on Wyandotte Street and continued over Broadway. Annie Chanute, the children, and their two servants watched the grand marshal and a corps of musicians ride by, followed by the mayor, members of the City Council, and chief engineer Chanute. Officers of local and eastern railroads, politicians, and reporters followed in coaches, while spectators and members of the bridge crew walked proudly behind. The throng proceeded to the bridge, which was decorated with patriotic bunting, and greeted with loud cheers the *Hannibal* locomotive as it crossed the bridge toward them from the north. The enthusiastic crowd then walked across and the celebration continued on the north side of the river. Several older citizens delivered speeches with typical frontier oratory and Chanute presented Joseph Tomlinson, the superintendent of superstructure, a gold watch in the name of all bridge company employees. As in the previous year, a balloon ascended in the afternoon from the levee near the Gillis House[25] amid the cheers of the crowd and the firing of a cannon.

More than 1,500 guests gathered in the evening at the Broadway Hotel for the Citizens' Banquet; Annie, wearing an elegant silk evening dress, joined her husband, the first man to erect a permanent structure across the Missouri. The hosts proposed many toasts, acknowledging everyone involved in the bridge project. In response to the toast to the chief engineer and his skill, Chanute stood and was greeted with waving handkerchiefs and enthusiastic applause. "With the modesty that always characterizes a true genius, he responded by taking no credit to himself, but said that he had been fortunate in securing faithful and skilled men,"[26] was the reporter's comment, published the next day. The celebration ended with spectacular fireworks late in the evening.

During the next week, newspapers around the country reported on the bridge opening with front-page headlines. The *New York Times* reprinted a report from a St. Joseph, Missouri, newspaper: "The inauguration today of the first railroad bridge over the Missouri is a triumph not for Kansas City alone; it demonstrates the practicality of the enterprise, it teaches the lesson that railroad systems may be independent of rivers." The write-up gave a lengthy account of the enterprise and closed with, "Mr. Chanute is the Engineer-in-Chief. The work is a splendid monument to his genius. It costs about $1,200,000, and has been two and a half years in building."[27]

Kansas Citians had shown how to celebrate the opening of a great bridge, with parades, speeches, a banquet, and fireworks. The citizens of St. Louis followed their example when the Illinois & St. Louis Bridge opened in 1874, and so did New Yorkers with their suspension bridge across the East River in 1883, colloquially called the "Brooklyn Bridge."

The structure of the Kansas City Bridge marked an epoch in bridge design, showing much of the romance and daring of the early development of America's West. Chanute and his staff had overcome the difficulties as they occurred, learning from each new experience. To share their acquired knowledge, Chanute and Morison coauthored a 140-page, oversized, cloth-bound book, *The Kansas City Bridge*, published by Van Nostrand. Chanute ordered 250 copies for the bridge company, including ten leather-bound books with fourteen photos of the construction.

Because many rivers were equally difficult to bridge, engineers from around the world were interested in all the pioneering details of the construction process. William Green was just one engineer who bought the book and used the information for his railway and road bridge across the Murray River at Echuca, in southeastern Australia.[28]

When the Western Society of Engineers (WSE) of Chicago celebrated its thirtieth anniversary in January 1900, Chanute shared a long-harbored secret: "Now that the Kansas City bridge is being mentioned it has brought to my mind the recollection of a confession which I have long wanted to make, but which through indolence I have postponed from time to time until I feared that I would pass off from this stage without having confessed the error—an error which shows how easy it is to blunder." Building the Kansas City Bridge, he was sure that all piers rested on rock. Seven years later, a diver examined the substructure and reported a three-foot-wide and four-foot-deep crevice under the pivot pier. The chief engineer at that time had a new caisson sunk around it. Twelve years later, a tornado damaged the superstructure; when examining the substructure, divers found the same caisson again undermined. "I instance this blunder and I make this confession simply so that we all may be on our guard against making similar mistakes."[29] Chanute had also learned that the only safe rule in dealing with this river was to never trust it and always expect that it will do the thing you do not want it to do, and then you will finally overcome all difficulties.

The completion of the bridge at Kansas City called for the construction of about four hundred miles of connecting roads, bringing urbanization to the Kansas frontier. The thirty-seven-year-old Chanute built this rail system and connected it with eastern railroads, bringing profit to both systems. During the first 230 days of operation, 5,263 locomotives had pulled their load across the bridge, and $5,706 had been collected in tolls from street traffic. The bridge with its connecting railroads was probably not a particularly novel idea, but allowing the north-south rail lines to cross the east-west lines gave the bridge a transfornational power, enabling conveniences and economies for modern commercial success. Historian Charles Glaab wrote that Chanute, having established himself as an engineer with a national reputation, had created for $1.2 million a bridge that stimulated the growth of a commercial emporium of the West.[30] The 1866 decision by Joy and the directors of the Chicago, Burlington & Quincy to extend across the Missouri River at Kansas City was just one event that contributed to the city's rapid growth. Had the bridge crossed elsewhere, Kansas City might have been just another town somewhere in the "undeveloped west."

The bridge at Kansas City survived earthquakes and floods. Workers demolished it in 1917 after the new Hannibal Bridge, crossing the river about two hundred feet upstream, opened for traffic.

Octave's Parents in France

While Octave Chanute worked on his career, his parents experienced unsettling times. Joseph, his father, had moved back to Paris in early 1851, but his estranged wife Elise was still not interested in sharing life with him. The fifty-five-year-old Joseph joined the Firmin-Didot Brothers Publishers as a staff writer, and his biographical research subsequently found publication in the *Nouvelle Biographie Générale*.[31] For additional income, Chanut became a professor at the *Académie de Dijon* in 1861.

In October 1854, almost four years after Joseph had returned to France, the forty-two-year-old Elise moved with her two younger sons, nineteen-year-old Leon and twenty-one-year-old Emile, to New Orleans. Leon died of yellow fever a year later, and Emile married a Creole belle, Emilie Fourchy, in 1859. Elise moved back to France in late 1862 to settle her mother's estate; still officially married, she needed legal authorization from Joseph to redeem her mother's investments. But the two were not on speaking terms. She tried to adjust, but the happy Parisian life, as she remembered it, was gone and the city seemed bleak to her. Feeling lonely, she put her belongings into storage and once more went back to New Orleans to live with her son Emile and his family.

Letters between father and son crossed the ocean several times a year, but each man was busy with his own life. Then Octave received a note that his seventy-three-year-old father had died on August 2, 1869, and had been buried in the Père Lachaise cemetery. When his friend Thomas Meyer wrote that he wanted to visit Joseph in Paris, Octave shared what little information he had: "The death of my father whom I had the misfortune to lose over a year ago, under such fanciful circumstances that it is only with great sorrow that I refer to it now. He was kicked by a horse in the street and this accident brought on a disease of the kidney, which caused him excruciating pains to which he succumbed in about two weeks, so that I had no chance to write. He had fortunately made the acquaintance of M. Peyronnet who replaced me in his last moments and who has since been good enough to arrange his affairs."[32] Years later, Jacques Peyronnet saw Chanute's name mentioned in Russian newspapers and wrote to ask for money. Thankful for Peyronnet helping his father thirty-five years earlier, Octave Chanute sent a draft of five hundred francs.

Hearing of her husband's death, Elise moved back to France a final time, settled his estate, cashed her mother's investments, bought a house in Ivry, and survived the brief Franco-Prussian war.

The "Border Tier Road"

A group of optimistic businessmen had chartered the Kansas & Neosho Valley Railroad in 1865 to run from Kansas City south through eastern Kansas and the Indian Territory (the future state of Oklahoma) toward Preston, Texas.[33] Within a few months, two other Kansas railroads also requested to lay tracks south through the Indian Territory. Naturally, the Indian tribes opposed the entrance of any railroad into their nation,[34] so the Interior Department declared that the first railroad to reach the border between Kansas and the Indian Territory in the valley of the Neosho River would be authorized to continue building its line south. A rail-laying race to the Indian Territory was about to begin.

In July 1866 the Cherokee Indians ceded their land in southeast Kansas, originally intended as "Neutral Land" between Indians and whites, off limits to both, to the federal government.[35] Secretary of the Interior James Harlan then sold the 800,000 acres of the ceded "Neutral Lands" to the American Emigrant Company just prior to leaving office, but the next secretary, Orville Browning, declared the sale illegal. Believing that any unoccupied land on the North American continent was open to preemption, white squatters began to move in.[36]

Meanwhile, James Joy had become interested in turning the Bostonians' railroad system from an east-west direction toward the south into a region with relatively quick access to the lucrative cattle market and little competition from other railroads.[37] Conveniently, on October 9, 1867, Secretary of the Interior Browning sold the former Cherokee Neutral Lands for $1 an acre to Joy, his brother-in-law. The Senate ratified the treaty in June 1868, over the vociferous protests of settlers, squatters, and local politicians.[38]

Even though all this politicking seemed somewhat removed, Annie and Octave, living in their comfortable Kansas City home with their young family, became concerned. Reading the local newspapers, and knowing Joy so well, Octave felt reasonably sure that the Bostonians' desire for expansion would soon become his next professional challenge. How right he was! In late July 1868, two members of the Boston Party, Joy and Nathaniel Thayer, bought the financially struggling Kansas & Neosho, placed themselves on its board, appointed Chanute to the chief engineer's position, and changed its name to Missouri River, Fort Scott & Gulf Railroad.[39] The locals soon called it the "Border Tier Road," because its tracks ran parallel to the Kansas-Missouri border.

Joy was aware of the railroad construction race to reach the border to the Indian Territory but at first concerned himself more with local competition. With railroad traffic centering in the bottomlands of Kansas City, Joy wanted a union depot. "The competition we can not escape, and I would rather have it done on the same ground, with proper regulations, than at half a dozen places."[40] Chanute accordingly bought forty-one acres adjacent to his machine shop near the bridge site and built a two-room depot.

While the bridge crew erected the piers across the Missouri River at Kansas City, Chanute began surveying the Border Tier Road to the south. Needing good people, he asked his mentor Henry Gardner if he knew of any young men interested in joining as assistant engineer at a pay of $100 to $125 per month. The company hired its former chief engineer, John Chapman, and John Runk as assistant engineers, and George Morison joined as a leveler.

In the meantime, Joy interviewed twenty-one-year-old Frank Firth from Boston and told him, "I will give you work in Kansas and you shall begin as engineers usually begin." Two weeks later, Firth walked up the bluff to Chanute's house in Kansas City and signed on at $2 a day, plus board. To introduce Firth to basic railroading, the two men walked twenty miles through the night over the graded roadbed, meeting the engineering party in Olathe the next morning. Chanute assigned Firth to assist Morison, and Firth quickly learned that "Mr. Chanute values discipline." He described his new life in a letter home: "We live in three tents, besides the kitchen, and have a springless wagon to jolt to and from work, and we move the 'city' from point to point. The officers are chief engineer, three assistant engineers, three non-professional assistant engineers, three axe-men, one cook, one teamster and one boy. There are a great many snakes here—no mosquitoes."[41]

After the Civil War, a hardy generation of settlers organized what became known as the "Granger" movement to protect themselves from some of the abuses of the railroads. Under pressure of the Grange, state legislatures enacted laws for the regulation of rates, called "Granger Railroad Laws." To avoid problems with the budding Grange organization, Chanute hired local contractors to supply ties and perform the grading; an estimated four hundred men worked on the first eighty miles south of Kansas City, receiving the relatively high wage of $4.50 a day.[42] In contrast, some Congressmen thought that $2.50 a day for the Union Pacific Railroad crew building the first transcontinental railroad was too high.

Agents for the Border Tier Road negotiated a right-of-way through Charles Bradshaw's land, and Chanute platted his first town in Kansas using his standard layout. Lenexa, now a suburb of Kansas City, went on record in August 1869.[43] Surveying and grading progressed steadily southward with a wood-burning locomotive from the Kansas Pacific Railway pulling the construction train over the well-ballasted roadbed, crossing streams on sturdy stone culverts and Howe trussed bridges, manufactured in Chicago.

In past projects, Chanute had learned that the expense of cuts and fills was justified even when earth had to be moved by pick, shovel, and wheelbarrow. As chief engineer he had to plan the route to minimize curves, because they were as deadly as steep grades to the efficiency of a locomotive hauling a string of cars. To bring the tracks through southern Johnson County, Kansas, into Spring Hill required considerable rock cutting, but the locals were not willing to supply funding for the construction. Responding to this situation, Chanute built the tracks one mile east of town and a station two miles north on land supplied by W. A. Ocheltree. He explained, "No place will prosper, unless we nurse it into life. The business of a Rail Road is to transport freight and passengers, not to coddle sickly towns. Hunt up trade and we will furnish accommodation."[44] Spring Hill citizens later gave $1,500 plus land for a sidetrack to connect their town with the railroad.

Farther south, the locating of the line presented several engineering challenges; the tracks had to cross the Marais des Cygnes River and Pottawatomie Creek, and long trestle approaches were required to cross the broad marsh of Goose Lake. This demanding construction effort would prove a good learning experience for Firth, so Chanute promoted him to Division Engineer, with a pay increase to $150 a month, and assigned this section to him.

In early July excessive rains flooded southeast Kansas[45] keeping Chanute fully occupied. The fast-flowing water washed out roadbeds, but also undermined the piers of Firth's first railroad bridge. The middle span collapsed on July 16, killing several workers. With input from Chanute, Firth rebuilt the bridge with two 140-foot-long Howe trussed spans and a deeper foundation.

Joy had approved the direction of the railroad, so Chanute acquired land for stations and towns. He wrote to a potential investor: "There will be some good speculation in land along the road which I can engage in, without interfering with the interest of the road and with the knowledge and consent of its proprietors."[46] Establishing a town site in the frontier typically marked the beginning of "civilization" for settlers, who usually appreciated Chanute's town layout, with its right-of-way and depot policies, with subtle changes tailored for each new community. Private parties could buy and sell land along the line but were required to give 50 percent of the proceeds to the road. Several investors had purchased land by quitclaim in the Mine Creek valley, and Chanute platted the future town site, recorded in the summer of 1869 as Pleasanton, Kansas. Newly arriving settlers—Chanute usually called them "immigrants"—opened a general store and a "commodious" hotel, and the town started to grow.

Needing water for his locomotives, Chanute installed Burnham's water tank at the next depot site, the future town of Osaga. Trying to turn over land quickly, Chanute wrote his friend Thomas Meyer about a potential investment opportunity in this new town: "Eighty acres were purchased for $1,600 and the rest being donations from the people to secure the Statute. One half of it is in the name of Mr. McDonald, a banker of Fort Scott, and the other half, now in my name, is designed for you. I do not suppose you will make your eternal fortune in this town, but I think a few hundreds, perhaps a thousand or two can be made out of it. You can take the full half upon the terms mentioned in my earlier letter, or I will retain a quarter and manage your interest for you."[47] There is no reference in Chanute's letters if Meyer accepted this offer.

Because the post office mixed up mail for Osaga and Osage Springs,[48] postal officials requested a name change. Local settlers suggested naming their town "Chanute." Octave did not approve, so they named it Fulton instead, which recorded a population of 185 citizens in 2000.

Joy had promised Fort Scott citizens that he would erect railroad shops in their town, but the former owner and now the general manager, Kersey Coates, unearthed an old contract stipulating that the first shops were to be built in his hometown, Kansas City. As the new owner, Joy simply told his chief engineer to ignore the old contract. Following orders, Chanute bought seventy-two acres in the northern part of Fort Scott and erected a brick depot, similar to the Chicago & Alton facilities, with stalls for eight engines and a blacksmith shop.[49]

Joy summed up progress in his report to the Chicago, Burlington & Quincy directors in June 1869: "The bridge at Kansas City, which is the south-western terminus of the Hannibal & St. Joseph Railroad, is now complete, and the running of regular trains is commencing. This opens an unbroken connection for trains from Chicago to all the Kansas roads. The Kansas Pacific runs directly west from Kansas City four hundred miles, and is soon to be opened to Denver. The Missouri River, Fort Scott & Gulf Railroad is nearly done to Fort Scott, and will soon be open to the southern border of the State, promising large increase to the traffic of this Company."[50]

Chanute's efforts prevailed. On November 25, the iron horse pulled into Fort Scott and "awakened the metropolis of Southern Kansas with its shrill notes, opening a new era of commerce."[51] A week later, the entire population celebrated in the Wilder House. Leading citizens proposed toasts to James Joy, the "Railroad King," and to "O. Chanute, the skillful and accomplished engineer. An enduring monument of whose genius now spans the broad waters of the Missouri, and which will descend to future times as a splendid triumph of engineering art and skill." Passenger train service began on December 9, 1869, taking six hours, thirty-five minutes to reach Kansas City.[52] Previously, the one hundred–mile trip by stagecoach took about twenty-four hours, and much longer during the rainy season when rivers had to be forded.

The Race to the Indian Territory:
Bleeding Kansas Erupts Again

South of Fort Scott, the Border Tier Road entered the former Cherokee Neutral Lands. Joy had transferred his land title to the directors of the Border Tier Road, who in December 1868 established a land office in Fort Scott. Settlers who had improved the land received legal notice to make entry of their land to prevent a sale to other purchasers. In an open letter, Joy wrote: "Nearly all the value which the newly established towns will have, will be given by the road, and it will cost us four or five million to build it. As we take all the risks, it will not be deemed unreasonable that we should retain for ourselves some of the advantages which the railroad will create in the country."[53]

To reach the border to the Indian Territory first, Joy now had to regard the construction race more seriously, as the Union Pacific Southern Branch, the later Missouri, Kansas & Texas Railway, or KATY, had made good progress with their 180-mile line from Emporia City, Kansas, toward the Neosho Valley. The third competitor, the Leavenworth, Lawrence & Galveston Railroad, had reached Ottawa, Kansas, but local bond problems had stopped construction.

Chanute hoped for few problems putting down tracks over the next sixty miles, but challenges soon arose. Fifteen years earlier, antislavery "Free-Soilers" and proslavery "Border Ruffians" fought for what they considered their rights in the Kansas Territory, turning the region into "Bleeding Kansas." Now the antislavery element was replaced by the antirailroad efforts of a squatter association. In true frontier style, these settlers organized the "Cherokee Neutral Land League," established a "death line" south of Fort Scott, and announced that any railroad man caught trespassing would receive severe treatment.[54]

Late in May, Kansas Governor James Harvey requested federal troops to control the hostilities south of Fort Scott. Much to Chanute's relief, the "Boys in Blue," four companies of the sixth U.S. Infantry, arrived in southeast Kansas on June 10, but the hostilities continued. Listening to the local politicians, Chanute wondered if they were the main troublemakers in the construction race. The Lawrence Journal reported Kansas Representative Clarke saying: "I do not advise violence to be used to prevent the construction of the Missouri River, Fort Scott & Gulf railroad through the lands occupied by settlers, but I do advise you to stand firm and united and no road can be built without your consent. Suppose the prairie grass should by some accident burn the ties and burn up wagons, tools and instruments of the engineers, could that be charged to you? I never did say that."[55] In the next few weeks, fires of unknown origin consumed about 26,000 railroad ties, and Leaguers attacked Chanute's crew twice, burning tools and tents, and stealing their two mules.[56]

To push construction ahead, the railroad company erected boarding shanties to house the workers and the troops. In mid-August, chief engineer Chanute wired a dispatch to the press: "The graders are following close upon our heels, and for their protection, I have them camp at the same place with the soldiery. We have not yet been molested, although, there is no question about the Leaguers intent, whenever they see a safe opportunity to make a raid, either upon my party or the laborers. For the purpose of having my men relieved from any undue apprehension from an attack by night, I have had a small area enclosed with a rude stone-wall banked up with earth, into which we can retire. It is hardly entitled to the name of 'Fort,' yet I think it would do us good service if our neighbors should be foolish enough to attempt driving us away."[57] Rightfully concerned about his own safety as well as that of his men, Chanute convinced Joy to hire private detectives for all of their protection. Returning home on weekends, there is little doubt that he had to assure his wife about his personal safety.

Then, related problems surfaced. Chanute's longtime mentor John Jervis had preached that the roadbed, usually the most neglected part in early railway construction, should be ballasted with broken stone about one to two feet deep to overcome the effect of lifting, owing to frost. Intimidated by the Leaguers, farmers were not willing to sell stone, so the roadbed was precariously short of this commodity. This upset Chanute, because he knew the consequences in terms of wear and tear.[58] Following the 1873 financial panic, management relearned the lesson, because the company lacked money to replace worn rails or roadbeds, resulting in derailments and slowed schedules.

Requiring immense supplies of ties and lumber for bridges in the construction process, Chanute wondered about future sources of hardwood as the road crossed sparsely forested eastern Kansas.[59] A recently passed Kansas legislative act encouraged the growing of timber, so Chanute suggested to the president of the Border Tier Road, Horatio Hunnewell, to plant a variety of trees in two 640-acre sections to determine the best and fastest growing. The company then collected a state-sponsored bounty and eventually supplied trees for telegraph poles and ties to all the prairie railroads.[60] These early concerns for natural resources could well have been the genesis for Chanute's timber preservation experiments and business a decade later.

Even though the troops provided limited protection, Leaguers continued harassing the railroad crew. "If the people in the Neutral Lands do not want the Railroad, we will not attempt to build it through a hostile population. Settlers may be left to go to market with their ox teams. If malicious men burn our property, we will wait till the people are ready for a road. What is right, just, and fair will prevail."[61] To circumvent the problems with the Leaguers, but still win the race to the Indian Territory, Joy acquired the competing Galveston Road (see "The 'Galveston Road'" later in this chapter) in June 1869 and told Chanute to move his men to build that line instead.

In the middle of February 1870, a legislative committee arrived in Fort Scott to verify if troops were still needed. The inspectors met the "Boys in Blue" at their quarters, established in a small valley below the 480-foot-long Drywood Bridge crossing the gorge 60 feet above them. "The appearance of such a bridge in a prairie country where there are no large streams and no timber in sight a moment before reaching it, has something novel about it, and we were more than pleased when our conductor gave us a chance to look around," commented one of the inspectors. On the return trip north "the train of two cars glided along at the rate of twenty miles an hour, seemingly as smooth as on the old roads of the east."[62] Inspectors reported that troops were still needed, and they remained in the area until 1873.

Eventually, Bostonians needed to decide on the future direction of the line south of Columbus. The original charter stipulated the road to run into the Neosho Valley to cross the border to the Indian Territory at Chetopa, but the citizens of Baxter Springs, about twenty miles to the east on the north-south military road into the Indian Territory, also wanted the tracks to come their way.

With the construction race heating up, Bostonians realized that their Galveston Road could not reach the Kansas border in a timely fashion, so they withdrew the line from the race in mid-March. Joy then instructed Chanute to transfer his men back to the Border Tier Road and to build it quickly into the Neosho Valley. Not wanting to work under the threat of the Leaguers, most workers quit and joined the competing KATY, making it difficult for Chanute to meet Joy's latest orders.

Hearing of Joy's decision to continue building the tracks in the original direction to Chetopa, unhappy Baxterites threatened to pull the surveyor's stakes unless the tracks pointed toward Baxter, but a more rational citizens committee met with Chanute and then Joy.[63] Next, a Baxter delegation traveled to Boston and offered the Bostonians $150,000 in bonds, or twice what Chetopa at the legal crossing point into the Indian Territory had promised, plus 1,100 acres of good land. They succeeded and Chanute received a note from Joy: "The conclusion of the Board is to build the Road by way of Baxter Springs, and all the forces must be at once transferred to that line and the work pushed with all powers. Do this instantly."[64]

Chanute felt sure that taking the railroad to Baxter into the Spring River valley would not allow them to win the construction race, so he proposed to build an extension to Baxter, while the main road would go into the Neosho Valley, but Joy was not interested. On April 2, the Baxter newspaper reported that "Col. Chanute of the M. R. F. S. & G. R. R. was in Baxter and located the depot, roundhouse and other R. R. Buildings on the J. S. Barnes addition. Work will commence next Monday."[65] The admiring and at times sarcastic public frequently assigned military titles to the engineers, which "Col. Chanute" did not appreciate.

Only a couple of weeks later, a most serious problem occurred. Fifteen armed Leaguers met Chanute's engineering party as they returned from the south and held a "squatter's court." The Leaguers burned the wagons, tents, blankets, surveying instruments, commissary stores, and so forth, and drove the subordinates away, with orders never to return in the employ of the railroad company under penalty of death. They then marched the two engineers Chapman and Runk several miles south, stripped off their coats, blindfolded them, and administered fifteen lashes to each man. They then ordered the two men to never return and never mention what had occurred.[66] The gang leader told the engineers that only one vote had saved them from hanging. Chapman and Runk managed their way to the railroad tracks and telegraphed Chanute, who rushed south with several armed guards accompanying the train.

Late in April, the locomotive of the Border Tier Road puffed into Columbus, and locals received free rides to Fort Scott the following week. Without interference from the Leaguers, Chanute established an office for himself and the railroad, but when John W. Davis tried to open a Land Office, Leaguers gave him notice—with pistols cocked—to leave town or face lynching.

Under the watchful eyes of the military, the construction of the Border Tier Road now progressed steadily toward Baxter Springs, but then heavy rains caused more work stoppages. Because there was uncertainty about the actual location of the border between Kansas and the Indian Territory, Chanute finalized the survey. One story reports that a spy from the competing KATY Railroad, dressed as an Indian, tried to mislead Chanute by showing him a pile of stones that he declared to be the official corner marker of the border.[67] Another story stated that an old settler showed the marker of the original 1837 survey, which he claimed to be the true starting point of the Cherokee Neutral Lands. These two markers were a few miles apart. What made this borderline confusing was the fact that the McCoy Survey Line of 1837 showed the parallel 36°30' north latitude as the southern boundary of the Kansas Territory. However, the map of Kansas approved by Congress in the Kansas-Nebraska Act of May 1854 showed the 37° parallel as its southern boundary.[68]

The construction train arrived in Baxter Springs on April 27, and the iron band reached the border to the Indian Territory on May 1, 1870, about 160 miles south of Kansas City. For the two thousand Baxterites, Thursday, May 12, was the day to celebrate the arrival of the railroad, and speakers representing both sides of the settler issue stirred the pulse of the crowd;[69] neither Joy nor Chanute attended the party.

The KATY railroad crew worked with the same goal as Chanute, to win the construction race to the Indian Territory. For their final twenty-four-mile stretch,

CHIEF ENGINEER'S OFFICE.
O. CHANUTE, Chief Engineer.

MISSOURI RIVER, FORT SCOTT & GULF RAILROAD,
LEAVENWORTH, LAWRENCE & GALVESTON RAILROAD.
Columbus, Kansas City, Mo., April 22d 1870

The Missouri River, Fort Scott & Gulf Railroad, Chief Engineer Chanute's letterhead, which needed to be manually corrected after moving his office to Columbus, Kansas, in early 1870. Chanute Papers, Manuscript Division, Library of Congress, Washington, D.C.

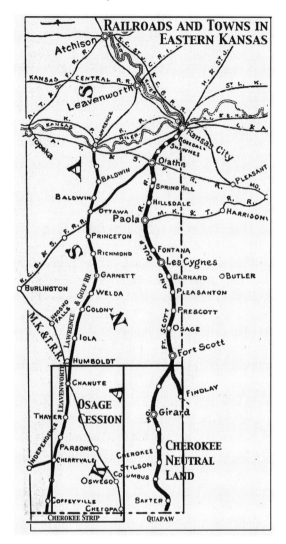

Railroad construction race to the Indian Territory. Railroads and towns in eastern Kansas. From *Annual Report, Kansas City, Fort Scott & Gulf Railroad* (1880).

Superintendent Robert Stevens ordered the rails laid without grading,[70] and the KATY's tracks reached the legal crossing point in the Neosho Valley one month after Chanute's Border Tier Road reached the border in the Spring River valley. Both railroads applied to the new secretary of the interior, General Jacob D. Cox, to continue laying tracks south.[71] Joy felt confident that his paid lobbyists would arrange a deal, even though the railhead at Baxter Springs lay twenty miles east of the legal crossing point.

On July 20 President Ulysses S. Grant announced that the KATY had the sole right to construct a line through the Indian Territory, "as this road had arrived first on the southern border of Kansas, designated by law as the point where one railroad was authorized to enter the Indian Territory."[72] The disappointed Chanute wrote to the ticket agent at the Hannibal & St. Joseph: "While I do not expect our Kansas Roads to repeat the splendorous success of the Illinois lines, I think those in the Joy interest will be worth all they have cost. The only question is whether the present owners will have the patience to wait."[73]

With construction completed, Chanute submitted his final report to the board:

> In accordance with the instructions received from the President of the Company, the Road was built as a first-class Road, and the wisdom of this policy is already apparent in the cheapness of its maintenance. The track is laid with fifty-seven pound rails, thoroughly fished at the joints and laid on about 2700 ties per mile, mainly of white and burr oak and black walnut. The culverts are of stone. The truss bridges are all of Howe truss pattern. The station arrangements are made ample to accommodate the present and prospective business of the line. Neat and substantial depot buildings were erected at most stations, and the Road is already better provided than is usual with new Western lines. The Water Stations, placed twenty miles apart, each consist of one of Burnham's frost-proof tanks, with a capacity of 50,000 gallons, and are all supplied from living streams or ponds.[74]

Ironically, the land grants, as promised by Congress, did not materialize, because the land belonged to the Indians and profit from freight and passenger traffic also failed to develop as envisioned. Perhaps Bostonians had foreseen a railroad terminating in Baxter to be more profitable than one extending through the Indian Territory. Built for just over six million dollars, the Border Tier Road was the best-paying line in Kansas during its first year of operation, with earnings exceeding $5,000 per mile.[75] Having huge cattle yards at its terminus, the rapidly growing Texas cattle trade grew even faster. Additionally, the railroad transported metal ores and coal from southeast Kansas as fast as it could be mined.

Long after Chanute had left eastern Kansas, the Missouri River, Fort Scott & Gulf Railroad further extended into Missouri and was later incorporated into the St. Louis & San Francisco Railroad.

Kansas City & Santa Fe Railroad

James Joy was an entrepreneur who manipulated the destiny of many miles of railroad, making decisions instantly and frequently consulting no one. Historian Craig Miner concluded that Joy trusted few people in the railroad business, but chief among his confidants was his engineer, Octave Chanute, whom Miner described as "a brilliant self-taught engineer who later doubled as liaison officer for the Joy interest. And he enjoyed the confidence of Joy's Boston backers."[76] Naturally, Joy wanted all railroads coming into Kansas City to pass over his new bridge, once it was finished. To accomplish this goal, Joy needed a thirty-one-mile feeder road to connect the Leavenworth, Lawrence & Galveston Railroad at Ottawa, Kansas, with the Missouri River, Fort Scott & Gulf Railroad at Olathe, allowing traffic from southern Kansas to flow directly into Kansas City for further distribution.

The Kansas City & Santa Fe Railroad and Telegraph Company received its charter in March 1868, with the financial support of local communities. Joy simply told his manager and chief engineer Chanute to build this line, meshing this with his other job assignments. From the hub in Kansas City, the North Missouri Railroad then moved goods to St. Louis, while the Hannibal & St. Joseph handled traffic toward Chicago and the east; all these were "Joy Roads."

The "Galveston Road"

The Leavenworth, Lawrence & Fort Gibson Railroad received its charter in February 1858. Senator James Lane acquired it in 1864 and changed the name to Leavenworth, Lawrence & Galveston Railroad, colloquially called the "Galveston Railroad." When Congress authorized the railroad to go through the Indian Territory, provided it reached the state line first, Joy showed his interest in purchasing it, but insisted on receiving every alternate odd-numbered section of land for ten miles on either side of its tracks. The commissioner of the General Land Office denied the claim, but Joy's brother-in-law, Secretary Browning, ruled that the road was entitled to the lands. Joy then acquired the Galveston Road to circumvent the problems with the Leaguers on the Border Tier Road and told Chanute to reassign his crew, so he could still win the construction race with the rights to reach the Gulf of Mexico.

At that time, Kansas railroad companies were both land and transportation companies. As land companies, they wanted to sell the granted land at the highest possible price with the least possible expense. As transportation companies, they sought to have the adjacent land developed speedily by settlers, who would provide traffic to increase the value of the land and the railroad company.

Taking his first trip over the newly acquired road in August 1869, Chanute reported to Joy that the line suffered from poor construction, lack of motive power, and no station houses or shops to make repairs to any equipment. As the first order of business, he rebuilt the twenty-seven-mile roadbed between Ottawa and Lawrence and ordered material for a bridge across the Marais des Cygnes River at Ottawa. To move supplies across the Kaw River at Lawrence, he first used flatboats but then erected a temporary bridge in November 1869.

Initially, Lawrence city fathers had agreed to supply land and funding for a shop, but Chanute considered Ottawa a better choice.[77] Ottawa, founded in 1864, had a population of four thousand and claimed to be the largest city of its age in Kansas. Its citizens voted overwhelmingly to aid, offering $60,000 in city bonds, and also supplying forty-two acres for a railroad shop.[78] Chanute

Bird's-eye view of Ottawa, the largest city of its age in Kansas. Howe Truss bridge across the Maria des Cygnes River, typical for other bridges erected by Chanute in Kansas and Nebraska, and the Leavenworth, Lawrence & Galveston Railroad shops with depot, water tank with windmill, and roundhouse, erected in 1871. Drawn by E. S. Glover. Map Division, Library of Congress, Washington, D.C. No. G4204.O8A3 1872 .G6.

then designed and erected a roundhouse with sidetracks, water tank, and station, which soon became a pride of Ottawa citizens. The agreement arranged by Chanute between the railroad and the city bound the railroad to preserve shops in town, and the succeeding Santa Fe had to make a substantial payment to terminate that contract in the 1940s.[79]

With the Galveston Road construction underway, Morison joined Chanute as the assistant engineer. Princeton, nine miles south of Ottawa, was the next depot site that soon became a shipping center for the district.[80] The railhead then pushed through unbroken wilderness to the southern part of Franklin County, and Chanute platted Richmond Station as the terminus for the year.[81]

A news release, most likely from Chanute, optimistically informed the public that crews would finish the Galveston Road to the southern border of Kansas by late 1870.[82] About five hundred men worked in the sparsely settled, treeless plains, where springs were scarce but well water was available at a depth of about twenty feet. Reading the *Journal of the Franklin Institute* regularly, Chanute saw an earlier article titled "Aerometry" in which the translator also discussed windmills and mentioned John Smeaton's paper on the natural powers of wind to turn mills. Chanute ordered a copy of Smeaton's article and studied how to efficiently position the rotating blades or "sails," and that "the center of gravity of each sail should travel in its own circle with the velocity of the wind."[83] He shared this information with John Burnham, who then designed a windmill to lift water out of the ground and into his water tank. Workers installed the first such tank, plus windmill, at Garnett, Kansas. Two years later, a windstorm "blew the windmills to atoms,"[84] destroying all equipment between Garnett and Ottawa but doing little damage farther south. Chanute again tried to understand the forces of the wind.

In March 1870 the competing KATY reached Humboldt, while the Galveston Road crew continued the construction about twenty miles to the north. Accord-

John Burnham's frost-proof water tank, with attached windmill. From Alfred R. Wolff, *The Windmill as a Prime Mover* (1885).

ingly, the Bostonians withdrew it from the construction race and Joy resigned as president. James M. Walker, a director of the Chicago, Burlington & Quincy, became the new president. Walker was a careful student of railroad affairs, always soliciting the views of his associates,[85] including Chanute, to whom he looked for a good working relationship.

With congressional authorization and with the Border Tier Road completed, Chanute redirected the Galveston Road later in 1870, to bring it closer to the head of the lucrative Texas cattle trail. The town of Welda on the gently rolling prairie became the next depot site, and newly arriving settlers named one of the roads Chanute Avenue. Arriving in the Neosho Valley, the tracks crossed Iola, the Allen County seat. The road entered Humboldt from the north on the east side of the Neosho River, while the KATY had arrived on the west bank several weeks earlier. A few miles farther south, three communities formed the later town of Chanute, Kansas (see the following section), and the tracks then entered the ceded land of the Osage Indians.

Continuing as an airline, Chanute established Cherryvale about fifteen miles south of Thayer, the previous year's terminus, and ten miles east of Independence, the new county seat. The citizens of Independence also wanted a railroad, so Chanute suggested they hire a contractor for the grading, build a trestle bridge across the Verdigris River, and supply ties for the ten-mile branch. The Galveston Road would then furnish rails, lay the track, and operate the line upon surrender of the city bonds.[86] Workers completed the feeder line two years later, as recommended by Chanute.

The Galveston Road, with its telegraph line, reached Coffeyville, 150 miles south of Lawrence, in September 1871. Earlier-arriving settlers had established Parker as a hoped-for terminus, but Chanute had acquired land toward the northwest and platted Coffeyville around the trading post of James Coffey. He erected corrals for cattle, driven up from Texas, a two-story banking facility, and a luxurious hotel, where guests could watch the activities in the stockyards from the lobby.

Years later, Chanute's oldest daughter Alice recalled her father taking her by train in the early 1870s from Lawrence south into the wilderness, even though the federal government permitted no railroad man to cross the border into the Indian Territory. As the Galveston Road crew worked on a preliminary and unauthorized survey, she and her father followed in shaky wagons, fording streams and often waiting for the men to clear the underbrush from their path. The twenty-one-year-old lady never forgot this trip.

The Leavenworth, Lawrence & Galveston Railroad, built to Chanute's high standards, was ready for big business that was slow coming.[87] Initially, the railroad only supplied settlers with manufactured goods, while locals consumed the produce grown by farmers along the line. The road was later incorporated into the Atchison, Topeka & Santa Fe Railway.

Leavenworth, Lawrence & Galveston Railroad letterhead (1872). Chanute Papers, Manuscript Division, Library of Congress, Washington, D.C.

Chanute, Kansas

Settlers began populating the northern part of Neosho County, Kansas, as early as 1856. A decade later, the news traveled fast that two railroad lines would create a junction point somewhere south of Humboldt. With "town-talk" rampant, settlers thought this an opportune time to establish villages along the future right-of-way, hoping to profit from the boom that inevitably would come with the opening of the new rail line.

The KATY Railroad arrived in April and the Galveston Road arrived in June 1870. Local settlers platted New Chicago between the tracks of the two roads and recorded the town on June 11 with hopes it would grow like its namesake. Within a few weeks there were fifty buildings and a two-story hotel either completed or under construction. By the end of the year, the community bustled and the KATY erected a depot and station, but lightning burned it soon thereafter. Tioga, recorded one day earlier, sat 1.5 miles northwest of New Chicago, and the Galveston Road listed it in their timetable as a station with a land office.[88] A third town, Chicago Junction, was platted on July 9 and settlers platted a fourth town in March 1872, recorded as Alliance. Naturally, each village wanted to be the center of business and political rivalry developed, with the two railroads encouraging jealousy as part of their own competition.

Late in 1872, Tioga and New Chicago city fathers approached Chanute, now general superintendent of the Galveston Road, to request a new depot. Chanute suggested, "Why do you not consolidate your towns and pull all together? Try it and then let us talk about the depot. There surely must be some equitable basis upon which you can jointly prosper." To stay one step ahead, Chanute contacted John Scott, in charge of the Railroad Land Office. Trying to refute the unfortunate belief that all railroad men were schemers, he wrote: "While I do not wish to ask you to remove against your own desires, I may say that I so much admire concord that I should like to promote it, if within our power. I should like to have your office moved, we would probably put a platform in front of it and have the passenger train stop there, so that the people could get on and off if desired. More than this, I should not like to promise without consulting Mr. Walker."[89]

The City Councils of the four communities then voted to consolidate; a new town, Chanute, Kansas, came into being on January 1, 1873,[90] honoring Octave Chanute's input and foresight. The Galveston Road moved their land office to the railroad junction and built a new depot, convenient for everyone. Combining New Chicago, Tioga, Chicago Junction, and Alliance, with a total population of eight hundred, brought new order and prosperity to all its citizens. Commenting on the honor of naming the town after him, Chanute wrote in 1907 to the Friends of the Chanute Library, "This honor, neither sought for nor expected, was a most valued mark of esteem."[91]

Moving Up in Management

With railroad construction coming to a close in southeast Kansas and foreseeing turbulent financial times, Chanute considered a career change and a move to Colorado or California. "The mountains are strewn with the remaining wrecks of capital, squandered for want of intelligent and skilled direction, while all the real outfits are doing work. The engineer can choose to work in some reduction or smelting works at a fair salary or about a furnace or mine in case he prefers the more useful metals."[92] During an eight-day trip to the Denver area, he obtained a fair glimpse of the country but decided not to shift into mining.

Looking for new challenges, particularly in management, Chanute considered an offer to become the general superintendent of the Kansas Pacific Railway, but declined, fearing that railroads were already built ahead of the means to make them profitable. Bostonians then promoted Chanute to the general superintendent position of the Galveston Road, but he should continue, as chief engineer, to bring the line to its terminus at the Kansas-Indian Territory border. The *Kansas City Journal of Commerce* reported that "Mr. Chanute has been called to surrender his interest in the work now nearly completed. We can only hope that his success in the future may be commensurate with his indefatigable zeal in the past."[93] In his new position, he handled the affairs of three railroad companies in southeast Kansas and a road from Atchison, Kansas, north into Nebraska, in addition to the construction of stockyards in Kansas City. He admitted to Walker, "I find I have to work about 14 hours a day in order to close up my old work and get the hang of the new one."[94]

Compiling his report for the directors of the Galveston Road in 1871 and noting its low earnings, Chanute explained to Walker: "We want to make the people living along the road feel that it is their road, built with their aid for their benefit, and it is in their interest to work with us to make it a success. Our charges for fares and freights are high, and yet the Road does not pay. I want to fill up the country and make it rich and prosperous, to let us reduce charges."[95]

On February 4, 1871, Nina Octavia, Chanute's sixth and last child, was born in Kansas City. After Nina's birth, Annie "was very near dying. For a month she was confined to bed and only within the last few days has been able to come out of her room. You can imagine how anxious I have been and how thankful I now feel that she has been preserved to me. My wife and her baby (it is a girl) are now spending much time in the open air and stay in the garden until late in the evening,"[96] Chanute wrote to his former coworker Joseph Tomlinson. Early in April, he requested passes for his wife and children to visit her family in Peoria for a few months.

As the new general superintendent, Chanute established an office at Massachusetts Avenue in Lawrence, Kansas. He liked the town, with its tree-lined streets, and bought the home of George Schweitzer for $8,000, just south of the Galveston Railroad depot. With Annie feeling better, the family moved to Lawrence in November 1871. To make the house, with its hearth and fireplace between the parlor and library, comfortable for family and guests, Annie selected a custom-made carpet, a status symbol for civil engineers living in the frontier. The Chanutes listed their Kansas City house for sale but then rented it to Edwin Bowen,[97] who had accepted the position at the Kansas Pacific that Chanute had turned down a few months earlier. Five years later, Chanute worked for Bowen when both held executive positions with the Erie Railway; the growing railroad industry was indeed a small world.

Locomotive engineers at the Kansas Pacific Railroad joined the "Brotherhood of Locomotive Engineers," a soon-to-be powerful railroad union. To retain good workers on his road, Chanute considered pecuniary inducements: "The locomotive engineer's post is one of dangers, and on their skill, judgment and fidelity, the safety of the public largely depends. As a class, they thoroughly appreciate the importance of the trusts that are confided to them, and not only do they uncomplainingly endure hardships, exposure and necessary overwork, but they have furnished many examples of self-devotion."[98] Walker and Joy agreed that the best performers should receive company-owned land or a bonus.

When his ticket agent needed to stay home to take care of his ailing child, Chanute stepped in and attended their convention at St. Louis, joining officials of other western railways to discuss the evils of free passes. James Walker reportedly once said, "I think the grant of a free pass to make one friend creates half a dozen enemies, but the friends were frequently very valuable." During the discussion, everyone agreed that only the president, or his representative, should authorize the issuance of free passes for business purposes only, while reciprocal annual passes should be given to the officers of the connecting lines.[99] This is what Chanute had advocated, but during this "Gilded Age," free passes were largely used for certain individuals to open doors to valuable favors. This railroad evil continued to plague the industry until in 1906 Congress finally passed a law prohibiting the issuing of free passes to anyone not connected with the operation of the railroad.

A frequently occurring problem consisted of cattle crossing roadbeds and being killed or injured; this created a loss to the farmers, suffering to the stricken animal, and danger to the traveling public. The Kansas legislature had provided an act in 1867 "that any person planting an osage orange within ten years from the passage of this act, and successfully growing and cultivating the same, shall receive an annual bounty of two dollars for every forty rods so planted and cultivated; the bounty to commence as soon as said fence shall be declared a lawful fence, and continue for eight years thereafter." Trying to maintain good relations with farmers and minimizing cattle suffering in advance of any legal problems,[100] Chanute encouraged farmers along the railroad to plant osage hedges adjacent to the right-of-way. The railroad paid $.35 per rod for planting and trimming the hedge, and the farmer could collect the additional bounty from the state.

In May 1871 the Galveston Road, with its depot south of the Kaw River at Lawrence, and the Kansas Pacific, with a depot on the north side, announced that they would erect a railroad bridge across the river "co-jointly." Chanute built this bridge with four 150-foot Howe truss spans on wooden piers at a cost of $70,000;[101] it opened for traffic in March 1872.

Foreseeing trouble ahead, Chanute wrote to one of the directors of the Kansas City Stock Yards, Charles F. Adams: "A Rail Road is so complicated a scheme. The main difficulty is that we have built the Roads in advance of the necessary population to make it pay from the beginning. This, time will cure, but we must build up the country to make it rich and prosperous, before we can look for any dividends. I am convinced that we built too many Rail Roads in the United States for the past three years, that a reaction is inevitable and that the sooner it comes the better it will be for existing interests."[102]

Atchison & Nebraska Railroad

When the state of Nebraska announced land grants to anyone constructing railroads in their state, the Atchison & Nebraska Railroad was chartered to run between Atchison, Kansas, and Lincoln, Nebraska, a sparsely populated but rich grain-growing region. Lacking capital and experience, the original investors approached James Joy to take over the unfinished railroad.

Building ahead of traffic was unprofitable, but to wait too long invited encroachment by rivals. "I know of nothing which in my opinion would be so unwise as for us to fold our hands and allow rival enterprises to occupy our domains, when we can supply the want and keep them out," Joy explained to John Green, a director of the Chicago, Burlington & Quincy.[103] In late 1870 the Bostonians took over the Atchison & Nebraska and placed themselves on the board, a common pattern in western railroad projects. Peter T. Abell remained president and Chanute was appointed general superintendent and chief engineer; Morison became the resident engineer and Frank Firth the assistant engineer.

Chanute visited Atchison several times to determine a location for a possible bridge across the Missouri. Traffic, however, amounted to only ten cars a day, thus the existing ferry service seemed satisfactory. In a letter to an inquiring stockholder, Chanute explained that traffic over the Kansas City Bridge averaged about 250 cars a day and that "we have 1,010 miles of rail road terminating in this city, against 140 miles at Atchison."[104]

On January 12, 1871, the crew finished grading the roadbed to the Kansas-Nebraska border, six weeks ahead of Joy's schedule. Early in March, the iron horse snorted from Atchison north to the state line. "We cannot hope to make it profitable until it is extended into the Nemaha Valley and I hope you will soon make arrangements that one shall have more than thirty-five miles to operate,"[105] Chanute wrote to Joy, reporting a total profit of only $180 in the first month.

The Atchison, Lincoln & Columbus Railroad was incorporated in April 1871 to extend from the Kansas-Nebraska state line through the Nemaha Valley to Columbus, Nebraska.[106] The incorporators were, among others, Chanute, Morison, and Andrew Cropsey, the former agent of the Eastern Extension in Illinois. Next, Joy acquired the Burlington & Southwestern and consolidated the three roads into the Atchison & Nebraska. Bostonians then ceremoniously laid the first rail into Nebraska in May 1871. One month later, Morison moved on to become chief engineer of the Detroit, Eel River & Illinois Railroad, so Chanute put Firth in charge of building the road and stations into Lincoln, Nebraska.

The Atchison & Nebraska reached Falls City, Nebraska, in July. Citizens had supplied land for the station and Chanute designed a thirty by sixty–foot building with a freight room on the west side, a neat little ticket office on the southeast corner, a ladies' waiting room on the northeast, and a gentlemen's waiting room in the center. There was a twelve-foot-wide by one hundred–foot-long platform along the front side of the building for everyone's convenience. About three hundred yards east of the depot sat a turntable, and another mile farther east was a water tank with a windmill for pumping water.[107]

The rapidity of laying tracks had been remarkable, and Chanute wrote to Firth, "I congratulate you upon your arrival at Falls City. You had better put on your trains at once and push beyond as fast as you can."[108] Even though the construction crew experienced an occasional shortage of iron, work progressed rapidly, because managers assigned all contracts to local teams interested in finishing the railroad quickly. The road reached Humboldt, Nebraska, about twenty miles west of Falls City, in early September and Table Rock in November, when winter suspended operations earlier than anticipated.

Building the railroad was not without its tragedies, some of which affected Chanute in his managerial position more than others. After a heavy rainstorm in June 1872, the twenty-five-year old Firth and his assistant, O. E. Allen, inspected a recently repaired bridge, sitting on the front end of the locomotive under the headlight for a better view. The locomotive had not gone on the bridge more than fifteen feet, when it and the tender plunged to the bottom of the ravine, while the passenger cars became disconnected and remained safely on the track. Both men were buried under the engine in the water.[109] The accident instantly killed Allen, and Firth died a few days later. Chanute took on the difficult assignment of notifying the families.

The 160-mile railroad from Atchison reached Lincoln, the capital of Nebraska, on September 1, 1872. Southeastern Nebraska now lay open for immigrants, and earlier settlers could become commercial farmers. But Chanute remained concerned and confided his gloom to Morison: "I am having an absorbing and ungrateful task in the endeavor to make a success of our Road. Emigration to Kansas has ceased and the people are being panicked by the collapse and exhaustion of the means they brought with them."[110] In 1888 the Atchison & Nebraska consolidated with the Chicago, Burlington & Quincy Railroad.[111]

Kansas City Union Stock Yards

Starting in early 1866, several Bostonians had acquired large shareholdings in the West Bottoms Land Company on the Kansas-Missouri state line, where a person could set one foot on "bleeding Kansas" and the other on "progressive Missouri." Seven railroads brought livestock into Kansas City for handling in separate yards in the West Bottoms, creating a chaotic situation similar to that of Chicago in the 1860s. Chanute had some ideas and telegraphed Joy: "When would it be convenient for you to have me call upon you. I wish to receive your instructions about the cattle yard and other matters. I go to Lincoln tonight and can come up to Detroit from there."[112]

To erect efficient Union Stock Yards, Chanute bought land in the spring of 1871 as a trustee and Joy supplied another thirty-five acres.[113] The articles of incorporation for the yards were executed, with James Walker appointed president and Chanute vice president for the first year. A stock certificate book ordered in Chicago later burned in the October 1871 fire, but the printer delivered a new book as soon as he was operational again.

Chanute designed the Kansas City Stock Yards to be as efficient as the Chicago yards, meshing the supervision of its construction with his other railroad duties. To him, it was of utmost importance to give absolute satisfaction to the railroads as well as the drovers. Each railroad received about five hundred feet of platform for receiving and shipping livestock, and the yards featured a "Drover's Paradise," an exchange restaurant with a barroom, a hay barn, a stable, and a wagon stall. Chanute also installed the recently introduced Fairbanks livestock scales with which workers could weigh a complete herd on the hoof, reducing labor costs and expediting the transfer of living freight. Pens had grass or sand covers, while others were floored and roofed to shelter the animals against the sun. Total cost to Boston financiers ran about $80,000. One of the partners of the Armour, Plankinton Meat-Packing Company from Chicago became the first tenant in the Kansas City yards, and Chanute made a "little" profit[114] selling them land for $200 per acre, with a right-of-way to the railroad tracks and a private switch.

During the first year of the stockyards' operation, about 20,000 head of cattle came north on the Border Tier Road alone for further distribution to the East. Business increased steadily, and by 1900 the Kansas City Union Stock Yards were the second busiest in the nation, next to Chicago, also designed by Chanute. A flood devastated the area in 1951 and the yards, with their associated businesses, did not recover; the gates closed in 1974, ending Kansas City's history as a cow town.

Expanding and Growing His Engineering Network

Living and working with men of the roughest sort on the frontier, Chanute looked for interaction with progressive-minded professionals. He joined the American Association for the Advancement of Science (AAAS) in 1868 and the Kansas Academy of Science in 1871. On a trip to Chicago in July 1869 to discuss strategies with Joy, Chanute attended a meeting of the just-formed Civil Engineers Club of the Northwest (later known as Western Society of Engineers). It was a good gathering, as most chief engineers of the railroads entering Chicago and other public works attended. Chanute joined as a charter member.

The American Society of Civil Engineers (ASCE) reorganized in 1867. Chanute's friends Meyer, Gardner, and Jervis signed his nomination for membership and he joined in February 1868. The 4th Annual ASCE Convention met in June 1872 in Chicago, the first such gathering outside the metropolitan walls of New York. Almost eighty members attended, including Chanute and many of his friends: Henry Gardner, now chief engineer for the Michigan Central; Matthias Forney, editor and part-owner of the *Railroad Gazette*, from New York City; William Shinn, the ticket agent from the Pittsburgh & Fort Wayne, now living in Pittsburgh; and Thomas Clarke, who had joined Samuel Reeves in a bridge design company in Philadelphia. Most out-of-town engineers boarded at the recently renovated Tremont House. With a general concern about the volatile railroad situation, there were plenty of opportunities to exchange insider information on railroad projects and the general financial outlook.

During that convention, Chanute joined the committees "On Publications" and "On Constitution"; he wanted meeting notices and abstracts of papers published in the engineering press, but he also wanted student memberships and local chapters. To become better known and more involved, he paid his fellowship fee of $250 and became an ASCE Fellow in July 1872. At the same time, James Joy became interested in joining the ASCE, but he could not become a member because he had not been "actively employed in the practice of the profession for five years, and had not been in charge of some work in the capacity of Superintending Engineer." So Joy paid his fellowship fee and members elected him a Fellow later in 1872.

In what little spare time he had, Chanute worked as city engineer for Kansas City. The municipality paid him in city warrants, which were not as desirable as cash, but he continued to learn, became better known, and could reach out to a network of other engineers. A horse railroad was just one project for which he provided a survey, and he submitted a charter of a Chicago company as guide.

In an effort to utilize the prevailing winds, Chanute gave entrepreneurial thoughts to using this readily available energy. Local farmers produced a surplus of grain and corn, thus flouring windmills would prove cost-effective for them. Chanute contacted the owner of the Wind Engine & Pump Company in Batavia, Illinois, where John Burnham was the general sales agent, hoping to receive and set up one mill on a trial basis.[115]

Navigating the water-depleted Missouri by steamboat during the late summer months was difficult, so Kansas Citians considered the concept of a shallow-draft barge line as an economical alternative. Chanute's neighbor Howard Holden, now the treasurer of the Kansas City Board of Trade, asked several engineers for their professional opinions whether grain could be shipped profitably by barge to St. Louis and beyond. Comparing the barge operation with the efficiency of the railroads,[116] Chanute determined that a barge line could be cost effective, but improvements to the riverbed should come first.

Kansas City needed waterworks for its growing population, a frequently discussed topic in the early 1870s. Chanute submitted his proposal on the construction cost to the Board of Trade,[117] and the city then contracted with the National Water Works Company of New York. Former associates Kersey Coates and Howard Holden were elected president and secretary-treasurer of the new waterworks that began operation in 1875, two years after Chanute had left Kansas.

Moving East and Up

For Kansas Citians, the late 1860s and early 1870s brought much progress, largely owing to the efforts of Chanute, who led several projects simultaneously. Starting in November 1867, rail traffic went north via the Kansas City & Cameron to connect with the Hannibal & St. Joseph Road. In July 1869 the bridge across the Missouri River opened. The Missouri River, Fort Scott & Gulf Railroad was completed to Baxter Springs in May 1871. The Kansas City & Santa Fe opened under a lease agreement with the Leavenworth, Lawrence & Galveston Railroad in August 1870. In September 1871, the Galveston Road reached Coffeyville, Kansas, and the Union Stock Yards opened late in 1871.

Working in the West was a rigorous learning process, but laying railroad tracks into unsettled regions also provided opportunities to learn about the social and economic impact of the expansion of America's frontier. In 1872, the total rail-road network consisted of more than 66,000 miles, and much of the nation lay within the sound of a locomotive whistle. But the public wondered why money was being wasted on running lines of iron through barren and unpopulated sections of the West. Having experienced the lack of income-generating com-merce on newly established railroads, Chanute agreed that some were indeed questionable investments.

Even though "Ev'rythin's up to date in Kansas City," Chanute wanted to leave the precarious West and become chief engineer of an eastern railroad. The fi-nancial aspects of the competing railroads offered intrigue, but he preferred to work in an engineering field with a future he thought he could control. Being concerned about the situation in the West, Chanute wrote to Charles F. Adams: "I suppose in Massachusetts, you could prevent the operation of a Railroad from becoming an enormous instrument of favoritism and corruption. I should have no such hope in the West, filled with adverse adventurers and broken down politicians. It would be too good an opportunity to plunder . . . Besides, what the people of the country need, is absolute fairness at all points in the management of the Railroad whether such points can be reached by competition or not."[118]

While in Chicago in June 1872, Chanute also met Hugh Riddle, vice president of the Chicago & Rock Island, who hoped to move up to its presidency; knowing Chanute and his capabilities, he offered him the general superintendent and chief engineer position. Looking for input Chanute then discussed the oppor-tunity with his president, James Walker: "As construction has ceased, you can undoubtedly secure a Superintendent for our Road who will do quite as well as myself, for less money than is being paid to me. To avoid all misunderstand-ings, I desire that this shall in no event be considered as an indirect method of getting an increase in salary. I believe however that our success in Kansas will be neither rapid nor easy. Loaded with debts as our Roads are, it may be many years before they become profitable while the hostility of the settlers threatens to take the very kernel from the nut. It would be more pleasant to myself to operate a road from Chicago than from Lawrence."[119] Even though Riddle's offer sounded interesting, Chanute did not accept and kept his eyes open.

Early in March 1873, having just passed his forty-first birthday, Chanute resigned from the Bostonians' enterprise and the Leavenworth, Lawrence & Galveston Railroad. He had accepted the chief engineer position with one of the largest eastern trunk lines, the Erie Railway, with an office in New York City. The editor of the *Railroad Gazette*, Matthias Forney, added his personal comments after the official announcement: "The Erie is to be congratulated on its good fortune."[120]

CHAPTER 4

At the Top

IN THE POST–CIVIL WAR YEARS, land grant railways were frequently not as profitable as their promoters had hoped. "We have built a good many more miles of railroad than the country will support for some years and many weak concerns must go to the well. In fact, I look for a magnificent smash at no distant day, when the investing public will awake from its folly, and the years 1869, 1870 & 1871 will be remembered as the English remember 1844–1850, the time of the Railway Mania,"[1] Chanute wrote to a director of the Leavenworth, Lawrence & Galveston Road in early 1873.

Perhaps to insulate himself from a "magnificent smash," particularly in the West, the thought of moving up into management at an established eastern railroad appealed to Chanute. Four major eastern trunk lines connected the populous East Coast with the West, and the Erie Railway was one of them. In the late 1860s, three financiers struggled for its control; Jay Gould eventually became the Erie's president and one of the most powerful robber barons in American railroad history. Directors ousted Gould in the spring of 1872 and charged him with stealing more than $10 million from the Erie's coffers. James Joy, going through difficult times with his allies in Boston, was the rumored choice to become the Erie's next president. Instead, directors elected Peter H. Watson, a patent attorney and a director of both the Lake Shore & Michigan Southern Railroad and the Standard Oil Company, as the Erie's new president in July 1872.

To take the Erie out of its embarrassment, the road had to spend less and earn more money, so Watson initiated radical changes in its upper management, administration, and operation.[2] To make the Erie profitable again required an extensive general overhaul: the line needed to be double tracked and possibly narrowed, it needed more efficient depots for passengers, modernized terminal facilities for handling freight, renewal of its antiquated equipment, and it needed extensions into Boston and Chicago. To accomplish all these changes required an estimated $50 million. British stockholders still had a good deal of confidence in the company, so Watson went to England to obtain funding.[3]

The Erie System. From *Erie Railway Timetable* (1887).

Besides money, the Erie also needed good people, especially a competent chief engineer informed on the latest technological developments and possessing the skills to organize and carry out projects efficiently. Watson wanted a professional problem-solver who could work under political and economic pressure. Through his network of railroading friends, Chanute had heard of Watson's improvement plans and showed his interest. He felt sure that joining the Erie would pose a challenge but that he would also learn more about the industry.

While the daily press discussed the frauds of Jay Gould, Erie management offered the chief engineer position to Chanute in late February 1873,[4] and he accepted the challenge. In mid-March he traveled 1,300 miles over three days on three different railroads from Kansas City to New York, traversing "Indiana, Ohio, Pennsylvania and New Jersey like a flash, rushing through towns with antique names, some of which had streets and car-tracks, but as yet no houses," as Jules Verne had described in his latest novel *Around the World in Eighty Days*. Perhaps this quote from Chanute's favorite author provided a poignant transition, leaving his old job of building a lofty rail system in the West and starting a new job rejuvenating a rail giant in the East.

Climbing the Corporate Ladder at the Erie

Arriving in New York full of energy and fresh ideas, Chanute was anxious to help make the Erie a profitable trunk line again. His first day on the new job was Monday, March 17, 1873. "My headquarters are at the Grand Opera House, the Erie office building, at the Corner of 23rd Street and 8th Avenue, and I room at 315 23rd Street. My family will still be in the West, where I propose to leave them until I have satisfied myself as to the result of my experiment in coming East,"[5] he wrote to his friend Thomas Clarke, now a partner of Clarke, Reeves & Company, a bridge building firm.

Some of Chanute's close friends wondered why he wanted to work for a railroad loaded with internal and external problems. When his mentor Henry Gardner inquired, Chanute replied:

> I came upon the Erie, because I felt it a matter of duty to the people of the West to improve their access to the seaboard. The plans of the Erie's managers are vast and generous. I see a large opportunity of being useful, coupled with so many elements of failure, extraneous to myself, that I doubt whether I can accomplish even a part of what seems desirable. The former management of the Erie Railroad was simply abominable, and plundered it in the most extraordinary way. The present management is entirely honest, but so beset with difficulties that it may prove anything but permanent. Meanwhile the Road has more business forced upon it than it can possibly do, and has so completely outgrown its conveniences that everything is being done at much greater cost than it should . . . I will try pretty hard to accomplish it.[6]

The well-groomed, forty-one-year-old Chanute may not have fully realized the political and economic realities of his move east. At the Erie, the chief engineer's office did not carry the authority and prestige he had experienced in the West. Here the superintendents, division general managers, and the chief engineer reported to the vice presidents, and only the vice presidents discussed business with the executive council and the president. So, Chanute started two steps below the president.

The slightly older William Pitt Shearman, husband of Annie's recently deceased cousin Mary James, had joined the Erie two months earlier as its corporate treasurer. The Erie's second vice president, George Blanchard, had worked with Chanute on the Ohio & Mississippi Railroad in 1863. They now collaborated in estimating resources and future traffic to beat the competition and increase the road's revenues. Because the Erie's connection between Buffalo and Philadelphia was shorter than any other road, Chanute proposed straightening the line to further cut operating costs. He also suggested resurveying several extensions to shorten them as well. The piers along the west bank of the Hudson River, leased from the city of New York, required much attention, and Chanute shared his knowledge of concrete work with the city engineer Charles Graham, but he also recommended painting the oak fenders above the waterline with "dead oil" to retard decay.

Some former coworkers followed Chanute to New York, even though he remained unconvinced if this was a prudent decision. Hiram W. Diggins, his trainmaster from Ottawa, Kansas, became the Erie's superintendent of second track work, but life in the East proved not to his liking, and he returned to Kansas late in 1874. George Morison had finished his engineering project in Detroit, so Chanute hired him in June 1873 as resident engineer, Eastern Division, but quickly promoted him to first assistant engineer to work directly with him. Morison had become interested in building bridges, and both engineers knew that all the wooden bridges on the Erie were deteriorating and should be replaced in iron.

Because bridges were a major part of the Erie's improvement scheme, Chanute needed a uniform basis for guiding bridge builders in submitting proposals so that he could compare bids realistically. After discussing this idea with his friend Thomas Clarke, Chanute and Morison then collaborated in writing specifications for new iron bridges of different lengths,[7] adopting specifications from Clarke's company. The Erie thus became the first railroad company with published bridge design guidelines. The two engineers also introduced rules for individual builders, incorporating their ideas as prominent bridge engineers.[8] Five years earlier, Chanute had let the contract for the Kansas City Bridge by buying wood and iron by weight; he introduced this scheme at the Erie, making the price of the material the chief determinant of cost. This practice of bridge contracting soon became widely accepted in the United States.[9]

Next, Erie management approved a drafting design office. Looking for a capable civil engineer to check bridge plans and calculate stress and strain in bridge members, a practice not generally carried out by railroads at that time, Chanute hired the thirty-four-year-old Charles Schneider later in 1873.[10]

Only half a year after joining the Erie, the business crisis that Chanute foresaw two years earlier began as a major economic reversal in Europe and reached the United States in the fall of 1873. The signal event in North America was the failure of Jay Cooke & Company, the country's preeminent investment banking concern, which had invested heavily in the second transcontinental railroad. Cooke's fall touched off a series of events that eventually encompassed the entire nation. In the aftermath, a bitter antagonism developed between workers and the leaders of railroad and manufacturing enterprises.

Dictated by the economy, the Erie cut wages and crews in March 1874, and the more than one thousand workers at the Susquehanna, Pennsylvania, depot went on strike. Erie officials restored order by paying the strikers and then dismissing them; Chanute was not convinced that the company should discharge the workers, because they needed money to feed their families. The Susquehanna depot reopened a few weeks later, and the Erie rehired everyone but the strike leaders. But labor tensions erupted again and again in the decade to come.

To put the Erie on equal footing with its eastern rivals, Erie management considered extending the line into Chicago and St. Louis, possibly even to the Pacific Coast. Chanute executed a preliminary survey and traveled in October 1873 with Erie officials to Chicago to identify a terminus.[11] Prior to any expansion however, the Erie had to narrow its gauge to four feet, and eight and one-half inches, the standard width used by most North American roads.

Coming from eastern Kansas, considered by New Yorkers as the "Wild West," Chanute's life in the big city took some adjusting, but he soon felt comfortable in his new position and moved his family from Peoria to a rented house in New York. Their elder son, seventeen-year-old Artie, entered the Irving Institute, a boarding school for teenagers, specializing in natural sciences, and Chanute became a patron of the school. Their oldest daughter, sixteen-year-old Alice, attended a boarding school for young ladies, at which the French language was part of the teaching curriculum. Both schools were located in Tarrytown, about twenty-five miles north of New York. A few years later, the two younger daughters also attended boarding schools in Tarrytown. Because Annie enjoyed spending the summer months with friends at Manitou Springs, Colorado, and Octave frequently traveled, the couple felt little need to own a house in New York City.

The Elements of Freight Cost

Settling into his new job, Chanute regularly attended meetings of the American Society of Civil Engineers (ASCE); its members were the "who's who" of the engineering profession. Becoming better acquainted with these individuals gave Chanute opportunities for interaction with like-minded professionals and personal improvement.

The general public was interested in cheap transportation, so Chanute researched the true cost of railroad freight traffic in his standard systematic way. The Erie had to beat the competition, but its rates needed to be fair to the public and reasonably profitable for the railroad. To determine the actual trends of costs and earnings, he gathered statistical data on seven New York state railroads and realized that the published tariffs were seldom the ones charged; rates and fees fluctuated greatly and were reduced for individual shippers whenever such an action was necessary to prevent traffic from going to a rival route. He also discovered that each road interpreted and published its data differently. "The first objection to a tariff, fixed by the State, is that the cost of transporting a ton a mile varies from 30% to 40% with the season on the same road. How, then, can an unvarying charge for all the months be a fair one? The second objection is the fact that the cost of transportation varies greatly on the different railways." Chanute concluded that the true cost was made up of seven general elements: roadway charges, general expenses, station service, track repairs, car service, train service, and insurance. But managers also needed to consider the indirect costs, such as the building of the road, its gradients related to the cost of fuel, wages paid, speed and volume of traffic, and finally, the competition.

Without a uniform baseline, Chanute wondered how any state legislature could determine the true cost for shipping, and summed up his research: "There is no term of comparison more fallacious to apply to individual cases or particular shipments, than the average cost per ton per mile." Tariffs, he felt, could not be uniform for different roads or for different circumstances of traffic; they had to be comparatively lower for long distances than for shorter ones.

Chanute read his paper "Elements of Cost of Railway Freight Traffic"[12] at the February 1874 ASCE meeting and distributed reprints widely. Other engineers analyzed this controversial topic concurrently, but Chanute's paper, with its comprehensive statistics, was the most thorough inquiry of the time, so railroad managers and state commissioners consulted it freely. When Congress needed to approve an appropriations bill for the army, they sought Chanute's opinion on what percentage of the railroad's tariff rates would constitute a fair charge for the government's use of roadbeds, switches, and depots belonging to private railroads. In his response, Chanute submitted a table with the actual trackage cost and earnings of several railroads, and concluded: "I am of the opinion that the percentage sought for will vary from 12 to 30 per cent of the 'tariff rates' upon newer road, such as the land grant railways, but that it should be ascertained separately for each road from time to time as the cost of maintenance and tariff rates vary."[13]

Traveling to Chicago on Erie business, Chanute again met his longtime friend Henry Gardner, who now worked for James Joy as chief engineer of the Michigan Central. Gardner shared his thoughts on transportation cost, but he also recalled an episode, attributed to Joy. The Baltimore & Ohio (B & O) wanted to establish the shortest line among its competitors from the east to Chicago. To achieve this, their crew had to lay its rails across the Michigan Central tracks in northwestern Indiana. Fearing competition, Joy ordered Gardner to prevent the building of the B & O tracks—one way or another. Soon thereafter, the Porter County sheriff arrested Gardner and several of his men and charged them as rioters for trying to disrupt the B & O construction efforts. Fearing prosecution, Gardner asked to meet with B & O officials and explained that "he only obeyed Mr. Joy's orders and could not have acted otherwise."[14] Hearing tales such as this must have confirmed Chanute's decision to leave Joy and to work for the Erie.

Only half a year later, the almost sixty-year-old Gardner died in Chicago. Chanute knew that he would sorely miss his mentor's wise input and thoughtfulness.

A New Phase at the Erie

Even though the Erie had a large traffic volume and income, financial problems continued because its antiquated procedures and equipment limited the road's profitability. Watson resigned from the Erie's presidency in June 1874 and stockholders elected Hugh Jewett as the new president under a ten-year contract. To familiarize himself with the state of affairs, Jewett asked his management team, including Chief Engineer Chanute, to join him on an inspection tour to determine how to increase traffic and reduce operating costs. Jewett then initiated another reorganization in September 1874 and hired the former superintendent of the Kansas Pacific, Edwin Bowen, as the new general superintendent. Having worked with Bowen in Kansas, Chanute looked forward to another good working relationship.

London accountants arrived in New York to investigate and verify that the railroad would pay dividends. Captain Henry Tyler then reported in October 1874[15] that the Erie's property stood in fairly good order, but recommended several improvements. Chanute was sympathetic to all of them as the great diversity of rails, roadbeds, and bridges needed his constant attention. As a case in point, bridge No. 60, just north of Jersey City, had burned a few months earlier. Sparks from a passing locomotive had caused the fire that stopped rail traffic for more than a week. Tyler also suggested avoiding speculation in coalfields. Chanute most likely smiled when he read this sentence, because the Chicago & Alton Railroad management had made a similar statement to him a decade earlier.

The First Major Improvements: Rails

With minimal funding available, Chanute used various approaches to increase the efficiency of the road. Because rails were a critical component, Chanute joined Ashbel Welch and Matthias Forney on an ASCE committee to determine the "best form of standard rail sections of this country; the proportions which the weight of rails should bear to the maximum loads carried on a single pair of wheels of locomotives or cars; the best method of manufacturing and testing rails; the endurance, or the 'life' of rails; the causes of the breaking of rails and the most effective way of preventing it." This presented a perfect opportunity for Chanute to learn, because Welch had designed a steel rail in the mid-1860s that had outperformed all earlier rail designs, and Forney had always been interested in developing more efficient locomotives that required reliable rails.

Few ASCE members shared their knowledge, so the first report, most likely initiated by Chanute, included a strident statement: "The Chairman of the Committee is instructed by his colleagues to give 'particular Jesse' to the members of the Society who have furnished no answers or information; he asks each recusant to consider himself the recipient of a tremendous poke, and keep the fear of a similar punishment before his eyes hereafter." Eventually, discussions at ASCE meetings brought results, and members of the American Institute of Mining Engineers (AIME) also shared useful information, opening an interdisciplinary communication. The committee report, presented at the ASCE Annual Convention in June 1874, was the most exhaustive treatise on rails published up to that time.[16]

Looking for a logical explanation of why rails broke, Chanute researched the contour of the rail as it wore by the action of the locomotive wheels. Placing carbon and tissue paper between the wheel and the rail, he documented the "footprint" of the locomotive, even though this procedure usually pulverized the paper in the actual contact area.[17] Studying many worn wheels and their corresponding rails, he observed that the point of bearing between the rail and the wheel became larger as the metal wore, increasing the resistance and ultimately breaking the rail.

Comparing the footprint of the locomotive's wheel with the shape of the worn rail, Chanute then determined a new railhead cross-section.[18] Believing that the metal ought to be where the wheel wore the rails the most, he placed a large amount of metal into the head and gave the rail a straight and slightly flaring, but thin, web and flange. Even though his design looked similar in cross-section to the standard sixty-six-pound rail,[19] the beveled "Chanute Head" did not wear as fast as the rails with smaller heads, much to the surprise of railroaders. Not lacking in sense of humor, Chanute reportedly told Henry Prout: "Looking at the rail on end, it is a very small affair, six or eight inches in area, but when looked at in longitude, we find that it is over 600,000 miles long, and every inch contains a blunder."[20]

The combination of unevenly worn wheels and rails also affected the locomotive's banking ability, especially under unfavorable conditions such as tight curves and imperfect roadbeds.[21] Chanute's rail design, in conjunction with proper curvature of the tracks and sufficient raising of the outside rail in the roadbed, allowed for better horizontal balancing and smoother banking of the locomotive. These findings influenced Chanute's thinking seventeen years later

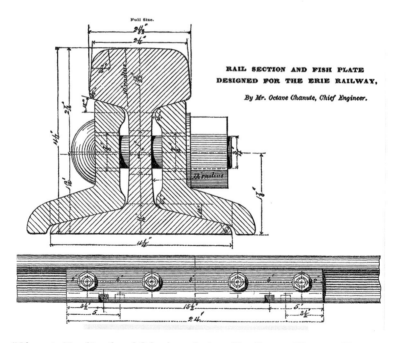

"Chanute Head" rail and fish plate, designed by Octave Chanute. From *Railroad Gazette*, February 6, 1875.

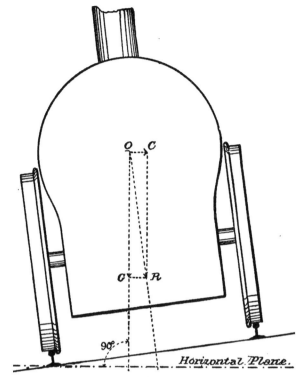

ELEVATION OF OUTER RAIL. CENTRIFUGAL FORCE.

Cross-section of a locomotive going into a curve on
an elevated outer rail. O is the center of gravity, O C
the centrifugal force acting horizontally, and O G the
direction of the weight owing to gravity acting vertically.
From W. B. Parsons, *Track, A Complete Manual of
Maintenance of Way* (1886).

when he tried to figure out a control system for a man-carrying flying machine
(see "Continuing Development: Louis-Pierre Mouillard" in chapter 7).

Replacing rails and connecting the sections usually proved troublesome. At
first, Chanute used Albert Fink's metal fastener to join the rail ends, but as he
learned more about rails, he designed a new "fishplate" that provided a more
solid connection between rails of different cross-section and weight.[22] Increased
weight of rolling stock, greater speed, and heavier traffic demanded reliable
tracks, thus the various aspects of rails were an often-discussed topic at ASCE
and AIME meetings. Alexander Holley, the recognized authority on the Bessemer process to produce quality steel rails, read his paper "On Rail Patterns"
at the annual AIME meeting in Philadelphia in 1881. In it he pointed out that
more than 60 percent of the current rails were fashioned after Chanute's design.

Receiving a preprint copy, Chanute submitted his written discussion: "I now unexpectedly find from Dr. Holley's paper that the considerations, which guided me, are thought worth enumerating, and that it furnishes a good text from which to preach a sermon to railroad men. I hope, however, to satisfy you that I am not entitled to as much notice as Dr. Holley has been pleased to give me."[23]

The large-headed and thin-flanged "Chanute Head" rail was the progenitor of what became the standard rail type in the United States. Starting later in the 1870s, most North American railroads adopted the design and put many thousands of tons into the tracks. Naturally, these rails were not without opposition and engineers continued to look for improved designs.

In the early 1890s Chanute donated his many railroad reports, drawings, and rail sections, collected during the past thirty years, to the National Museum. "In themselves, they are of little value, but may become useful to such as you may want to refer to the history of our railroad development."[24]

The Floods of 1875

Every spring the Delaware and Susquehanna Rivers flooded their respective valleys, but early in 1875, the excursions of the rivers became alarming. The first serious trouble occurred at Port Jervis, New York, about seventy miles northwest of New York City on the Delaware River. A large ice jam threatened the area, and the mayor of Port Jervis requested urgent help from the Erie: "The company's property and this place are both in danger from the ice gorge in the river. Please send some one competent to examine the gorge, and advise us what is best to do at once."[25] Chief Engineer Chanute rushed west, only to watch the bridge upstream of Port Jervis collapse. Its remnants, together with large chunks of ice, then pushed against the trussed spans of the Erie's recently erected Barrett Bridge and lifted its spans from their foundations like reeds.

The built-up ice had to be broken, so Chanute's first thought was to melt the ice. He ordered fifteen barrels of naphtha from Binghamton, New York, poured the fluid into depressed places on the ice, and ignited it. This did not bring the anticipated result.[26] Next, Chanute had five holes drilled into the ice and filled them with gunpowder, which resulted in a spectacular explosion, but removed only three or four square feet of ice. Something else had to be tried. Alfred Nobel's patent had been reissued and ASCE members had discussed the nitroglycerin explosive at recent smokers. Mixing an inert material with the nitroglycerin allowed the safe shaping of the mixture by hand that could then be stuffed with a fuse into predrilled holes. So Chanute ordered three hundred pounds of nitroglycerine and electric blasting caps. "These experiments were more successful and a huge ridge of ice which formed the dam was blown partially away. When the barrier was broken, the water above it burst through the small opening with

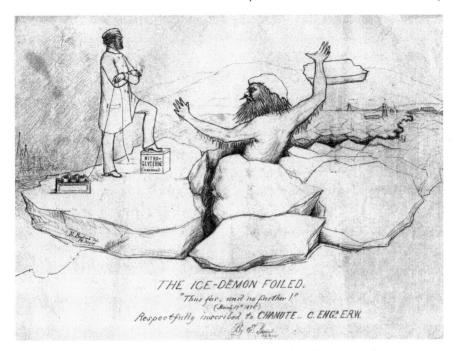

THE ICE-DEMON FOILED.
"Thus far, and no farther!"
(March 11th 1875)
Respectfully inscribed to CHANUTE_ C. ENGᴿ E.R.W.

"The Ice Demon Foiled." A grateful citizen presented this drawing to Chanute for rescuing Port Jervis, New York, from the devastating flood. Courtesy of the Chanute Family.

irresistible force and great volume, saving the town from ruin."[27] Colorful reports of Chanute "dissolving the ice" were published nationwide.

With one problem solved, Chanute immediately had to handle the next one, because the Erie needed a new bridge to replace their destroyed Barrett Bridge across the Delaware. A reporter from *Pomeroy's Democrat* described the situation: "The new bridge at Port Jervis is certainly a model of engineering skill and mechanical ingenuity. At 7:30 AM the three spans of the old bridge were taken away and the bridge was lost, and at 2 PM the work of rebuilding the bridge began. O. Chanute, chief engineer, and E. S. Bowen, general superintendent, were on hand, and it was through their labor and constant supervision that the work was so well done and so soon completed. Mr. Chanute is one of the best engineers in the United States, and Mr. Bowen combines railroad experience and knowledge of a mechanical engineer."[28] It took Chanute's men ten days and nights to rebuild the bridge in iron, then traffic rolled again.

Williamsport, Pennsylvania, along the western branch of the Susquehanna River reported a similar ice build-up the following week, so Chanute suggested the nitroglycerin blasting technique to Philadelphia & Erie management. The explosion blew ice and mud a hundred feet high, allowing the water to flow. The *New York Times* report concluded: "It has not yet been decided whether this method of clearing the river from ice will be adopted by the Erie Railroad in the future."[29] This blasting method was far from perfect, so engineers around the country discussed its merits, while still looking for better ideas.

Bridge No. 16: The Portage Bridge

The next disaster happened only six weeks later. Just before daybreak on May 6, 1875, the Portage Bridge, or Bridge No. 16 on the Buffalo Division, burned after a train had passed.[30] This 850-foot-long and 234-foot-high timber trestle bridge had opened for traffic in 1852 and provided a fast connection from the mainline to Buffalo.

Bridge No. 16 across the Genesee River, erected in 1852.
From letchworthparkhistory.com.

Chief Engineer Chanute and Assistant Engineer Morison took the next train west to evaluate the damage, but the destruction of the bridge across the Genesee River north of Portageville, New York, was complete. Traveling back to New York City, the two engineers developed a plan. The railroad company could temporarily guide traffic over the mainline via Erie, Pennsylvania, to Buffalo, New York, so it would be most efficient to rebuild the bridge in iron using the original layout.[31] Erie management accepted Chanute's proposal, and the Watson Manufacturing Company of Paterson, New Jersey, received the contract for the ironwork one week after the disaster.

At the next ASCE meeting, Chanute and Morison described the damage to the bridge and the picturesque gorge (modern-day Letchworth State Park) where the Genesee River, after passing the remnants of the bridge, flowed through a gorge with spectacular perpendicular walls and plunged more than

Bridge No. 16 burned in May 1875. From letchworthparkhistory.com.

one hundred feet straight down. Half a mile downstream, the river took another
ninety-foot plunge into a small pool before continuing in its windings. As part
of the discussion, one member suggested they fly in a balloon for a better view
of the area, similar to Jules Verne's description in his latest adventure article,
published in *Scribner's Monthly*.

Erie workers cleared the site and prepared for the masonry repair. Chanute
suggested that Morison encase the heavily damaged stone piers with *béton*, as
the French called the "artificial stone" that later became known as concrete.

Bridge No. 16 was rebuilt in iron after the 1875 fire. In 1880 the uprights
and girders were strengthened and piers rebuilt. From Chanute, *ASCE
Transaction* No.227, "Repairs of Masonry" (1881).

Morison supervised the construction of the replacement bridge, now with ten 50-foot spans, two 100–foot spans, one 118-foot span, and girders in the simple Pratt truss pattern. The new "High Bridge" opened for rail traffic on July 30, only 86 days after the fire; railroad officials, employees, and tourists stood along the cliffs, witnessing the official engineering test.[32]

Three years later, just when Chanute thought everything was in good working order for the winter, he received a wire that the rapidly flowing river had undermined the Portage Bridge piers. To save the viaduct from self-destruction, repairs had to be done right away, even though it was late in the season. Chanute suggested applying an artificial riverbed bottom of concrete to stabilize the masonry piers. Following additional engineering calculations, the crew also strengthened uprights and girders, making the bridge better than new.

Dark Clouds with Silver Linings: Trip to Europe in 1875

Business normally picked up in spring, but in early 1875 it was slow. Two weeks after the burning of the Portage Bridge, Jewett reported to Erie stockholders the utterly hopeless financial condition of the road. One week later, the Erie declared bankruptcy and President Jewett was appointed the receiver.

Everything seemed to collapse around the forty-three-year-old Chanute. With problems developing everywhere, he was at the end of his energies. Not knowing which way to turn, he made an appointment with his physician, who recommended a lengthy vacation far away from the hectic business in New York. Following doctor's orders, Chanute decided to take an extended leave of absence and go to Europe, just as some of his friends had done in the past. He took Annie and their three younger children to Peoria, where they would stay with members of her family and be safe. Traveling back to New York, he started a diary:

> Parted from my dear wife and children with great reluctance. . . . In the present state of depression in business, my estate would not be sufficient for their education and maintenance until they were able to take care of themselves. Would it not be wiser to remain with them and endeavor to recover my shattered health by idling a few months at home away from business? On the other hand, can I remain in this country and forget business and recover the mental tone and balance necessary for future success and the earning of a living for the loved ones? Is it not better to run the risk of disaster for the certainty of complete recovery from overwork? These cruel doubts have harassed me for weeks.[33]

Artie, Chanute's oldest son studying at Yale, met his father at the wharves of the Inman lines and gave him a good send-off on June 6. Ten days after leaving New York, the steamer *City of Chester* arrived in Liverpool. Being curious, Chanute introduced himself to George Lyster of the Mersey Docks, who gave him an extended tour of their docks and warehouses, cranes, and flat belts that seemed more efficient than the Erie's docks in New York City.

After spending a few weeks in England, Chanute traveled to Paris to visit his mother and other family members. Feeling a bit more like his real self, he began his "walking tour" in late July, also using railroads and steamboats. His first stop was Brussels, then up the Rhine River to Basel, and by train through Switzerland. At Ragaz, he visited a spa, still feeling tired but also concerned. "Am horribly anxious, no news from America and have only enough money to return to Paris."[34]

During the last week of his therapeutic vacation, he reacquainted himself with working life and met representatives of the Erie's British mortgage holders in London. After spending three months in Europe, Chanute embarked on September 7 on the steamship *Bothnia*, the newest in the Cunard fleet. Arriving in Boston, Chanute had to pay $42.89 duty for his purchased presents, "which left me with $4 and a ticket to New York, whence I arrived on Sunday morning with $0.59 in my pocket and so my vacation is at an end." This was the final entry in his diary.

Octave Chanute and his mother, Elise de Bonnaire Chanut, at her home in Ivry, France (August 1875). Courtesy of the Chanute Family.

Reenergized and full of ideas, Chanute successfully picked up his job where he had left it; life and work now proceeded at a more normal pace. He rented a house at 128 High Street in Brooklyn and brought his family back to New York.

Life Continues

Because the Erie had introduced fewer improvements during 1875, its financial situation was under control. Later that fall, Chanute received a promotion to assistant general superintendent, only two and a half years after joining the Erie.

In November 1875 Chanute regretfully accepted Morison's resignation from the Erie, as he wanted to join another engineer and form "Morison, Field & Company," a private bridge contractor. Five years later, Morison sold out and moved to Chicago to design and build major steel bridges across the Mississippi, Missouri, and Ohio Rivers. In the next decade, their paths would cross periodically because they were members of the same engineering societies.

Knowing that safety on the Erie should never face compromise, Chanute read with interest Richard Rapier's paper on fixed signals of railways, published by the British Institution of Civil Engineers (ICE). This comprehensive paper prompted the ASCE to appoint a committee, consisting of J. Dutton Steele, Chanute, and Charles Fisher, to research the same subject in North America.[35] Chanute then introduced fixed signals more rigorously on crossings and drawbridges on the Erie, providing a safer railroad.

Another recurring problem concerned the longevity of parts used in bridge construction. Having watched fellow civil engineer Onward Bates inspect and test various pieces for the approaches in the St. Charles Bridge in the late 1860s, Chanute initiated regular testing of pieces at the Erie shops as well. For him, eyebars with their pins were the critical links holding bridge members together, thus these tests determined the correct proportions of the heads and pins. "For the weight of the structure, for the effect of the wind and of changes of temperature, a factor of safety of 3 shall be adopted, while for the live load the factor of safety shall be 8."

As the behavior of metal under stress and strain became better understood, the factors of safety needed reassessment; engineers now looked beyond breaking strength and focused on the metal's elasticity limits, recommending "a series of experiments upon compression on parts of various shapes, length and materials needed to be made. The results of such experiments may lead to important changes in our current bridge building practice, and materially add to the safety of our bridges."[36]

In December 1875, Chanute accepted an appointment as one of three consulting engineers to recommend a plan for the New York & Long Island Bridge across the East River at Blackwell's Island. Because none of the submitted proposals provided construction details or cost, Chanute prepared specifications for this bridge, incorporating many of the unpublished guidelines of the Erie. Receiving the next set of proposals, he asked to hire Charles Schneider, the Erie's chief draftsman, to verify the calculations. The consulting engineers reported the three best proposals in February 1877,[37] and each consulting engineer received $12,500 from New York state, but the bridge company put the project on hold.

With his additional appointment as the Erie's superintendent of motive power and rolling stock, Chanute turned his attention to potential fuel economy by streamlining the locomotive and cars. He reviewed his notes from 1871, in which he had calculated wind forces, and now tried to determine the most efficient shape offering the least resistance. He compared different-shaped locomotives with various loads, speeds, tracks, and atmospheric conditions.[38] Knowing that crosswinds and headwinds created resistance, he assembled tables showing the ratio of increased atmospheric drag to speed, but felt unsure about their interpretation. Trying to grasp the interplay of forces, he shared his data with other engineers, hoping for meaningful feedback.

Even though little was known about the longevity of concrete, Chanute used it more and more in various repair projects. "When freshly made it looks like a loose, incoherent heap of damp sand, slightly coated with lime, and a more unpromising mortar can scarcely be imagined. When put in place, however, it forms a plastic mass, capable of being rammed into crevices of all masonry. When set, it becomes a hard and imperishable stone." The tunnels on the Buffalo Division, built in the early 1850s, had started to disintegrate and the Warsaw culvert presented serious problems in late 1875. Building a new 146-foot tunnel would cost a great deal, because it ran under a heavy embankment, so Chanute decided to have the inside lined with concrete. With winter fast approaching, he suggested speeding the curing process by heating the sand and water prior to mixing. Crews completed the project for $2,200, instead of an estimated $36,000 for a rebuilt tunnel. Railroad management harbored misgivings about the potential waste of money on this unique repair, but an inspection next spring revealed no water percolating through the embankment. After another cold season, and the winter of 1877–78 was a severe one, the inspectors found no weak points or defects; "it was as perfect as originally built."[39]

In the spring of 1877 another unbudgeted expenditure required immediate attention; the sandstone inside the culvert at Clifton, New Jersey, had become dislodged and hampered traffic flow. The culvert could be replaced for $6,000 or repaired using Chanute's scheme. Workers smoothed the sandstone, lined the inside of the tunnel with concrete, and restored the buttresses at both ends at a cost of $600.

Then the Blauvelt culvert, about thirty miles north of Jersey City, fell in. This section of the Erie was known as the "Milky Way," because it went through the rich dairy region that supplied milk for New York City. Rail traffic had to be detoured, increasing expense to the Erie and the milk sellers. In December 1877 Chanute had the culvert boarded up and a repair crew went into "winter quarters." Two months later, rail traffic again passed through the newly concrete-lined culvert, lowering the cost of transportation for everyone.

Economies and Technology: The Battle of the Gauges

The narrowing of the Erie's original six-foot-wide gauge provided an ongoing subject of discussion. With Chanute in charge of motive power, Erie directors approved the laying of a third rail from the mainline to Buffalo, a distance of 157 miles;[40] this allowed the Erie's native equipment to use the original six-foot-wide gauge while the newly purchased equipment and rolling stock of other roads used the "standard" or "narrow" gauge. But financial negotiations with English bondholders broke down, again, and the editor of the "Financial Affairs" column of the *New York Times* summed up the situation: "Stock holders who bought the stock at $23 on the faith of the statement in regard to the third rail are naturally beginning to wonder when the laying of the rail will be commenced. One day there are positive assurances that the third rail will be laid and the stock goes up, and inevitably on the following day it is reported that there is a hitch in the negotiations, which will prevent the laying of the rail, and, presto, the stock goes down. The point today was that the rail would not be laid, and accordingly the Erie dropped to 20⅛."[41]

Jewett and the board then decided to narrow the gauge with or without stockholder approval, and in April 1876 workers installed the third rail between Suspension Bridge 4, northwest of Buffalo, and Waverly on the mainline. Now there was a uniform gauge between Chicago and Philadelphia, via the North and South Shore Lines, the Erie, and the Lehigh Valley Lines. By joint agreement, the trunk lines increased the through-traffic rate between New York City and Chicago, and the Erie began passenger service over leased tracks using newly purchased sleepers and drawing-room coaches with standard-gauge wheelbase.

But the freight and passenger rate wars soon flared again, and one reduction followed the other, as each railroad tried to cut the other's throat. Rates soon fell below the actual cost of transportation, but the Erie, still in receivership, kept the cost of a passenger ticket $1 below its competitors. The one-way ticket between Chicago and the east coast sank from $30 to $12.

The Erie's pricing paid off; many visitors traveled to the Centennial Exposition in Philadelphia in 1876, and the Erie bragged that it carried an estimated five million passengers during the traveling season without a single passenger being injured or a piece of luggage lost.[42] Even though the passenger business operated at a loss, the increased total traffic brought profit, and for the first time in many years, the Erie Railway paid its employees promptly on November 15, 1876, including back pay. With lower debt payments over the past sixteen months, the road could afford to take care of its people, stated the editorial comment in the *New York Times.*[43] To Chanute, the future started to look brighter.

The First Major Railroad Strike: 1877

The last quarter of the nineteenth century was a critical time period for North American railroads, as the expansion phase gave way to the organization phase. Cornelius Vanderbilt, the president of the New York Central, sought a uniform rate to allow profitable freight carriage by all railroads, but he also wanted a central board to establish rules and tariffs.[44] Late in July 1874, representatives of the principal eastern railroads met at Saratoga, New York, but its results raised a storm of opposition, as the public thought that the railroads were meeting to create a monopoly.

Three years later, the eastern trunk lines formed the "Joint Executive Committee of the Trunk Line Association," and fellow ASCE member Albert Fink, who had turned his interest from a bridge-designing career to railroad economics, accepted the role of commissioner to supervise traffic flow and apportion the total competitive traffic among the trunk lines. Chanute regarded Fink as the perfect person for this difficult job[45] in these trying times. The five major eastern railroads then agreed on a mutually acceptable freight rate and the share of the trade each would have. They lowered the rate on eastbound freight traffic, while maintaining the westbound rates.

With a general decline in business in 1877, Erie management announced a 10 percent wage cut effective July 1, 1877; Chanute and Bowen journeyed to Hornell, New York, the southern terminus on the Buffalo Division, to explain the situation to representatives of the various trade organizations. Not willing to accept pay and crew cuts, workers threatened to strike,[46] tear up the tracks, and burn the company's property.

Other railroads also introduced wage cuts, and on July 16 workers on the B&O initiated a strike that spread rapidly from Baltimore west, increasing in brutality and intensity as it expanded. A number of local authorities sympathized with the strikers, because they viewed their action as a protest against the rapid mechanization of the American Society, while some urban citizens simply felt frustrated with the arrogance of the powerful railroads.[47]

Meanwhile, the Chanute family spent the summer in the hill country near Suffern, New York, about thirty miles north of Jersey City, renting a house along the Erie tracks.[48] Because there was no immediate indication that Erie workers would strike, Chanute joined his family for the weekend.

However early on Friday, July 20, Erie workers struck at Hornell,[49] disrupting traffic and pushing incoming passenger trains onto sidetracks. Erie Headquarters received the news shortly after midnight, and Jewett ordered all division supervisors and his superintendents to Hornell. General Superintendent Bowen sent a special train to Suffern early Saturday morning to fetch Assistant Superintendent Chanute, whose experience in dealing with labor issues on the western roads would prove valuable, as the difficult situation developed. Annie quickly put her husband's clothes into his always ready double valise, and Octave left for Jersey City. Chanute's daughter Lizzie recalled that later the same morning, a special train came by, with her father standing on the rear platform of the last car. Seeing his children, he tossed a newspaper from the slowed train, which they collected and took to their mother. An attached penciled note stated that he was on his way to break the strike and that he would contact them as soon as he could.[50] At Corning, Chanute and Bowen picked up the Steuben County sheriff with his deputies and continued to Hornell, arriving at 9:30 P.M.[51]

Next morning, on July 22, members of the National Guard arrived to clear the rail yard of strikers. Jewett wanted his trains to go through,[52] so the Erie's westbound express train arrived as scheduled in Hornell at 9 A.M. with two passenger coaches, a baggage car, and a mail car. The stranded passengers of the previous day boarded, and Chanute and Bowen watched the *Pacific Express* leave the depot, with soldiers standing on the platforms and the locomotive.

The railroad tracks out of Hornell climbed over one of the steepest grades on the line to Tip Top Summit, an elevation of 1,760 feet. To prevent the train from going west, strikers had covered the rails on the incline with soft soap. As soon as the locomotive struck the soaped tracks, its wheels started to slip. The strikers then overtook the soldiers and uncoupled the passenger cars from the slowly moving train, but let the locomotive with the mail car continue.[53] Bowen and Chanute were furious but powerless.

Receiving the latest report, Jewett ordered another train assembled and taken out via the Buffalo Division, where the grade was not as steep. As soon as this train left the depot, strikers forced their way into the cab of the locomotive, threw the firemen from the engine, and threatened to kill the engineer. Later the same day, the 23rd Regiment of Brooklyn arrived, and Bowen tried to take this train east on the mainline, but strikers overturned the slowly moving locomotive. Chanute had seen much ugliness in eastern Kansas where settlers fought the railroad, but this mob violence was incredibly fiercer and uglier.

Seeing about 1,200 soldiers at Hornell, the strikers may have realized that they were starting to lose control. The strike leader, Barney Donahue, who had been discharged from the Erie a few weeks earlier, requested a meeting with railroad officials just before midnight. Bowen, Chanute, and other supervisory personal met the strikers in the superintendent's car and listened to their ultimatum. Bowen responded quietly that the company's earlier proposed settlement was unchanged, but he also reminded the strikers that the bankrupt Erie remained in court custody.

Next day, New York governor Lucius Robinson, who was also a director of the Erie, declared the strike a riot, and receiver Jewett forwarded a warrant to the Steuben County sheriff to arrest Donohue.[54] The Erie and the strikers then reached an agreement: the men would resume work at the reduced pay and the company would discharge no one on account of the strike, except the strike leaders. With the strike settled, Chanute remained in charge at Hornell, directing the reopening of business, while Bowen went to East Buffalo, where strikers had occupied the Erie rail yard. With no prospect of an early return, Chanute shipped his valise with his dirty clothes and a note to his wife that he was well, but that he would like fresh clothes.

These first major strikes of railroad workers in the United States—historians soon called these the "Great Strikes of 1877"—were a response to wage cuts and other grievances, including the dominance of the railroad as part of American life and commerce. It tested Chanute's management and labor relation skills, and he remembered his days in Hornell vividly in years to come. During this time of economic uncertainty and labor unrest, Chanute progressed in his career with the Erie, and his skills provided a benefit to his employer and himself.

Using Funds Wisely

With funding still tight everywhere, railroad managers had to cut operating costs to a point previously thought unattainable, and civil engineers, already trained on sparse funding, helped them. Chanute, collaborating with Bowen and Blanchard, introduced technical advances, better equipment maintenance, and more efficient organization of labor and traffic. Whenever funding was available, he improved the roadbed's grades and curves and strengthened wooden bridges with additional masonry support and iron stringers until workers could replace them in iron. He initiated the purchase of two gravel pits in western New York, so ballast could be replenished as needed to make the roadbed more stable.[55] As a result of all these improvements, the length of trains grew from eighteen cars in the early 1870s to thirty-five cars a decade later, and some locomotive engineers reported back to Chanute that they had clocked speeds of eighty miles an hour on smooth and straight tracks.

In August 1878, the line between Binghamton and Sayre, Pennsylvania, a distance of about forty miles, and the extension from Susquehanna to Jersey City, received a third rail, allowing passengers to travel west on express trains and in Pullman sleeping coaches without changing cars.[56]

Iron structures replaced wooden bridges, but the physical characteristics of iron and steel were still not completely understood. Chanute commented: "Until we know more about the product, it is probable that the chemist, the manufacturer, and the testing engineer will have to proceed simultaneously hand in hand, until the composition of the metal, its manufacture, and its capacities are thoroughly understood, and the uniformity of its product assured." Updating the general specifications for future iron bridges on the Erie,[57] Chanute thought that the manufacturer should verify the quality of the structural members and supply reliable, ductile, and safe steel for the same price as wrought iron. "Parties proposing to substitute steel for particular parts will be required to furnish evidence of its strength, elasticity, uniformity in production, and adaptability to the intended purpose."[58] Bidders also had to submit stress calculations before starting the job; Chanute then personally checked material and workmanship in the field, making the Erie possibly the first American road to introduce regular testing and inspecting.

Early in the summer of 1879, Chanute had designed a larger, more efficient roundhouse with space for forty-one locomotives to update the terminal facilities in Buffalo. The first section of the roof went up without trouble, but the second one collapsed, injuring several carpenters and ironworkers.[59] The construction crew then used more care in erecting, and the roundhouse was in full operation two months later, serving the company's needs and protecting the equipment from the weather, especially the Buffalo winters.[60]

Another challenging repair in the spring of 1880 was the double-tracked West Paterson Bridge across the Passaic River, which was "cracking open and falling asunder." The two bridge piers had settled, and the constant vibration from the trains had split them apart. Discussing the problem with John Goodridge, the president of the New York Stone Contracting Company, Chanute suggested something novel: he had the piers wrapped entirely in a shell of concrete from four to twelve inches thick, with three iron reinforcement rods, or "rebar," em-

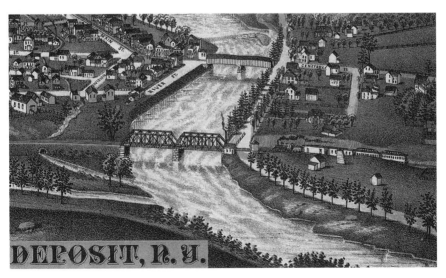

Deposit, N.Y. 1887, bird's-eye-view. The rebuilt iron bridge is typical for the many bridges erected by Chanute during his time at the Erie. Drawn by L. R. Burleigh. Library of Congress, Map Division, Washington, D.C. No. G3804.D53A3 1887.B8.

bedded on either side.[61] A century later, it is common knowledge that reinforced concrete combines the compressional strength of concrete with the tensile, or bendable, strength of metal to withstand heavy loads. Chanute then reported: "It is confidently believed that the piers are now good for many years," as he had saved time and money while minimizing interruption to traffic flow.

When Chanute took over the chairmanship of the ASCE Timber Preservation Committee early in 1882, he shipped hemlock ties, the cheapest and softest wood available, to the St. Louis Wood Preserving Company, operated by Joseph P. Card, and placed the treated ties into the Bradford, Pennsylvania, extension. Chanute's relationship with Card would be a lasting one (see chapter 6).

More Financial Turbulence

The aftereffects of the 1873 panic and the 1877 labor unrest, with its continuing depression, created a troublesome situation for many railroads, and the bankrupt Erie still did not see profit, so shareholders ordered the road sold. Receiver Jewett assembled a complete inventory of the property, and on April 24, 1878, the road was auctioned off for $6 million to the former Erie director and politician, Edwin D. Morgan. The board elected Jewett president of the new company that entered history as the New York, Lake Erie & Western Railroad.[62]

Meanwhile, the demand for tariff regulation gained strength, for several reasons. Industries complained about the railroad rebate, giving preferential treatment to large shippers, and certain shippers were charged the published rate, with part of the payment returned later. This discrimination in rates between individuals, locales, and class of freight forced merchants, farmers, and shippers to agitate for government regulation. Fink and Chanute believed that the federal government should set freight and passenger charges, not state governments, or the lines themselves. In their opinion, the railroads were quasi-public institutions, because they were built with the aid and resources of the people.

To investigate freight rates and their rebates, the state of New York established the Hepburn Committee in June 1879.[63] As part of the state's investigation, committee members also inspected the Erie's facilities, with Chanute as the main contact. A lawsuit followed, with the People of New York State as plaintiff. Jewett of the Erie and Vanderbilt of the New York Central explained in court their freight rates and why there were special rates for certain manufacturers. After a few days of hearings, it became clear that the roads managed their affairs like other private corporations, extending favors or denying them, advancing rates or cutting them, according to what it would repay them.

While the Hepburn Committee sought answers, the Erie requested the assessment reduced on its Long Dock property in Weehawken, New Jersey. The state commissioner of railroad taxation claimed that the site should be assessed for its value as a railroad terminal, while the counsel for the Erie insisted that the assessment should be reduced on account of its general depreciation of value. On August 4, the court summoned Chanute to testify in Trenton. "At this point, a gentleman whose face very strongly suggested the model of Wm. Page's bust of Shakespeare, was called forward by Mr. Parker and sworn as Octave Chanute, the Chief Engineer of the Erie Railway," reported the newspaper. In cross-examination, Chanute stated that the original cost of the docks was $1,743,492, but its value should align with the current assessment at $420,000, because the property had depreciated. The judge differed in opinion, and ordered the Erie to pay back taxes.[64]

A couple of months later, Chanute appeared in court again, this time as an expert witness. The Hepburn Committee wanted details about rate structure and the actual cost of transportation.[65] Chanute explained that the general cost of railroad operation had been reduced by 20 to 30 percent since 1873, but a proposed flat rate of $10 per one hundred pounds of freight from Chicago to New York would probably pay only for the "direct" charges, leaving very little profit to any railroad.

Mastering Technology:
Uniformity in Railway Rolling Stock

The almost fifty-year-old Erie Railway used as many designs for locomotives as there were shops in the country. There were more, in fact, as each shop experimented at the expense of its patrons, introducing new designs as an improvement on its former plans. To increase the efficiency of transporting freight, Chanute wanted a more powerful engine to pull a longer string of cars. Years later Chanute humorously recalled the story of a John W. Keely, who had invented a "motor" that reportedly produced a "tremendous power." Keely had experimented with different substances, and on one occasion, he had poured a mixture into a vessel that suddenly exploded into his face; his "motor" was the result.[66] Hearing these details, Chanute eliminated Keely's invention.

Seeing the Consolidation-type locomotive in action on the B & O convinced Chanute of its economic advantages.[67] Alexander Mitchell designed this locomotive in the mid-1860s for slow, heavy pusher service but did not patent his invention, so every major American railroad eventually adopted it. Chanute recommended to the Erie board to introduce these "monster locomotives," with eight instead of only four wheels, which he thought more advantageous on steep grades. Erie shops built four units, and two outside companies then received the contract to build thirty additional locomotives. In writing the contract, Chanute tried to introduce the principle of "interchangeability of parts" for easier maintenance and decreased future operating costs.

Even though each contractor received exactly the same drawings, each used different screw threads and placed bolt holes in slightly different positions. Spare parts did not make a good fit without filing and chipping, and there was no interchangeability. Chanute later stated: "Few engineers realize how much annoyance and increased cost result in the operating and repairing of railway rolling stock from lack of uniformity in their construction, nor how much efficiency and economy are promoted by building them with absolutely interchangeable parts, so that duplicates can be kept on hand, which will be sure to fit, in case of breakages."[68]

Frustrated, he mentioned his findings to his two friends Matthias Forney and Robert Thurston. With their support, the ASCE formed a committee to consider the "standard system of screw threads, and the best method of maintaining exact sizes of screws, so that bolts and nuts may be interchangeable." The importance of adopting standard forms and sizes for hardware and machinery was obvious, thus committee members determined the absolute essentials for each manufactured part: screws must have a given number of threads per inch, threads must have the form and proportions as designated, and the diameters of the screws must conform to the size specification. The Pratt & Whitney Company joined the effort and spent large amounts of money on equipment to meet the desired accuracy in producing taps and dies.[69] A standardized system for the production of parts, combined with a taste of innovation, helped build the foundation for economic strength in the United States,[70] and Chanute was one of the engineers pushing its introduction.

The new locomotives needed coal more quickly and economically, so Chanute designed coal chutes with a storage capacity of about 330 tons. The Erie erected the first such facility in Waverly, New York, on the Susquehanna Division, with eleven pockets and chutes on a 110-foot-long storage platform.[71] The next problem was not expected, because the railcars of the longer trains pulled apart on the steep inclines in western New York and Pennsylvania. Using a scientific approach, Chanute suggested using a dynamometer to record the pulling forces[72] to determine the ultimate load capacity of the Consolidation locomotives. Improved couplers, followed by the introduction of Westinghouse air brakes, soon created a more efficient system.

With most mechanical problems solved, labor problems emerged, because the Erie's locomotive engineers did not want these "monster locomotives" on their road. When asked, the men told Chanute privately that they were afraid of losing their jobs, reminding him that the B & O had doubled the length of its freight trains and then downsized the crew after introducing the Consolidation. As the superintendent for rolling stock, Chanute made sure that there was no reduction in the workforce on the Erie. Soon the increased freight traffic pushed men and equipment to their limits, surpassing any previous experience of the road; the Consolidation, with its "limitless" power, moved loaded trains, twice as heavy, for half the cost of its earlier brethren.

Next, the Erie's bridge specifications needed revision to accommodate the increased load and length of trains, so Chanute hired his former chief draftsman

Charles Schneider as a consultant. A few years later, Schneider made a name for himself when he designed and erected a double-tracked cantilever bridge across the Niagara River; having learned the advantages of concrete under Chanute, he erected the girders on reinforced piers.[73]

Chanute and his staff had achieved some degree of success.[74] In the early 1870s the average cost of repairs in the Erie shops was 9.17 cents per operating mile, but by the late 1870s it had dropped to 4.33 cents, yielding a saving of $675,000 annually. Some savings came from other reforms, but the beginnings of standardization, inspection, and interchangeability were major factors. For the Erie's 1880 fiscal year report, Chanute stated that the cost of haulage had been reduced from 1 cent to 0.5 cent per ton-mile, and the treasurer reported net earnings of $7 million, an increase of almost 48 percent.[75]

From Broad to Standard Gauge

In June 1880 the Erie finally submitted to the "battle of the gauges." Upper management, and especially Chief Engineer Chanute, had spent weeks in preparation for the complete narrowing of the tracks. Between Dunkirk, New York, and Hornell, a distance of 198 miles, the ties were marked and spikes were laid out for the move. The actual work began at 4:30 A.M. on June 22, and a crew of eight hundred men reduced the line to standard gauge in three and one-half hours. Then, one inspection train started from Dunkirk and another from Hornell, meeting in Olean. At 8 A.M., the news flashed over the wires that the work was completed, and trains started to roll again.

The narrowing of the gauge, with the interim use of a third rail, fascinated the traveling public, and many people were curious about the "why" and "how." Chanute gave an interview to a reporter from the *New York Tribune*, reprinted in several papers. He explained: "When the Erie was built, the six-foot or broad gauge was chosen, because management believed that more power could be gained by having a broad base to the boiler, and greater security would be insured by a broader gauge. But these principles have since been proved to be fallacious."

When asked about new rolling stock for the narrower gauge, Chanute replied, "We have ordered 30 new engines and 3,000 new freight cars. The present rolling stock will not be altered but will be replaced as fast as worn out by those of narrower gauge. It would cost about half a million to change all the cars, but more than three times that amount would be necessary to alter locomotives, as new boilers would be required." He also explained that the Erie could now connect with and run its new cars over the tracks of almost any other road.[76]

Later the same year, the Erie finally reached Chicago. Jewett had acquired the Atlantic & Great Western Road in January 1880, with its terminus at Marion, Ohio. Chanute then surveyed the Chicago & Atlantic Railroad between Marion and Hammond, Indiana, and Erie traffic went for the last few miles on leased tracks into Chicago. Soon there were four daily through-trains, requiring twenty-six hours, thirty-five minutes for the trip from New York to Chicago.

Then, yet another rate war began. The one-way, first-class ticket from New York to Chicago was reduced to $9, and immigrant rates from Castle Garden, the entrance point at the New York City harbor, were much lower.[77] By the end of October, the passenger rate war was almost over, but no one knew for how long.

For the 1881 Stockholders Report, Chanute reported a further decrease in the operating cost. The use of steel rails, the adoption of standard gauge, the utilization of faster and more powerful locomotives, and a host of other technical advances created a much more efficient system.

A Spectacular Bridge: The Kinzua Viaduct

In 1873 and 1874 Peter Watson, at the time president of the Erie, had purchased more than 27,000 acres in Delaware, Elk, Jefferson, and McKean counties in Pennsylvania. A decade later, coal, oil, and timber from these locales provided significant income, but to ship the "black diamonds" economically and profitably to the populous eastern cities required the railroad tracks to cross the mountainous Pennsylvania countryside.

To connect Bradford, Pennsylvania, with the mainline in the north across the Kinzua Creek's deep ravine required a two-thousand-foot-long bridge at the extreme height of more than three hundred feet above the valley floor. But building railroad tracks across the sky also required knowledge of the wind forces through a deep, narrow valley and of the material to build the structure. Chanute had discussed the shape of bridge members at a recent ASCE meeting. "We all know that a convex surface presents less and a concave surface more resistance than a flat plane to the wind. This knowledge should be applied to bridge members."[78]

Most people considered bridges as utilitarian structures, but to Chanute their appearance was also important. He explained:

> The remark is sometimes heard that bridges and viaducts are great eyesores to the artistic public, and that engineers are rectangular utilitarians who pay no attention to looks. . . . How far an engineer will be justified in spending other people's money on ornament will depend upon the circumstances, but as wealth and taste have increased, there seems to be a disposition on the part of capital to take a more liberal view of this matter. Engineers, in designing new structures may profitably inquire how bridges will look from an aesthetic point of view, and artistic effect will follow, by simply arranging the outline in accordance with the lines of maximum strains.[79]

Thomas Clarke, now a senior partner of the Phoenix Bridge Company, also liked challenges; he agreed to build a viaduct, higher than any other in the world, at a much lower cost than any other crossing. The novel construction began on May 10, 1882; a crew of about one hundred men erected steel towers first and then placed the girders in position, with the help of a traveling scaffold, projecting about eighty feet. "No staging of any kind was used, not even ladders, as the men were climbing up the diagonal rods of the piers, as a cat would run up a tree."[80] There were twenty-one spans, each sixty feet long, and the steel columns rested on masonry piers, fastened to iron shoes, embedded in the coping. Each column was anchored to its base by two large bolts, secured by an iron plate and nut to counteract the forces of the wind.[81]

The viaduct's construction proceeded rapidly. Workers completed the project on August 29 for less than $250,000, and a plaque, placed conspicuously on the southern approach, gave the year of erection, the names of the bridge builders, the chief engineer, and the officer of the railroad who accepted the work. This 302-foot-tall viaduct, with its single track, was Chanute's most spectacular bridge. Its bold design was of fascinating beauty, and the airy structure, with its elegant uprights, did not look as strong as it actually was. To watch a fully loaded coal train move over this "unsafe" looking viaduct was a curious spectacle, telling a triumphant tale of modern engineering.

Freight trains carrying coal, lumber, and oil, and excursion trains with sightseers crossed the bridge daily. To keep up with the increasing volume and weight of the coal traffic, the bridge was rebuilt in 1900 by the then-chief engineer of the Erie, Carl W. Buchholz. Because the original masonry foundation piers

The Kinzua Viaduct, in progress of erection, July 22, 1882. From Clarke, Reeves & Co., *Album of the Designs of the Phoenix Bridge Company* (1888).

remained solid, Buchholz only redesigned the superstructure.[82] The bridge
was well maintained until 1959, when it no longer passed safety inspections
and subsequently closed for commercial freight traffic. However, lighter-weight
trains with sightseers continued to roll across.

The surrounding area became the Kinzua Bridge State Park in 1963. In July
2003, a tornado with wind speeds in excess of one hundred miles an hour tore
eleven towers from their concrete bases.[83] Investigators found that the anchor
bolts, installed under Chanute's supervision, had rusted over the past 120 years.
The towers, either standing upright or lying on the ground, are silent reminders
of the former spectacular bridge.

More Improvements

Only half a year after the Kinzua Bridge opened for traffic, it was front-page
New York Times news that Chanute took charge of building another branch line
through the remote regions of eastern Pennsylvania, a country full of abrupt
hills and deep chasms.[84] The sixty-eight-mile-long Erie & Wyoming Valley Road
then carried the "black diamonds" from the Scranton coal district across the
Moosic Mountains to the mainline at Port Jervis.

To remain competitive, the Erie decided to establish its own stockyards after
becoming full owner of the Weehawken, New Jersey, docks on the west bank of
the Hudson. Chanute designed a cold-storage warehouse and constructed two
piers with warehouses, providing facilities for economical handling of livestock
and easy transfer to any point on either side of the harbor.[85] As part of the ASCE's
meeting in January 1882, Chanute invited members for a boat ride across the
Hudson to inspect and examine the work in progress. The yards opened later
that year.

To profitably supply the eastern cities with meat, shippers began experimenting with refrigerated cars, because it was more economical to ship dressed meat
by rail than livestock on the hoof in freight cars. Nelson Morris generally receives
credit for shipping the first refrigerated carload out of Chicago in 1874, but it
was Swift & Company, putting their improved cars on the rails in 1878, that
made the shipment of dressed beef by rail a financial success. Chanute wanted
to introduce refrigerated cars to the Erie's fleet, so he looked for the most efficient refrigeration system. The field was large and the patented De La Vergne
system seemed to be the best on the market.

After meeting the owner, John De La Vergne, Chanute decided to become
their consulting engineer in July 1883, working for a retainer of $300 a month
plus expenses, while remaining employed by the Erie. He liked his cool beer
in the summer and he knew who needed refrigeration equipment, so he visited
brewers in Denver, Milwaukee, and St. Louis, listened to the needs of prospective customers, and shared his ideas on the different systems.[86] In his standard

fashion, he wrote a pamphlet on "Refrigeration Machines"[87] and suggested improvements, which the owner may not have appreciated. Half a year later, without prior warning, De La Vergne cut Chanute's compensation even though they had signed a ten-year contract. Frustrated, Chanute terminated the contract.

Moving from Job to Job

Possibly reflecting on his own life, Chanute assembled a table in 1880 showing how the American railroad system had grown from two thousand miles in 1838 (when he arrived as a child in Louisiana), to almost six thousand miles (when he began his career with the Hudson River Railroad) in 1848, to almost ninety thousand miles in 1880. His railroading career had spanned the era from design by empirical "rules of thumb" to the use of mathematical and analytical techniques, introduction of pioneering innovations in bridge design, materials testing, and inspection and the development of construction specifications. Underscoring his contribution to make the Erie profitable again, Chanute revised the table and submitted it for the 1881 Stockholders Report, showing a steady increase in the road's earnings since he joined the Erie in 1873.

The late 1870s marked the end of the Reconstruction period after the Civil War and the beginning of the Industrial Revolution, with its epic battles between labor and capital. This was Chanute's time as the Erie's chief engineer, with additional duties as the assistant general superintendent and superintendent of motive power and rolling stock, weathering corruption, labor disturbance, rate discrimination, bankruptcies, and ever-increasing regulation in an atmosphere of cutthroat competition.

Chanute told fellow ASCE members in 1880, "If Engineers desire to take a higher rank than they now occupy in this country, they must study new paths for themselves, and be no longer men of routine, ordered hither and thither by promoters of schemes and the magnates of Wall Street." It was time for him to move on. On September 2, 1883, Chanute resigned his high-paying position at the New York, Lake Erie & Western Railroad Company to go into private consulting practice, but agreed to remain a consultant for the next year.

In his final report to General Superintendent Bowen, Chanute recorded all the modernizations he had introduced:

> The first and most important one was the reducing of the gauge of the entire road, with all its sidings, yards, and branch lines from the broad gauge to a standard gauge of four feet eight and one-half inches. The magnitude of this work might be dwelt upon almost indefinitely. Such of the equipment as was worth the expense, was changed from broad to standard gauge, which meant the changing of some 10,417 cars and 404 engines. After the engine came out of the shop it was practically a new piece of machinery. The main line is now all double track and laid with steel rails of a standard pattern for the entire

distance, a total of 1,513.5 miles, every rail and fastening being of uniform pattern and interchangeable format. Of the 335 bridges on the road, 272 were replaced with modern structures of improved design and iron construction, and abutments and piers for all iron bridges have been built or rebuilt with first-class masonry. Sixty-nine water-tanks were replaced with frost-proof tanks of Burnham's design. Brick passenger and freight houses were erected along the entire road for handling and caring of property and passengers.[88]

A few months later, George Blanchard resigned in early 1884, and Edwin Bowen earned promotion to the second vice president position. Suffering from a nervous breakdown, Bowen then decided to transcribe Chanute's final report to him and submitted it as his report to the stockholders, and also resigned. A few weeks later, at the annual directors' meeting in November 1884, Jewett resigned, as his contract had ended. Ongoing improvements came to a standstill, and so did the road's earnings. The New York, Lake Erie & Western Railroad declared bankruptcy in May 1893 and reorganized two years later.

CHAPTER 5

Self Realization

AS AN IMPATIENT TEENAGER, the eighteen-year-old Octave had written to his father: "Although I hope to became wealthy in a certain—or rather an uncertain—number of years, I do not think that making money should be the only goal a person should have. Money is only precious because of the pleasures it can get for us. Why should we give up those pleasures for 20 or 30 years in order to buy them at great cost later, when we cannot enjoy them any more? Consequently, I believe that the only way to achieve happiness is to save all one can but without depriving oneself of the many pleasures of one's age and position."[1] In due time, Chanute learned that hard work and good intentions were not always enough to earn a fortune in the rough American business arena.

Chanute was a striking figure, about five feet, six inches tall, weighing about 150 pounds, and careful about his health; he was a well-mannered man of high character, meticulously honest, and displayed an energetic style that commanded attention and respect. But Chanute also had a certain fondness for adventure, mingled with a self-preserving sense of caution that did not always show through.

To achieve results with limited manpower and funding, he had developed communication skills and familiarity with a broad array of scientific, mathematical, and engineering knowledge, an absolute requirement in his chosen career. More than once he had demonstrated his ability to solve complex engineering problems by analyzing and defining technical difficulties in a systematic manner. With the industrial age under way, he also believed that "Progress in civilization may fairly be said to be dependent upon the facilities for men to go upon their business with other men, not only in order to supply their mutual needs, but to learn from each other their wants, their discoveries and their inventions."[2]

The engineering profession had given him administrative experience, personal contacts, and status, but he strived for higher goals and wanted to emulate European civil engineers, who did not just design public works but sought new challenges and possessed the energy to fight for innovation. They were expected

to be businessmen and originators of new enterprises. But as a practicing professional, Chanute at times felt constrained by his employer, and sometimes had to
choose between following policies contrary to his best judgment or resign from
his high-paying position. He was highly aware that several fellow ASCE members
who had gone into private practice were happy in their new chosen roles.

Envisioning his career, Chanute wanted freedom to realize his personal capabilities. He wished to solve problems, to attract clients who would seek his
advice as the authoritative voice on special projects, and to advise in lawsuits
as an expert engineering witness. And Chanute sought recognition and respect
from his peers.

American Society of Civil Engineers

The Society of Civil Engineers and Architects, formed in 1852 in New York City,
used the original charter of the British Institution of Civil Engineers (ICE):
"Civil Engineering is the art of directing the Great Sources of Power in Nature
for the use and convenience of man as applied in the construction of roads,
bridges, aqueducts, canals, river navigation and docks, and in the art of navigation by artificial power for the purposes of commerce, and in the construction
and adaptation of machinery."[3] This was a wide-open and growing field and its
boundaries were limited only by progress in science and technology.

The Society of Civil Engineers reorganized in 1867 as the American Society of
Civil Engineers (ASCE). A year later, in 1868, Chanute was elected a member.
At that time, membership required the recommendation of at least two active
ASCE members, and the candidate had to submit a statement outlining why
he wanted to become a member.

The post–Civil War years were a period of vast railroad expansion and the construction of bridges, tunnels, and other systems were vital to American westward
settlement and prosperity. Many prominent nineteenth-century civil engineers
entered the profession by apprenticing on public works projects or the railroads.
They now wanted some kind of licensing program to prove their qualifications[4]
and to allow the use of "C.E.," as the British had established. In the early 1870s
the ASCE board appointed a committee, with Chanute as one member, to examine how to determine the qualifications of a civil engineer. The committee
resolved that any candidate who wanted to join the ASCE needed to have been
"active in practice as such for seven years (or has graduated as Civil Engineer
and been in practice for five years), and has had responsible charge of work as
Resident or Superintending Engineer for not less than one year; not as a skillful
workman merely, but as one qualified to design as well as to direct public work."[5]

With many engineers living outside the greater New York City area, the ASCE
chose Chicago for its 1872 annual convention, far away from headquarters. A
Chicago Tribune reporter colorfully described its members: "They were all men

of more or less mark, with fine square heads, the bump of mathematics being largely developed in each individual cranium. It was a great group; there were men who had constructed mighty works that will long live after them. They were generally a jolly whole-souled party, bent on unbending themselves, and their humor was enjoyable although sines, cosines, tangents, cube root, the unknown quantity and other astronomical geometrical, trigonometrical, geological and algebraic materials entered into its composition."[6]

After moving to New York in the spring of 1873, Chanute regularly attended the ASCE's twice-monthly meetings. Here he sat among his peers, men who influenced the industrial, political, and financial future of the United States, eager to lend the civil engineering profession the same status they already enjoyed as individuals. The meetings were as much social as they were professional, with members being easily excited by engineering challenges. Less-formal "smokers," or smoking room discussions, in this "Gentlemen's Club" encouraged investigations and spirited debates that broadened Chanute's mental horizons and helped extend his personal and business network. Only half a year after arriving in New York, he was elected to the ASCE Board of Direction; he held this position for three years, then served two years as vice president, starting in 1879, and members again elected him a director for 1885. Chanute was elected president for 1891.

Chanute quickly became friends with the slightly younger Robert Thurston, who had graduated with a civil engineering degree from Brown University in 1859 and who now taught mechanical engineering at Stevens Institute of Technology in Hoboken, New Jersey.[7] Both men had similar interests and enjoyed taking mental excursions into the "borderland of science."

Bridge construction, with all its nuances, was a frequently discussed topic. To better understand the behavior of metal from the scientific and engineering aspect, Thurston favored regular testing of metal parts, because the method of manufacturing affected the metal's characteristics and performance. Even though Chanute had introduced regular testing of parts at the Erie shops, he found this different approach fascinating. He joined Thurston on an ASCE committee to organize a testing laboratory,[8] which was later established at the Stevens Institute. Discussing iron and steel a few years later, Chanute commented: "I can say little more than to add to the general acknowledgments of ignorance and make one of those confessions which are thought to be good for the soul. Having had some experience in the erection of bridges during the past years, I am aware that we yet need much information in order to proportion bridges to the best advantage. We still lack knowledge in the behavior of steel, the proportions of compression members and the influence of the size of a bar upon its strength per square inch."[9]

In 1882 Chanute chaired another committee to investigate how full-size parts, furnished by railroad bridge builders and manufacturers, could be tested on a government-sponsored testing machine.[10] The three main engineering societies (ASCE, AIME, and ASME) collaborated and then submitted their request to the army chief of ordnance to obtain congressional funding for a testing machine at the Watertown Arsenal, outside Boston. In the ensuing years, tests conducted at the arsenal played an important role in establishing engineering standards for construction materials; this was arguably the beginning of materials science research.

Rapid Transit and Safety in Transportation

Of the many problems occupying the urban American mind, few received more careful study than mass transportation.[11] As traffic became more congested in New York City, the demand for rapid transit increased and the public began to complain louder. To help New Yorkers, the ASCE established a committee in September 1874 to first investigate how to improve rapid transit within the city and then to recommend a transportation system for passengers and low-cost delivery of goods. Looking for an easier commute between his home, his office in the Opera House in Manhattan, and the Erie's terminal facility across the Hudson in Jersey City, Chanute became the chairman of the ASCE's Rapid Transit and Terminal Freight Facilities Committee and Matthias Forney, editor of the *Railroad Gazette*, acted as its secretary. The other committee members were Ashbel Welch of the Pennsylvania Railroad, Charles Graham of the New York City Department of Docks, and Francis Collingwood of the East River Suspension Bridge Company.

The five engineers conferred with public bodies and private citizens, seeking opinions on the best way to improve urban transportation. Before finalizing their report, Chanute requested feedback from the mayor of New York City. With his input and approval, the engineers presented their committee report to the ASCE membership,[12] concluding that an elevated, as opposed to underground, rapid transit system had the most merit. Legislation and funding followed almost immediately, and the resulting four elevated rail lines, operated by locomotives, eased the commuting problem during the next decade. The Rapid Transit Committee also suggested erecting warehouses adjacent to stations as the best method of distributing goods, but this scheme was soon forgotten. Twenty years later, another engineer proposed this as his idea, much to the chagrin of Forney and Chanute.[13]

With the trains rattling noisily above city streets, New Yorkers soon complained that the elevated structures were ugly and wanted the trains moved underground. In August 1891, the city hired four engineering experts—Chanute, John Bogart, Theodore Cooper, and Joseph Wilson—to make further suggestions. In his report, Chanute commented on the complexity of this engineering problem and suggested that the Rapid Transit Board should lay down general requirements, allowing latitude in design but not in safety.[14] After several public hearings, the Board of Rapid Transit Commissioners decided that building underground railroads presented an engineering problem of the first magnitude and was a costly project. The city then discharged the consultants, who received $1,500 each for their service.

Nothing happened in the next three years. Then, another committee, composed of Chanute, Clarke, Abraham Hewitt, Charles Sooysmith, and William Burr reexamined the issue. This time the commission decided that an effective solution required constructing underground railways and they approved the construction and funding.

This did not mark the end of Chanute's involvement with New York's urban transit. Even though the city of New York had agreed to the construction of the tracks, property owners did not consent, and the New York Supreme Court appointed three commissioners to report whether the elevated line should be built. Frederick Coudert, a leading figure in New York's legal, social, and diplomatic circles, was selected chairman of the Supreme Court Commission. Now the civil engineer Chanute, working as a consultant for the Rapid Transit Commission, and his former classmate, Fred Coudert, the attorney for the state of New York, collaborated to solve the problem so vital to New Yorkers. In March 1896, the commission unanimously agreed that the road should be built beneath the city. However, the political battle continued.

The future rapid transit in New York, as proposed by a board of civil engineers, including Chanute. From *Railroad Gazette*, January 1, 1886.

For more than two decades, Chanute was involved with New York City's urban transit and proved influential in shaping the system as it evolved from surface transportation to elevated trains and finally to a subway system. In his book *The Marvels of Modern Mechanism*, Jerome Crabtree summarized the situation a few years later: "All the available streets having been covered with surface lines and elevated roads, and the flying machine not yet perfected, the insistent demands for rapid transit forced the engineer to emulate the mole. By going underground they could use space not available for other purposes and propel trains at a higher speed than would be safe in streets used in common with vehicles and pedestrians."[15]

Reinvigorated

Returning to New York after spending three therapeutic months in Europe in the summer of 1875, Chanute joined the Century Association, just as his father had done three decades earlier. Meeting free-spirited thinkers offered a means of relaxation to avoid another burnout episode.

For the upcoming Centennial Fair in Philadelphia in 1876, Chanute chaired the ASCE committee on bridges, and announced: "American engineers have long enjoyed the prestige of having built the longest, boldest and most important wooden bridges in the world. During the past fifteen years, the attention had turned toward iron bridges. Some engineers claim to be in advance of European practice, in regard to economy of materials, facility of erection and possible length of spans, of iron and steel bridges. This fact, if it is a fact, the committee desires to establish."[16] Chanute asked fellow engineers to submit an album of their best bridge design for evaluation in an informal competition; he submitted his Kansas City Bridge book, coauthored with George Morison.

Continuing to research the elements of cost for railway freight traffic, William Shinn and Chanute presented a report, "On Uniform Accounts and Returns of Railroad Corporations"[17] at the 25th ASCE meeting in November 1877. The two engineers had convinced the commissioner of Massachusetts, Charles F. Adams, whom Chanute knew from his time in Kansas, to adopt the practice of uniform reporting, which if it proved successful in Massachusetts might encourage other states to adopt similar systems.

Chanute's reputation also led others to consult with him on the setting of appropriate freight rates. In late 1880, Chanute and William Muir, the former general manager of the Canada Southern Railway, were retained by the Northern Railway of Canada to set and arbitrate rates between the Northern Railway and the Credit Valley Railway in Toronto.[18]

As an elected ASCE director, Chanute frequently presided over meetings and learned more of its internal politics. As part of the modernization process, Chanute recommended adding the statement to the ASCE's constitution: "Its object shall be the professional improvement of its members, the encouragement of social intercourse among men of practical science, the advancement of engineering in its several branches, and the establishment of a central point of reference and union for its members." In November 1879, members elected Chanute a vice president; accepting this office allowed him to influence the direction and scope of the society. He had many goals, but he especially wanted new members to join and he wanted to promote open communication among its members so that all could learn from the successes and failures of others. He also thought that editors of engineering journals, with their broad view and knowledge, should receive invitations to join ASCE meetings. Not everyone favored all of these changes; resistance came especially from conservative old-timers who did not want a mass membership. Chanute also wanted an expanded ASCE library that should "lack nothing required of a library by the student or the accomplished engineer seeking professional knowledge," so he asked fellow members to donate some of their personal books and journals for everyone to use.[19]

Inter-Oceanic Transportation Projects

Even though the first transcontinental railroad had been fully operational between Omaha and San Francisco since 1869, many American businessmen still preferred traveling for two months from New York to San Francisco around Cape Horn. To cut the sailing time of the 14,000-mile all-ocean route in half would require a convenient passage across Central America.

After completing the Suez Canal between the Mediterranean and the Red Sea in November 1869, Monsieur Le Comte Ferdinand de Lesseps, a French financial promoter of the project, believed that the construction of a canal between the Atlantic and Pacific Oceans, across the continental divide in Central America, was just as easy. American engineers differed in opinion; they considered the one hundred–mile-long Suez Canal a ditch dug through a sandy desert, while a canal across Central America would prove far more challenging. Eads, Shaler Smith, and Chanute, among others, believed that a railway, transporting a steamer across the fifty-mile stretch of land, would be faster to build than a canal and would pay handsomely.

To interest American investors in his canal scheme, De Lesseps arrived with his family and a party of engineers in February 1880 in New York. Several ASCE members agreed to show their city's engineering highlights, with the Erie supplying carriages. Chanute, representing the ASCE and the Erie, acted at times as interpreter.[20]

De Lesseps and his engineering party accepted the invitation to the ASCE meeting on February 26. Vice President Chanute escorted the French visitors, greeted by the attending engineers with thunderous applause, to their seats on the platform and welcomed everyone.[21] De Lesseps read a paper on the "Inter-Oceanic Canal Project" in French, and several ASCE members read short papers on the same subject. In the discussion to follow, Chanute set the tone with his first questions: "What are the estimates of the probable cost of the undertaking, and the probable returns," and "Why go to the enormous expense of making a sea-level canal across the Isthmus when a canal with two or three locks on each side of the summit could be made so much more cheaply?" De Lesseps firmly believed in the practicality of his proposal, but he also reminded his listeners that he did not come to beg for money from American capitalists, he only came to explain his scheme. After the meeting, the group went to the gaily decorated Union League Club, where de Lesseps could personally meet the many civil engineers in attendance.

While the engineers discussed their various pet projects, the wives and daughters socialized with the Comtesse de Lesseps. Chanute's oldest daughter Alice was part of the reception committee,[22] because she wanted to practice her French language and show New York from her perspective. A performance in the Opera House was one of the visits she suggested.

One member in the engineering party of de Lesseps was Ernest Pontzen, who had joined the ASCE in 1876 as a "Corresponding Member." Hearing of Pontzen's interest in American railways, Chanute shared his experiences on the western roads and described the improvements he had introduced on the Erie. This was the beginning of a lasting friendship, in which each helped the other with future projects.

De Lesseps and his entourage then traveled to Washington, Chicago, and San Francisco and returned to France on April 1. Work on the Panama Canal began two years later, but the French Company ran out of money in 1889 with less than half the distance excavated. Theodore Roosevelt, at that time the assistant secretary of the navy, and in 1901 elected president of the United States, believed that a United States–controlled canal across Central America served a vital strategic interest, and this massive project restarted in 1904; the Panama Canal opened for business on August 15, 1914.

Following the visit of de Lesseps, engineers gave more thought to lowering the transportation cost on internal canals and introducing a more efficient power source for barges. To stimulate progress, the Erie Canal Company sponsored an engineering competition for an engine and propeller combination for use on canal boats. Chanute also cogitated on how to replace animal power. He envisioned an engine that was safe against explosion, unlikely to break down, quick to start, and operable with little skill by the mule drivers, propelling the boat forward with greater speeds for improved economy.[23] But progress did not come easily.

Annual Address: Progress in Engineering

According to the bylaws, the ASCE president needed to deliver an address on the progress of engineering at the annual convention. President Albert Fink, who was the commissioner of the "Joint Executive Committee of the Trunk Lines" in his working life, had scheduled a trip to England, so he asked for a volunteer to prepare and present the paper in his place. Everyone pointed to one member, always full of suggestions and inquiries, Vice President Chanute. Preparing this paper gave Chanute the opportunity for some personal soul searching:

- What had been accomplished in the United States?
- What had he, personally, accomplished as a professional civil engineer?
- What lay ahead in the engineering field?
- What would the future bring in his personal engineering career?
- What could be changed and improved in the engineering field?
- How could he contribute or should he change the direction of his career?

More than one hundred engineers from throughout the country attended the 12th ASCE Convention at Washington University in St. Louis in late May 1880. At the evening session, Secretary John Bogart announced that President Fink was in Europe and that Vice President Chanute had just sprained his ankle, which was quite painful.[24] Thus Bogart read Chanute's address.

"It has sometimes been said that, when in after centuries historians write of the period in which we now live, they will probably characterize it as the age of material and mechanical development. We are part of the movement, and perhaps can not well be judges over ourselves, but we can realize at least that from the end of the eighteenth century dates the time when man began to utilize the powers of nature and made our world a more comfortable abiding place." In his comprehensive review, Chanute discussed water works and hydraulics, canals, railways, and rapid transit, stressing that engineers should introduce economies developed in the United States in foreign countries.

Preservation of timber and development of lighthouses were two splendid fields in need of improvement. In marine engineering, Chanute stated, "the canal still awaits that improvement in its motive power, and steam is yet to be successfully applied to the propelling of canal boats. Let them [the American engineers] accomplish for the iron and steel steamer what has been done for the railway, the locomotive, and the iron bridge." The telegraph and telephone, introduced by Alexander Graham Bell only four years earlier, electricity, gas engineering and the manufacture of illuminating gas, metallurgy, and mining were ripe for improvement.

Agricultural engineering was the next topic, in which Chanute noted that farmers, especially those living in the Old World, still worked with the same tools as when Christ was born. Many machines that helped farmers were not the work of professional engineers, yet they were true engineering works. Here the most important invention was the cast iron plow, but the "plow of the future," utilizing a steam engine, had not yet been invented: "This problem is a field for the inventive engineer . . . where labor saving inventions could help human society."

In describing the future of engineering, Chanute aimed to raise self-awareness within his listeners: "The professional standard and position of the engineer will need to be raised, a closer union of civil engineers throughout this vast country is needed with a greater concert of action. All the men, eminent or rising, in the several branches of the profession should become an influence on society and truly useful to the country. Let us think how to raise even higher the standards of technical education, of working practice, and of professional honor."[25]

Chanute's philosophical message came through loud and clear; in his mind, only the civil engineer could lift mankind out of the dark ages. But he also wanted engineers to become better known and respected in society and receive full credit for their achievements. The audience listened with enthusiasm and moved to extend a vote of thanks to the author.

Interests in other Engineering and Scientific Societies

Starting in the late 1870s, the development of engineering education opened the path to closer collaboration between engineers and scientists. Several civil engineers attended the American Association for the Advancement of Science (AAAS) Convention in Nashville, Tennessee, in 1877, where Chanute was elected a "Fellow," a member professionally engaged in science and who had aided the advancement of science. Thurston then urged Chanute to join the just-established "Committee on the Relation of Science to the Industrial Arts." He wrote a brief communication,[26] but this topic did not really interest him.

The growth of transportation was much more to Chanute's liking, so he submitted a lengthy discussion in October 1879 to the ICE on William Carson's paper, "The Passenger Steamers of the Thames, the Mersey, and the Clyde," a topic closely related to river commerce in North America.[27] The production of iron and steel for rails was also of continuing interest, so Chanute joined the American Institute of Mining Engineers (AIME), formed as an offspring of the ASCE in 1871, and attended their annual meeting in Pittsburgh in May 1879. Fellow ASCE member William Shinn, with whom Chanute had collaborated on past projects, served as president of this group.

To further improve the image of the ASCE and its "usefulness," ten members, including Vice President Chanute, donated $500 each for permanent headquarters to provide space for meeting rooms and a library. Committee members then purchased a house at 127 East 23rd Street, and Chanute presided at the first meeting in the new headquarters in April 1881.

To draw engineers into professional societies and open communication, Chanute helped form the "Association of Engineering Societies" (AES) to publicize the activities of local and regional engineering groups. The association was chartered in December 1880, initially representing the Boston, Chicago, Cleveland, and St. Louis engineering societies. The board hired Henry Prout as manager and editor of their publication, *Journal of the Association of Engineering Societies*, and its first issue appeared in November 1881.

While Chanute tried to promote the civil engineering profession, his two friends Holley and Thurston were the driving forces in forming the American Society of Mechanical Engineers (ASME) in April 1880. The newly formed group held its first convention six months later and, appreciating the statements made by Chanute earlier, Thurston described the society's goals in his inaugural address: "When we shall have become sufficiently intelligent and good to exhibit 'perfect conduct' in all our relations, we shall find that the promotion of mutual welfare is consistent with the highest, the most perfect system of mutual co-operation and mutual aid in all the truest and highest aims of life. This is the fundamental rule to guide the engineers of this Society in every effort to aid the profession and to benefit the world."[28]

With the slowing of railroad expansion in the 1870s, American steel manufacturers had reduced their production, and Alexander Holley pushed to find other uses for steel. Unfortunately, Holley died prematurely in January 1882, and members of the three major engineering societies, including Chanute, collaborated in preparing his memoir.[29] This was probably the first instance in which engineers of the different societies worked together to honor one from their ranks.

In the past, Fink and Chanute had contemplated how the various engineering societies could work together, but the state of the ASCE in the 1880s made this difficult. To overcome the traditional attitude of many civil engineers, believing that they were superior to other engineers, took effort. In a letter to William Metcalf, a fellow engineer from Pittsburgh, Chanute expressed his concerns: "I am afraid that nothing but cold water could be expected from the American Society in New York, as it seems to be entering upon a reactionary policy."[30]

While the ASCE was in the midst of internal turmoil, Chanute was surprised to receive its nomination for president in 1888. He was flattered but declined, because his consulting business absorbed his time, but he also thought that he could not turn the society affairs around or change the general attitude of its members. However, he agreed to help pull the society "out of its present rut."

Infinitely Curious . . . Aeronautics

In the Middle Ages, the general public usually considered flying as witchcraft, with caricatures of witches flying on broomsticks a favorite in fairy tales and legends. Engineers in the late nineteenth century knew that witches did not have the power to fly, but birds demonstrated this skill daily. For Chanute, manned flight had always posed an unsolved engineering problem,[31] and he surely did not agree with people like Mark Twain that "in this age of inventive wonders all men have come to believe that in some genius' brain sleeps the solution of the grand problem of aerial navigation."[32]

More than once, Chanute had studied the forces of the wind blowing over the open prairies and against obstacles like high bridges, various shaped roofs of buildings, or fast-traveling trains. He had calculated and tabulated the reactive forces, but there was always some factor that did not fit into his calculation. After a while, this searching for a solution grew absorbing and interfered with regular duties, so he rolled up all his accumulated aeronautical material in 1874, tied red tape around it, and decided that it should remain that way until he could take up the subject without detriment to his career. However, he kept up his interest and continued saving clippings.

Trying to recuperate from his nervous breakdown, the forty-three-year-old Chanute sailed for Europe in June 1875. Members of the American Rifle Team traveled on the same boat and demonstrated the use of their Chinese bird kites to estimate the force and direction of the wind during shooting matches. Chanute found the kite's construction ingenious, watching the thin paper, attached to the wings, move under the action of the wind, simulating the flapping of bird wings.

After a brief stop in Liverpool, Chanute continued on to London, sightseeing and spending time on "other items of interest." He had heard of the Aeronautical Society of Great Britain holding its tenth annual meeting on June 23, with James Glaisher presiding over the well-attended gathering. Thomas Moy read a paper on aeronautical progress and remarked in closing: "I believe in work and in making use of the present time, and, as I told you in 1868, so I tell you now, that thought, work, and money can and will achieve aerial navigation."[33] Several engineers from overseas were in attendance, with Chanute most likely one of them.

To talk about flying with like-minded engineers was a somewhat accepted practice, but outside the engineering profession, many considered aeronautics terribly unorthodox, and a young man who cared to stand well with good people found it safest to say nothing about it. Reviewing the "Rise and Progress of Civil Engineering" in his address at the 1876 Centennial Exhibition, ASCE Vice President Theodore Ellis appears to be the first presiding engineer who mentioned air travel in an annual address, while looking into the future. He envisioned bridges of a one-mile span, solar heat as source of power, and that "the navigation of the air would become practicable."[34]

After the Erie Railway declared bankruptcy in April 1878, Chanute became concerned about another stress-related breakdown and sought mental relaxation. Reading colorful reports of Charles Ritchell's recently patented "flying machine" made him compose his own thoughts on mechanical flight: "Every few years a new flying machine is invented, tried and found a failure, each fresh attempt proving by the attention it excites how deep a hold the problem has upon the popular imagination. Inventors constantly retry old experiments, and in their isolated efforts they neither seek to ascertain what has already failed, nor the principles upon which success is possible."[35] He sketched a cloverleaf and placed the letters "D F O U" in each leaf on the top of the first page of his manuscript, however we do not know what message Chanute wanted to relay with this symbol and the four letters. He left the second page blank, and the third page contained calculations on lifting power. He then stopped, as he was not yet ready to delve deep into the topic of mechanical flight.

Always advocating progress in any transportation problem, Chanute mentioned in his 1880 ASCE address, "I do not believe that we are on the end of the great industrial movement which began with the steam engine. I think rather that we stand on the threshold of greater undertakings." The gasoline engine had recently shown success, and Chanute theorized that with this new motive power might come the solution of the last transportation problem. "I suppose you will smile when I say that the atmosphere yet remains to be conquered; but wildly improbable as my remarks may now seem, there may be engineers in this room who will yet see men safely sailing through the air."[35] At the next regular meeting, fellow engineers scoffed at him and told him that it was undignified and unprofessional for an officer of an engineering association to make a public statement hinting of the possibility of air travel and for any man to fly.[36] Abstracts of Chanute's address received wide publication, but his predictions on air travel were usually not included.

Almost a decade earlier, in March 1871, a tornado destroyed homes, twisted bridge abutments, and overturned railway cars in East St. Louis. Concerned about future bridge designs, Charles Shaler Smith, chief engineer of the St. Charles (Missouri) Bridge, and Chanute met the next day to evaluate the damage. At the December 1880 ASCE meeting, Shaler Smith discussed wind pressure on bridges, giving examples of destruction by the most powerful storms, including the St. Louis tornado. In the discussion, Chanute wondered why no one had mentioned the influence of the shape of bridge members. "We all know that a convex surface presents less and a concave surface more resistance than a flat plane to the wind. I dare say that some of us have ascertained this fact experimentally with an umbrella. The umbrella offers a resistance of 76%, if turned with its convex side to the wind, while the concave side offers a resistance of 1.9 times that of the plane. I believe that Sir George Cayley found that a sphere offered a resistance of but 42% of that of a flat plane. If we apply this knowledge to bridge members, we see at once that convex objects, like round ties and round closed columns, will offer less resistance than plane surfaces."[37] Chanute's continuing studies of the wind against fixed structures helped shape his understanding on lift and resistance when applied to aeronautical structures a decade later.

Slowly, engineers began to discuss the possibility of air travel more openly. At the second ASME convention in November 1881, following up what Chanute had stated the previous year, Thurston marveled over future inventions, made possible by three great inventors. "The first is he who will show us how to produce the electric current; the second is the man who will teach us to reproduce the beautiful light of the glow-worm and the firefly, producing energy without a waste." And the third great genius would be the man to build a steam-powered chariot and fly on wide-waving wings through the air. "We have not yet learned to fly like Daedalus, and thus have escaped the fate of Icarus, but the navigation of the air is on the point of real advancement. We are apparently approaching this far goal, and progress in the science of aeronautics is being observed year by year, and there is no department of engineering in which the art of the mechanic has opportunity for greater achievement."[38] This last statement could well have been part of the friendly rivalry between the civil engineer Chanute and the mechanical engineer Thurston. But Chanute fully agreed with his friend, as humans with all their intelligence and powers were beaten daily by every bird and insect. Had the time come to design a flying machine?

The Family

At smokers in the spring of 1882, Chanute and other engineers expressed concern how the civil engineer could fit better into the current social order. Ashbel Welch presented his thoughts in his ASCE presidential address: "Man has ceased to be a unit, and become only an atom of a mass. With the disappearance of the things themselves, the dear old phrases 'family fireside,' and 'domestic hearth,' are rapidly disappearing."[39] In order to survive as a professional, Welch suggested that the civil engineer should remove himself from large bureaucratic corporations and work as an independent consultant. This trend came into vogue in the early 1880s.

   Chanute, in his 1880 ASCE address, had told fellow members, "If Engineers desire to take a higher rank than they now occupy in this country, they must study out new paths for themselves, and be no longer men of routine, ordered hither and thither by promoters of schemes, and the magnates of Wall street." In September 1883 the fifty-one-year-old Chanute decided to take a new path, resigning from the Erie and moving his family back to Kansas City, renting a house at 1236 Broadway. For immediate income to supplement his relatively good financial standing, Chanute bought the Kansas City Transfer Company, a trucking firm with one hundred horses and seventy wagons that hauled merchandise between businesses and freight depots on either side of the Kaw River. His brother-in-law Charles James became the treasurer and general manager. Accustomed to quick verbal communication, Chanute had a "speaking telephone," No. 819, installed. Five years later, Chanute's company earned recognition as "one of the greatest public conveniences in the city, and efficient in all details of the transfer business."[40]

   Octave's and Annie's first-born, Artie, had graduated in 1880 from the Sheffield Scientific School of Yale College, the undergraduate school of what later became Yale University, with a "Bachelor of Philosophy" degree in chemistry.

His first job was as a chemist in Leadville, Colorado. In 1883 he married Mary Vanderbilt Lockwood of Tarrytown, New York.

Chanute's younger son, Charles Debonnaire, felt unsure about his future. Vacationing in Colorado, he watched his brother work as a mining engineer and decided to follow in his footsteps. In the fall of 1883, the almost sixteen-year-old Charley entered the Phillips Academy at Andover, Massachusetts, to become a mining engineer and chemist, with minor studies in economics and business.[41] He soon developed health problems and wanted to leave school to become a cowboy, so the concerned parents looked for an apprenticeship in the cattle business. Charley spent a year on a ranch near Pueblo, Colorado, working hard outdoors with ranchers and cowboys; he then reconsidered his career choice and asked to go back to school.

Octave's and Annie's three daughters attended preparatory boarding schools in Tarrytown. Alice, the oldest daughter, married George Boyd in 1884 and moved with her husband to New York City.[42] The two younger daughters, Nina and Lizzie, moved back to their parents' home after graduation.

In 1887, between his various consulting assignments, Chanute considered going to Europe and Annie and the daughters wanted to spend the summer with the Moores, Annie's baby sister Virginia and her husband Latimer. So, they jointly decided to cancel the lease on their rented house and instead board in a Kansas City hotel. The family sold furniture and other belongings or put it in storage; but after only one week at the Woodland Hotel, everyone agreed that this lifestyle was unpleasant. They took the furniture out of storage and moved to a house at 444 West 15th Street.

Octave and Annie Chanute (late 1880s). Courtesy of the Chanute Family.

One Sunday afternoon sitting around the kitchen table, the daughters insisted on learning more about the family of both their parents. Lizzie especially never grew tired listening to her mother talking about her Virginia heritage, but her father did not share memories about his family or childhood. Knowing that he had changed his family name when he became an American citizen, Lizzie decided to stop at the public library when visiting her older sister in New York. She found several entries in the various biographical encyclopedias on Chanut. There was a Pierre Chanut, French ambassador to the court of Christina of Sweden and a friend of the philosopher René Descartes, and Pierre's son, the clergyman Pierre-Martial Chanut. There was also Antoine Chanut, the director of the parochial school at the Saint Sulpice cathedral in Paris. She transcribed all the information and then asked her father if she was related to these men; he hesitantly replied that he thought they were, but he also stated, "It is folly for an American to take pride in ancestry. It does not matter who any person's ancestors are. In America, everyone stands on their own merit and should make a name for himself. If there is any ancestor worship in this household, let your mother continue to brag about those Virginia kin folks."[43]

Slowly, Chanute gave in and began recalling memories of his childhood days, his parents, and his maternal grandmother; he also contacted family members in France, asking for background information. Writing letters was a favorite pastime; almost every outgoing letter, whether business, pleasure, or family, he recorded and copied into letterpress books, neatly shelved and ready for use as reference. Sitting at his big flat desk, surrounded by densely filled bookcases, Chanute could smoke his favorite cigars and think about any subject that caught his fancy. In later years, his daughters sometimes made mild fun of their father, because he never discarded a piece of paper unless it was all "scribbled up,"[44] and even then Chanute reportedly considered saving it.

Consulting Engineer

Life as an independent consulting engineer was different. Now Chanute had to think of projects that were profitable and yet fun, but he also had to concern himself with collecting consulting fees, keeping track of expenses, and paying bills. For additional income, he still sold real estate in Illinois and Kansas and bought property when it looked financially promising. Not long into this process, Chanute became ill with malaria and could not work for several months; as a future precaution, he bought a health insurance policy for each family member.

The cost of carrying freight on railroads, essential for economic growth, continued to rise, and Chanute revisited the subject for the 1883 ASCE annual meeting.[45] With continuing rate wars threatening commerce, the general public wanted the government to regulate the rates. Chanute submitted a lengthy paper on the true cost of handling freight, hoping that wider knowledge of the actual cost would give railroad managers a better understanding of the equities of rate making, but also the often-doubtful desirability of cultivating one class of traffic over another.[46] The *Railway Review* published his research in two consecutive issues and as a pamphlet.

Civil-minded engineers promoted urbanization and civic improvements, but they also helped cure the evils that resulted from too rapid growth. Street pavements, lighting, clean water for human consumption, and sewage systems were important topics for every community. Chanute became involved with almost all of them, but held particular interest in sewage and storm water, which were usually combined and then discharged into adjacent rivers, lakes, or tidal waters. Because these same waters had to supply clean water for human consumption, the offensive matter needed to be purified. This interested him not only as a civil engineer but also as a taxpayer. He felt that it would save him and other citizens considerable money to have separate instead of a combined systems adopted in his hometown.[47] He read his paper to members of the Kansas City Academy of Science in November 1883, suggesting that city fathers should resist combining storm water and refuse from homes. Because Kansas City was built on a hill, open channels and street gutters could collect storm water and drain it downhill into the Missouri River. The widely read *Engineering News* of New York republished Chanute's paper, with the author's note inviting criticism. Robert Moore, ex-sewer commissioner of St. Louis, did just that and argued that the combined system offered a safer and healthier choice for any community and the space for open gutters could be used for other purposes.[48] Several other engineers also submitted thoughtful comments on the sanitary and economic aspects of each system.

After reviewing the competing arguments, Kansas City officials adopted Chanute's proposal. They hired Walter S. Dickey, an Irish immigrant, who set up the Dickey Clay Company to manufacture pipes for transporting household waste to sewage treatment plants. But political bickering continued, as some citizens felt that the concrete used for the pipes was of bad craftsmanship, and sued the city. Remaining involved in seemingly all aspects of the sewer issue, Chanute then served as an expert witness, explaining the characteristics of concrete in May 1885.

Street paving was another project for Kansas City, as the growing population demanded paved streets, especially during the rainy season. One of its citizens suggested studying whether stone- or woodmen should lay the pavement, and he wanted "an engineer like Mr. Chanute come here and spend a week or two, making a personal examination of our streets. He could give our people practical advice that would ensure first-class streets at the lowest cost."[49] Chanute accepted this project, and the city paved the streets with bricks.

When the James Street Bridge across the Kaw River between Kansas City, Missouri, and Wyandotte, Kansas, neared collapse, Chanute accepted the assignment to have the bridge repaired. The Wyandotte County clerk, Mr. W. E. Connelly, recalled years later that Chanute was modest in his charges and that the repaired bridge stood for more than thirty years, carrying ever-growing traffic. In an interview in the mid 1920s, Connelly described Chanute: "I found him a very plain, unassuming gentleman. He impressed me as having in reserve many resources, which he was not using on the job in hand. . . . He was a very sociable man and enjoyed conversation with an appreciative listener. He had a fine head, covered with curly hair, which he allowed to grow to a considerable length, and he dressed with individuality and had not much regard to fashion. If the day was warm, he invariably took off his coat and vest on entering my office. But he liked to have his shoes shiny and it vexed him if he stepped into the mud under the bridge."[50]

As an independent consultant, Chanute occasionally declined projects. The Washington Aqueduct Tunnel in New York City was one of these projects to which he responded: "I would be happy to act on the aqueduct tunnel, but for two trifling circumstances, 1) I know nothing whatever about aqueducts, and 2) I know nothing whatever about tunnels. Otherwise, I think I am alright."[51]

One of Chanute's next consulting assignments presented multiple challenges. Recognizing that Galveston was the largest and, hoped to be, the most important city in Texas, its citizens sought a deep-water entrance into their harbor. James Eads, who was familiar with the regimen of the Mississippi, agreed with Chanute in the mid-1870s that excavating the river and moving the silt to either side of the shore would offer the easiest way to deepen the entrance. In 1877 Congress authorized the dredging of a one hundred–foot-wide and twelve-foot-deep channel; the army Corps of Engineers then built jetties into the Gulf of Mexico to maintain the channel, but the waves soon refilled it with silt. Many years later, Carl Schurz told fellow senators that for the past thirty-seven years, the military engineers had been "scratching and scraping at the mouth of the Mississippi, and today the depth of the water is no greater than it was then."[52]

In the early 1880s, several civil engineers thought themselves better qualified than federally trained military engineers to supervise federally financed projects like the Galveston jetties, even though the government would carry out the work. In December 1885, 168 engineers from the various engineering branches, with Chanute as an ASCE officer, signed a document as the "Council of Engineering Societies on National Public Works" and submitted it to the U.S. Senate and House of Representatives.[53] Congress eventually authorized the appointment of independent engineers to serve as advisors on various commissions.

Galvestonians were not willing to give up on their deep-water entrance, and one of their wealthy citizens contacted Chanute. After inspecting the area, he thought of modifying the original work. "I saw clearly how success could be accomplished," he wrote confidentially to a fellow civil engineer who was involved with the original work.[54] Chanute proposed to position the jetties in a slightly different location and to install a "tidal gate," a design he also submitted to the Patent Office. But technical problems plagued the work, and Chanute gave up; the Gulf won again.

Another Galveston project was waterworks for its forty thousand citizens. Chanute and his associates submitted a bid to supply fresh water, pumped from Sweetwater Lake and several nearby artesian wells, and install three hundred hydrants at a cost of $150 each. The city accepted Chanute's proposal, amended the city charter to include a water tax, and erected waterworks in 1888–89.

Bridge Reform Act

The art of building railway bridges, an integral part of the expanding railroad system, developed from using wood and masonry to wrought iron and then steel. Even though Congress regulated the design of bridges across major navigable rivers, bridges collapsed for various reasons, and newspapers highlighted each new bridge disaster, usually blaming civil engineers for the accidents. Considering the impulsiveness of travelers and the many thousand miles of rails with all their curves, culverts, and bridges, Chanute considered disasters hardly avoidable.

In 1873 a ten-member ASCE committee, which included Chanute, investigated ways to avert bridge accidents. In their final report, the committee concluded that they could classify disastrous accidents into three groups.[55] First, some bridges were erected by incompetent or corrupt builders and then accepted by incompetent or corrupt officials. Second, bridges of good design and material failed from neglect on the part of their owners or from injury to the material during transportation or erection. And last, bridges were destroyed by derailment caused by trains being driven too fast or by wear from heavier traffic than that for which the bridge was designed. The committee also recommended mounting a plaque conspicuously on the bridge abutment, listing the names of the engineers responsible for the works.

A decade later, many people remained vague on the main principles of bridge designing and of the necessity for safe structures. Some engineers believed that the practice of bridge building was a matter of federal interest and should apply uniformly to all states, while others felt that each state should hire an independent engineer to look after the safety of their bridges. Members of the Engineers' Club of Kansas City assembled facts of the so-called "evils" in the methods of bridge building and appointed a committee to recommend bridge building legislation. Having participated in each step in the developing art of bridge building, Chanute accepted the chairmanship. After much discussion, the committee determined that American bridges were generally safe and bridge engineers did not deserve the sweeping criticism that the public had recently advocated. "Engineers have evolved within the past few years with new methods of designing and building bridges, by which greater strength and safety are obtained from a given amount of materials. Bridges, built by experts, seldom fail from errors in design. The trouble is that all bridges are not so built, and that bridge owners do not realize what risks they are incurring by letting contracts without professional assistance." In the committee's opinion, with which Chanute agreed, the building of bridges should be placed in the hands of expert bridge engineers, who would hold responsibility for the final product.[56]

Next, Chanute compiled a draft for "An Act to Promote the Safety of Bridges" and submitted it to the congressional committee "On Bridge Legislation." The draft was embedded into Senate Bill No. 2330, "An Act to Authorize the President of the United States to invite the International Congress of Navigation to hold its Ninth Session in Washington," and civil engineers should communicate with the chairman of the Committee on Foreign Affairs.[57] Congress did not enact a comprehensive bridge regulatory act until 1906, and by that time, a complete revision of every aspect of bridge construction was warranted.

Chicago, Burlington & Northern Railroad

With improved water and rail transportation, cities in the upper Midwest began to grow and prosper. The region around St. Paul and Minneapolis, Minnesota, developed surprisingly quickly, and Boston Party investors saw advantages in connecting the Chicago, Burlington & Quincy with the Twin Cities. The Chicago, Burlington & Northern Railroad received a charter in the fall of 1885, and Albert Tonzalin accepted its presidency. Naturally, Bostonians wanted the railroad to be completed quickly,[58] so Tonzalin contacted Chanute to accept the construction of the forty-two bridges on this new line.[59] This was Chanute's first major consulting job.

Much of the grading of the three hundred–mile road along the eastern bank of the Mississippi was done before the harsh winter season began.[60] The Union Bridge Company, formed in 1884 by Thomas Clarke and associates, won the contract for the three major bridges, and Chanute discussed quality requirements with his friend Clarke, just as they had done in the past.

While in Minneapolis, Chanute served as an expert witness on grade crossings and explained his opposition in court. "Two roads meeting would wish to hurry their trains through and would perhaps be antagonistic to each other; they would also be governed by different rules and regulations."[61] It would lead to better and safer results to entail the extra expense for either a bridge or a tunnel to avoid grade crossings at all times.

The bridge across the Galena (Illinois) River and the laying of the tracks going north to the Wisconsin state line encountered political problems right away. The Illinois Central, with Edward T. Jeffery as its general superintendent, claimed that the Chicago, Burlington & Northern rails were being laid on its right-of-way. The lawsuit that followed was quickly resolved, because the United

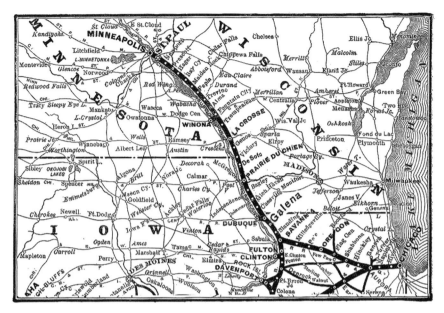

Chicago, Burlington & Northern Railroad. Chanute accepted the contract to erect forty-two bridges between Galena, Illinois, and Minneapolis, Minnesota. From *Grain Dealers Journal* (1891).

States Circuit Court dismissed it, but the court action delayed track work. The construction of the drawbridge across the Galena River was not delayed, being the first bridge finished. The news release announced, "The structure was pronounced one of the handsomest in this section. It has three fixed spans of 200 feet each and a draw-span of 200 feet."[62] The design of this bridge was novel, having railroad tracks on one side and a wide sidewalk for pedestrians and carriage traffic on the other. Six years later, a major flood washed out many bridges. The Burlington's bridge at Galena, designed by Chanute, not only withstood the flood-swollen river but also split the largest section of the Illinois Central's bridge that washed downstream against it.

Bridges crossing the marshy country around the Grant and Platte Rivers in southwest Wisconsin went up quickly.[63] The Wisconsin River Bridge at Prairie du Chien presented a few problems owing to quicksand scouring around the piers, and Chanute suggested locations for rip-rap protection. The North La Crosse Viaduct, with its long approaches and five spans, presented more engineering challenges, as it had to cross the tracks of several roads. With winter arriving early, the wide, shallow Chippewa River, about seventy miles south of the Twin Cities, froze quickly and Chanute miscalculated the depth of the underlying sand. A rapid thaw and early spring rains undermined the just-erected piers, causing the Union Bridge Company to increase their charge by $10,000 to make the pier foundations deeper. The St. Croix River Bridge at Prescott, Wisconsin, a few miles southeast of St. Paul, required a long trestle approach on the Minnesota side, but the first passenger train from St. Paul crossed the bridge on May 30, 1886, with curious spectators from miles around watching this historic event.

Managers pushed as hard as possible to finish the tracks and bridges for the Burlington road. On August 9, 1887, an excursion train with railroad officials ran over the road, and two months later, the first scheduled passenger train left Chicago for the Twin Cities without fanfare. Building the forty-two bridges provided a wide variety of adventures and challenges; Chanute reported them in good order and well maintained when he inspected them two years later.[64]

Chicago, Santa Fe & California Railway

In the 1880s, the Atchison, Topeka & Santa Fe Railroad (the public soon abbreviated the road's name to Santa Fe) ran as a through-car line from Kansas City west to California and south to the Gulf of Mexico. To compete with eastern

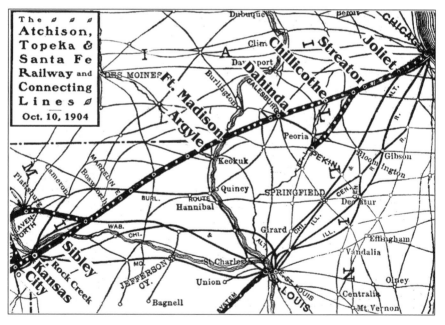

Kansas City to Chicago Airline (1887–1889). Chanute's assignment included the bridges at Rock Creek, Sibley, Argyle, Fort Madison, Dahinda, Chillicothe, Streator, and Joliet. From *The Atchison, Topeka & Santa Fe Railway and Connecting Lines* timetable, October 10, 1904.

trunk lines expanding west, William B. Strong, the president of the Santa Fe, researched the feasibility of building tracks between Kansas City and Chicago, the center of westward traffic operations. To make a profit, the shortest route had to connect the two cities;[65] Chanute's friend, Chief Engineer Albert Robinson, reportedly took a ruler at an early planning meeting and drew a line between the two cities to establish the route. Whether this story is true or not, the fact remains that the Santa Fe's airline was as straight as a railroad line could be.

The Chicago, Santa Fe & California Railway Company received its charter in December 1886, and Strong instructed Robinson to have the line between Kansas City and Chicago in full operation by January 1, 1888.[66] Having known Chanute since 1871 when they both lived in Kansas City, Robinson offered him the construction contract for the bridges. This included obtaining approval from the secretary of war for the two major bridges across navigable rivers, procuring of bids, designing and specifications, and general overseeing of the work. The compensation was $10,000, plus travel expenses. Chanute accepted this project, even though his other consulting contract with the Burlington was not yet finished.

Having just testified in Minneapolis on the evils of railroad grade crossings, Chanute favored Robinson's approach that the tracks of other roads should be crossed on an elevated level to avoid traffic hazards and commercial conflicts.[67] The first bridge on this airline crossed the drainage canal in Joliet, Illinois. Chanute ordered a through-girder bridge from the Joliet Bridge & Iron Company that also supplied the bridge across the Vermillion River in Streator. To cross the Illinois River with its low level banks, two miles east of Chillicothe and only fifteen miles away from his very first bridge at Peoria, Chanute designed a Howe truss channel structure with three 150-foot spans and a 302-foot-long swing span, bringing the overall length with its approaches to 1,412 feet. The area east of Dahinda, Illinois, on the Spoon River was rough country with deep cuts, requiring heavy grades and substantial embankment with trestlework. The bridge across the Des Moines River, about twenty miles northwest of Fort Madison, Iowa, ran about nine hundred feet in length, and the Grand River Bridge between Bosworth and Rothville, Missouri, was a 460-foot-long, low level bridge. The 268-foot iron viaduct at Rock Creek, Missouri, coming over the bluff just east of Kansas City, bridged the tracks of the Missouri Pacific and the Chicago & Alton with four spans.[68]

The Santa Fe's Bridge at Fort Madison, Iowa

The Santa Fe's plan to run straight from Kansas City to Chicago required a bridge to cross the Mississippi somewhere near Fort Madison. Under an original 1872 charter, the Fort Madison Bridge was authorized either as a high or a low bridge. After checking the site with its flood levels, Chanute suggested that the Santa Fe build a low bridge with a pivot span. To expedite the project, he made several trips to Washington to discuss matters with the secretary of war, William C. Endicott, and others. Congressional approval came in March 1887.

After discussing this bridge with Thomas Clarke, Chanute and the Santa Fe then awarded the Union Bridge Company the contract for this 1,925-foot-long massive railroad and wagon-bridge. The bridge featured eight spans and a four hundred–foot draw, as it extended from Fort Madison across the Mississippi and the wide sand bars on the Illinois side. Having learned from Pier No. 4 of the Kansas City Bridge that had been undermined twice, Chanute designed a new form of "coffer-dam" to erect the pivot pier. Here, the crew built the enclosure and lowered it about nineteen feet below the waterline to create a dry work environment and stabilize the preliminary structure. Workers drove the first pile in the permanent work on March 16, 1887, and on December 7, the first passenger train crossed the bridge, erected at a cost of just over $580,000.

Resident Engineer Walter W. Curtis worked closely with Chanute, who now encouraged him to compile a detailed paper on the experiences in building this bridge. In the meantime, Chanute had also nominated Curtis for membership in the ASCE, but the budding engineer did not have the required time in the field, so his membership was declined. Curtis then submitted his paper with Chanute's input to *Engineering News*,[69] a leading civil engineering periodical. Three months later, with the additional substantiation, Curtis was elected an ASCE member.

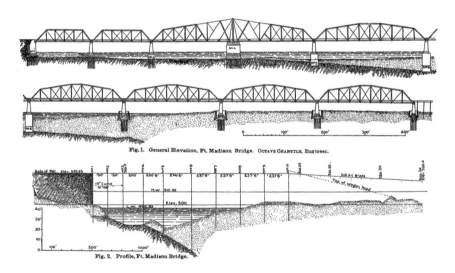

Fort Madison Bridge across the Mississippi River, designed by Octave Chanute, consulting engineer, assisted by Walter W. Curtis, resident engineer. From *Engineering News*, June 2, 1888.

Hoping that the arrival of the railroad and the bridge across the mighty Mississippi would entice people to move to Fort Madison, Chanute helped develop this town of about eight thousand mostly German citizens by contacting several architects. He also bought real estate, especially in the area of the future center of town, including a large corner lot, where he erected a building to house five stores, a bank, and apartments above, envisioning that each store would pay about $125 a month for rent. Income from this property later supported his youngest son Charley and his wife Emily.

The Santa Fe's Bridge at Sibley, Missouri

Congressional approval for a bridge across the Missouri at Sibley, about twenty-five miles east of Kansas City, had been granted in 1872 to the Kansas City, Topeka & Western Railroad and the Santa Fe acquired these rights in 1884. Even though the Missouri was relatively narrow at Sibley, the bridge construction presented several engineering challenges, bringing back memories of building the Kansas City Bridge. Chanute believed the flooding to be similar to that in Kansas City,[70] but local engineers questioned the flood level of thirty-seven feet in 1844; he then located several high-water-marks at Sibley that corresponded with the marks at Kansas City and determined the most suitable location for the bridge; the Santa Fe's chief engineer Robinson approved the plan in February 1887 and crews broke ground one month later.

Because Chanute's longtime friend Thomas Clarke had decided to leave the Union Bridge Company later in 1887, Chanute considered other bridge builders. The firm of Sooysmith eventually received the contract for the substructure, foundation and masonry piers, and the Edgemoor Iron Company won the contract for the superstructure. The 1,900-foot-long iron trestle viaduct was supported on twenty-one iron towers, resting on masonry pedestal piers[71] with three 400-foot-long trussed spans and five shorter ones.

The Santa Fe's William Breithaupt served as resident engineer in charge of the superstructure, and Chanute recommended hiring John F. Wallace, who had just completed his assignment on the Union Pacific, to oversee the substructure and the improvement work upriver.

In a personal discussion, Wallace mentioned his philosophy to the twenty-year older Chanute: "Cultivate your friends, as the measure of your success depends on your friendships. Next, cultivate a willingness to serve others, regardless of time, place or circumstances, not measuring your services pound for pound, as for compensation, but let the measure of what you give be overflowing."[72] Sharing the same general philosophy, the two engineers maintained their friendship, especially after both moved to Chicago and were members of the same engineering societies.

Bridge across the Missouri River at Sibley, Missouri. From O. Chanute,
J. F. Wallace, and W. H. Breithaupt, "The Sibley Bridge." From *ASCE Transaction* (September 1889).

On July 30, 1887, in place of a regular meeting, members of the Engineers'
Club of Kansas City, including ladies and invited guests, took an excursion to
Sibley. Because the Santa Fe tracks were not completed, the Wabash, St. Louis
& Pacific Railway supplied a car for the sixty-one people in the party. After an
excellent lunch, Wallace and Chanute explained the works, and a few adventurous members descended into the caissons, from which they emerged rather
quickly. The contractor for the substructure, Charles Sooysmith, then offered
a steamboat ride on the Missouri, and the group returned to the Kansas City
Union Depot late in the evening.

Much to Chanute's frustration, the first winter blizzard arrived earlier than
expected, stopping work for a week. Management wanted the bridge completed,
and so did he. Aware of the urgency, Wallace decided not to go home for the
holidays, and the appreciative Chanute sent him a box of cigars and a few bottles
of wine to help pass the Christmas time away from his family. Continuing snow
and ice storms with extremely cold weather delayed the bridge construction
and prevented the Santa Fe from opening the route on January 1, 1888. The
first freight train crossed the Sibley Bridge three weeks later, even though one
of the center spans was not quite completed,[73] and income-generating freight
and passenger traffic began in mid-February.

The Sibley Bridge was the most imposing bridge on the Santa Fe at that time.
The company completed it for just over $770,000 and in less time than any other
bridge of similar size. The three engineers Chanute, Wallace, and Breithaupt,
coauthored a paper on their experiences in erecting this bridge, read in June

1889 and published in the ASCE *Transactions*.[74] It received the "Thomas Fitch Rowland Prize" in 1890, a $50 cash award given annually to a paper describing accomplished works of construction, cost, and errors in design and execution. The ASCE Award Committee also recommended that the masonry specifications of the Sibley Bridge piers should be used as industry guidelines as the current best practices.

The story of the Santa Fe's five hundred–mile airline between Chicago and Kansas City clearly shows how careful planning, efficient handling of workmen, and engineering skills achieved success. Chanute's two large bridges that required congressional approval, three major bridges without federal approval, and several smaller ones, were built in just over one year. William Strong, president of the Santa Fe, gave his competitors a run for their money when the finest trains initiated passenger service on April 29, 1888, racing between Kansas City and Dearborn Station in Chicago on a schedule of thirteen hours, forty-five minutes.

Other Consulting Work for the
Atchison, Topeka & Santa Fe

Chanute continued working for Robinson as the Santa Fe expanded, while keeping up his obligations as part-owner of the Chicago Tie Preserving Company (see chapter 6). In 1887 he surveyed the Chicago, Kansas & Western Railroad to connect with the Union Pacific at Benedict Junction, Kansas, building the tracks along the section of Verdigris River that frequently overflowed its banks. Under Chanute's supervision, the crews raised tracks in several places on trestles and the bridge across Simmons Creek[75] had to be made significantly longer than first anticipated. With amazement, Chanute saw the many changes since he had put the "Joy Roads" down twenty years earlier.

Many of Chanute's consulting contracts with the Santa Fe ran smoothly, but errors happened when construction was rushed. The Leavenworth, Northern & Southern Railway, incorporated in October 1885, connected Leavenworth, Kansas, with Atchison in the north and Wilder in the south.[76] Robinson considered Chanute's design of the Maple Avenue Bridge across Three Mile Creek at Leavenworth a "botch." After visiting the site, Chanute assured Robinson that he would have the problem fixed. "There will be no loss to the railroad company, save the great annoyance of not having things exactly right in the first place."[77]

The Santa Fe expected good business from its ten-mile-long Pueblo & Arkansas Valley Railroad, bringing valuable coal to market. Unfortunately, Chanute erred in his initial design of the two-spur bridge near Pueblo; workers needed to add extra piers in the late stages of construction, increasing the overall cost and delaying the road's opening. Another costly error was the belt line bridge at the Corwith rail yard in Chicago. The bridge floor was designed too low and did not make a smooth connection with the remaining tracks. Although all these errors were clearly disclosed, they probably strained relations between the Santa Fe, particularly Robinson, and Chanute.

Robinson, who was now also general manager, needed to cut expenses, so he checked with Chanute how much additional money the Santa Fe would need to pay him. Not wanting to overcharge, but also not to lose money, Chanute consulted other independent engineers; the general consensus was to charge either a flat, monthly fee or charge between 2 and 5 percent of the final cost. So Chanute wrote to Robinson: "I shall be satisfied with the latter figure, and from this you may deduct whatever has been the expense of making good the mistakes or omissions made by my assistants, say at Corwith, and bridges at Pueblo and at Leavenworth."[78]

The Santa Fe declared bankruptcy in 1893 and Robinson joined the Mexican Central Railroad as president. Keeping up their friendship and showing his appreciation, Robinson mailed Chanute for many years an annual railroad pass with his greetings for the holidays.

The Next Step in Life and Career

Throughout his life, Chanute had worked hard to gain respect for himself and his chosen profession, but he observed that some engineers did not always act as professionals. Therefore, he wrote an editorial for *Engineering News* on "The Ethics of Good Consulting Practice," realizing that "a Code of Ethics, or of professional good manners and fair dealings, is necessarily a matter of slow growth. General acceptance of any rule of conduct has to be universally accepted. Engineering is a new profession, therefore it can not be said to have any such rigidly established rules of practice."[79] He felt strongly that elementary rules of good manners had to be observed within the engineering profession, just as they were observed at home.

Envisioning his future career (others might refer to this as retirement), Chanute sought a business that would supply income, challenge, and fun. Preservation of timber met these criteria, as few people cared about preserving natural resources in the early 1880s. To persuade railway managers about the benefits they would derive from longer-lasting wood would prove challenging and it would be fun to travel the country to meet personally with railroad officers, his long-term associates, to discuss his newest ideas.

Everything seemed to fall into place later in 1888. Annie's brother Charles James did not want to manage the Kansas City Transfer Company anymore, because he had bought a house at 2418 Fifth Street in San Diego, California. Meanwhile, Chanute had set his mind on wood preservation, being already part owner of the Chicago Tie Preserving Company and wanted to move to Chicago. So Charles and Octave decided to sell the Transfer Company; they finally found a buyer who offered time payment. Charley Chanute had just graduated from the Phillips Academy and agreed to move to Kansas City to manage the Transfer Company until the business would sell completely. Now the Chanute family could relocate to Chicago, Charles James could move to San Diego, and son Charley had a job in Kansas City.

Over the next two decades, Chanute continued to accept short-term consulting contracts, but no long-term bridge contracts. He spent his time creating a small industry to preserve timber, which made him money. However, at the same time, he spent money on his newest "side issue," aeronautics. The next three chapters will discuss each of these endeavors separately.

CHAPTER 6

A New Industry

GROWING UP, Chanute's father Joseph stressed that energy and persever-ance were necessary to achieve success. After becoming part of the workforce and watching his coworkers, the budding young engineer wondered why some of his acquaintances had worked hard throughout their lives without becoming wealthy. There must be something else he did not know, and the eighteen-year-old Octave theorized to his father: "I can see only one reason—it is how clever or cunning the person is. Moreover, circumstances are all important; they elevate men of little wealth and hold back many of talent." In due time Octave learned that a person had to meet the people and capitalize on the circumstances that opened the doors to success and wealth; and, sometimes the mysterious quality of "personality" also made things happen—or not.

Trying to capitalize on the circumstances in the early 1880s, and knowing that a first-class railroad required one wooden tie for every two feet of track, or 2,600 ties for each mile, Chanute gave more thought to the necessity of preserving wood. The hurried railroad construction into the outskirts of civilization had quickly depleted the once-plentiful timber supply, as trees were cut for railroad ties, poles, and bridges faster than nature could replenish them. Placing ties into the ground sped the natural decaying process, thus workers needed to replace them every few years. This presented an enormous waste, and a related ques-tion was even more thought provoking: "In the country known as the American desert, about 500 miles west of the Missouri, rainfall drops down to 6 to 9 inches a year. Here the soil is abundantly rich, but nothing can be grown without more moisture. It is of the greatest importance for us to know whether the cutting away of forests does diminish rainfall, and if by planting trees we can increase it."[1] Preserving natural resources deserved much more attention than it had previously received.

Having used a variety of wood in bridges and tracks, Chanute thought he was familiar with its idiosyncrasies, but he soon learned better. Robert Thurston, professor at Stevens Institute, had included timber in his lecture on construction materials, and De Volson Wood had incorporated Thurston's lecture notes in his textbook on the resistance of materials.[2] This provided Chanute a first source of factual information on wood and the causes for its decay.

As part of the 1880 federal census, the government initiated a survey of the nation's forest resources, including their economic properties and uses. Initial results forecast a rapidly diminishing timber supply, so it appeared of vital necessity to prevent its wholesale destruction. With this precept, the American Society of Civil Engineers (ASCE) appointed a committee in 1879 to investigate the best method to retard decay in timber. Secretary Bogart prepared a circular letter and mailed it to railroad managers and chief engineers. "Several processes had been in use in Europe for many years with satisfactory results, but the same result had not been the rule in the United States. Whether these failures have been due to imperfect preparation, or because we have blindly followed the foreign methods, we are not prepared to say; but the whole question is so important that a report of our failures even more than that of our successes, would be of great value to engineers."[3] This mailing received hardly a response, because most railroad men believed that ties would wear out mechanically before they would rot and that it was cheaper to renew ties than to spend money on preservation chemicals.[4]

Finding the topic of wood preservation intriguing, Chanute asked in January 1882 to take over the chairmanship of the ASCE's Preservation of Timber Committee. He prepared another circular letter and added with his characteristic dry humor: "Please keep this Circular and Card in a conspicuous place until you have answered it." The ASCE mailed about one thousand letters with a postcard, self-addressed to Chanute at the Erie, to practically everyone who may have had any experience with the subject. This time, there was a good response.

One of the first people to respond was Hugh Riddle, who had recently advanced to the presidency of the Chicago, Rock Island & Pacific Railroad. His roadmaster had informed him that their road had treated ties in the 1860s, but to ensure success, the work had to be done with care. Several other American roads had also used treated ties, recognizing that they wore longer and held the spike better than untreated ties, but their chief engineers were frequently unsure when and how their ties had been treated and how long they had been in the tracks.

Another report negatively impressed the committee members. After the Civil War, the Memphis & Charleston Railroad purchased a patented process using a mixture of arsenic and salt for $50,000, but "the poisonous nature of the ingredients soon brought disaster. Some shingles were treated for a railroad freight house at East St. Louis, and all the carpenters who put them on were taken very ill and one of them died. Then ties were treated for the road; cattle came and licked them for the sake of the salt and died, so that the track for ten miles was strewed with dead cattle."[5] This was clearly not acceptable; any chemical used for timber treating had to be safe for the environment.

Studying the available literature and patents, the eight members of the Preservation of Timber Committee realized that the published information at times contradicted the underlying data. To better understand the facts, Chanute began a systematic investigation, asking specific questions and opening communication to bring people together to discuss this far-reaching topic.[6]

To increase awareness of preserving timber, Chanute's committee arranged a display of treated and green timber for the National Exposition of Railway Appliances in Chicago in May 1883,[7] including micrographs of wood surfaces and wood cell structures. Members of the Western Society of Engineers arranged the exhibit, which was subsequently donated to the University of Illinois at Champaign.

The Preservation of Timber Committee submitted its final report for the ASCE Convention in June 1885,[8] pointing out that nature's once bountiful timber supply was indeed dwindling and that railroad managers should look at simple economics: lengthening the life of ties would decrease the yearly demand for renewals, slowing the decline of the growing forest. In this comprehensive report, Chanute collated and discussed the almost 150 reported experiments, grouping them under the names of the best-known processes and explaining the merits of each.

> **Open Vat Process** — Timber was steeped in a mercury chloride solution, probably the most powerful antiseptic next to creosote.
>
> **Burnett Process** — Timber was steeped for several weeks in a zinc chloride solution, as developed by Sir Joseph Burnett. This process was subsequently improved by forcing the preserving liquid under pressure (using a partial vacuum) into the wood in sealed cylinders, as developed by William Wellhouse.
>
> **Bethell Process** — Timber was injected with hot creosote in a closed cylinder under pressure, as developed by John Bethell. The success of this process had been established worldwide.

In the final section, Chanute described the lesser-known processes using various chemicals and also included the letter reports from each contributor, along with a translation of an article from a German railroad journal. The committee concluded that creosote, commonly used in Europe, provided the most effective protection for timber, but this chemical was expensive and not readily available in the United States, so they recommended burnettizing as their top choice.

By presenting past practices, the timber preservation report, authored by Chanute, provided guidelines for wood preservers and railroad officials. Major engineering publications discussed it favorably; its lengthy abstract in the *Railroad Gazette* closed with the statement that "railroad corporations and others interested would do well to give careful study to this report which contained more practical information on the subject than has ever before been published."[9]

It was arguably the most comprehensive publication available and laid the foundation for a new industry.

Opportunities

Europeans, with their limited timber supply, had perfected the processes to preserve wood starting in the 1840s, but such an industry did not exist in North America. In the early 1880s, when Chanute became more interested, timber remained plentiful and cheap. Untreated oak ties cost about $0.70 each and lasted about seven years in the roadbed, but the much softer hemlock ties cost only $0.25 and could be burnettized for an additional $0.25, making them last twice as long as the untreated oak tie.[10] This provided a 65 percent savings, but most railroad managers did not recognize the economic advantage.

Joseph P. Card moved to St. Louis after the Civil War and established the St. Louis Wood Preserving Company in 1879 to treat wooden blocks for the St. Louis Bridge across the Mississippi and for the city of St. Louis. These commonly used wooden street paving blocks, sawed from large timber and resembling bricks, cost less than stone and the chemical treatment made them almost as hard and durable. Hearing of the ASCE committee's efforts, Card shared his knowledge, as he had done extensive experimenting and improved the patented Wellhouse treating process.[11] Several railroad engineers, including Chanute from the Erie, became interested and shipped a few ties to Card for treatment as an experiment.

To grow his business and become better known, Card joined the ASCE as an associate member in 1883. Realizing that expertise and promotion were necessary components for a successful business, he mentioned a partnership to Chanute. The two men could complement each other: Card knew the business of timber treating, loved experimenting, and held patent rights for the Wellhouse process, while Chanute knew the right people, how to handle the promotion of the business, and how to convince railroad men to conserve natural resources. Developing a working relationship, Chanute suggested to Card, "If you are compelled to reduce the royalty, the contract had better be made on the basis of 1 cent a tie and a rebate made yearly. Will see Mr. Robinson as I am leaving for Denver tomorrow where I will also meet some of the Denver & Rio Grande people, some of whom are my personal friends."[12]

The two men formed "Card & Chanute" in 1884, and Chanute indeed contacted his many railroad friends. "Have become interested in burnettizing timber and ties which is yielding excellent results and which I desire to introduce. Please advise me whether I may make you a proposition, and if so, how many ties you would have treated per annum. It is desirable that as many shall be gathered at one place as possible, to save moving the works,"[13] he wrote to the general superintendent of the Pennsylvania Railroad.

Because Chanute had agreed to continue working as a consultant for the Erie under a one-year contract, he also contacted Edwin Bowen, his former associate, who related the Erie's interest in erecting a tie treating plant at Horseheads or Elmira, New York. Bowen stressed, however, that other roads should participate to lower the overall cost.[14] The Erie decided against building a plant, as it went through major management changes in the fall of 1884.

First Permanent Tie Treatment Plant:
Las Vegas, New Mexico

The Atchison, Topeka & Santa Fe Railroad, with Albert Robinson as chief engineer, was arguably the first railroad in the United States to respond to the timber crisis as forecast in the 1880 government survey. Struggling with the short lifespan of the mountain pine and the high shipping cost for ties from the East, Robinson shipped 384 pine ties to the St. Louis Wood Preserving Company for treatment by their Wellhouse process, as an experiment. Workers inserted the treated ties into the Santa Fe's tracks in 1881 and 1882 near La Junta, Colorado, and Topeka, Kansas;[15] a small plate was nailed to each tie for future identification. Four years later, Chanute and Card walked over the roadbed, and the little tag helped identify the treated ties. At the 1905 Railway Engineering Association (AREA) meeting, Chanute recalled how this record keeping turned bad: "One day, the roadmaster said to a new section foreman, 'I want you to take good care of those tags.' 'I will, sir,' said the foreman; and the next day he took them all off and sent them into the office, thus destroying future records."[16] By removing the tags, a concerned, but ignorant, worker had destroyed any record of the life of the ties.

Robinson was pleased with the results of the first batch of treated ties, so he took the next step. Knowing of Chanute's latest endeavor, he asked Card & Chanute for a proposal in late 1884 to erect a plant. Eventually, they reached an agreement to build the first permanent North American tie treatment plant at Las Vegas, New Mexico, about forty miles east of Santa Fe. The plant, with its two cylinders, was built for $30,000 and began treating ties using the patented Wellhouse process in July 1885.[17] Samuel Rowe, working for Robinson as resident engineer, became the manager and quickly learned the plant's operating procedures.[18] Because reliable statistics would prove useful, Chanute asked him to document and record every step in the operation.

The relatively simple treating process took about nine hours, at a cost of $0.15 to $0.20 per tie. The six-foot diameter and one hundred–foot-long treating cylinder had a door on one end, and a narrow-gauge track ran inside over its entire length. Workers placed ties on specially designed flatbed retort cars that were run into the steel cylinder; a door sealed the chamber. Next, they introduced a vacuum that removed much of the air and sap from within the

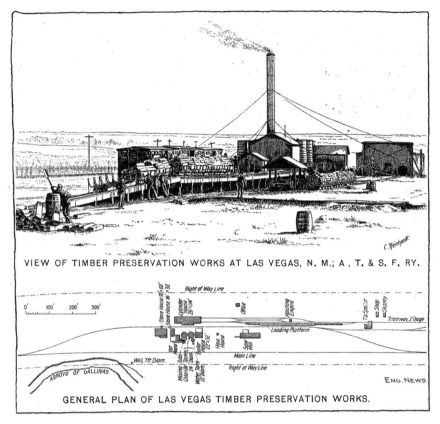

VIEW OF TIMBER PRESERVATION WORKS AT LAS VEGAS, N. M.; A, T. & S. F. RY.

GENERAL PLAN OF LAS VEGAS TIMBER PRESERVATION WORKS.

View and general plan of the timber preservation works at Las Vegas, New Mexico. From *Engineering News*, September 13, 1894.

ties. The vacuum was then released and the ties treated under pressure with zinc chloride, then with tannic acid and gelatin. After the treatment, the retort cars were pulled out and the ties moved to a storage area to dry.

To give a better understanding of the cost involved, Robinson reported to stockholders in 1892 that 2,431,622 new ties were placed into the tracks, at a total cost of $920,224. The overall operating budget of the Santa Fe for the fiscal year 1891–92 was $4,954,113, thus the cost of new ties amounted to almost 20 percent of the road's total operating expense. Fewer than 10 percent of ties received chemical treatment, because they were only used in areas where timber decayed quickly.

The Santa Fe declared bankruptcy in 1893; royalty payments of $0.01 per treated tie were slow in coming, and in June 1896 the Wellhouse patent rights expired. In 1897, Chanute estimated the average life of a treated tie on the Santa Fe's line as 10.7 years.[19]

Union Pacific Timber Preserving Works:
Laramie, Wyoming Territory

The Union Pacific Railroad (UP) had treated cottonwood ties, a very soft and perishable material, with zinc chloride for about one year near Omaha in the mid-1860s. The treatment was not particularly successful, but the ties held up long enough to enable the UP to build their tracks across the Rocky Mountains. When the treated ties became brittle, the company replaced them with untreated oak ties.[20]

Twenty years later, Chanute submitted a proposal to the general manager of the Union Pacific, S. R. Callaway, to erect a tie treatment plant along their line for $25,000, exclusive of grounds, tracks, and water supply,[21] operating under the Wellhouse patent and paying a royalty of $0.01 per tie to Wellhouse, Card, and Chanute. At a planning meeting in New York, Chanute was pleased to meet two people he knew well. Charles F. Adams, the former director of the Kansas City Stock Yards, who was now president of the Union Pacific, and Chanute's former classmate Frederick Coudert, who was their newly elected government director. The contract for erecting works at Laramie in southeastern Wyoming was signed during the winter of 1885–86.

Meanwhile, Chanute also finalized a contract with the Chicago, Rock Island & Pacific Railway to erect a plant in Chicago while he was still working as a consultant for the Burlington to erect bridges between Galena, Illinois, and the Twin Cities in Minnesota. Spending much time in Chicago, he also tried to attend the monthly WSE meetings.

Card purchased the equipment for the Union Pacific and shipped it to Laramie. In May 1886 Chanute went west to erect the plant with the help of the railroad crew and local contractors. Because this was his first installation, he really wanted his partner to come out and make sure that everything was done correctly, but Card was busy with their just-erected Chicago plant.

On July 26, 1886, the two-cylinder works at Laramie began to treat mountain pine ties that were floated down the big Laramie River and lifted with a tie boom out of the water. Chanute submitted a full description of the plant operation to the *Railroad Gazette*.[22] After spending three months in Laramie, Chanute finally returned to Kansas City. He not only needed to look after his other consulting businesses, but he also had to prepare for the American Association for the Advancement of Science (AAAS) meeting, at which he needed to deliver an address on the progress in mechanical sciences in Buffalo, New York (see "The 1886 Buffalo Meeting" in chapter 7).

To cut operating cost, UP management shut down the Laramie plant in 1887, after treating more than 200,000 ties; the works burned later the same year. Losing income from royalties, the concerned Chanute wrote to the Union Pacific's president in June 1888: "If the financial necessity for this suspension still continues, nothing more is to be said; but if it does not, I am more than ever convinced that it would be a measure of true economy to resume treating ties at Laramie. Although only two years have elapsed since ties treated at the works were laid down, perhaps an examination of the earlier ties laid, which if untreated would begin to decay, will furnish some indication of what can be expected."[23]

Ten years later the Union Pacific's general manager ordered an examination of all ties in the tracks; the treated ties were all still sound, verifying Chanute's earlier statement.[24] Chief Engineer John B. Berry then asked for a proposal to erect a new plant, which Chanute gladly supplied; the company erected a portable plant in Rawlins, Wyoming, in 1902 and workers rebuilt the Laramie plant in 1903.

Chicago Tie Preserving Company

Management of the Chicago, Rock Island & Pacific Railway Company knew of the economic advantages of treated timber, and their newly elected president, Ransom R. Cable, requested a cost estimate for a plant in 1885. The initial contract specified 100,000 hemlock ties treated annually for five years, so Joseph Card moved his works and family from St. Louis north to Chicago in the spring of 1886. He erected the Card & Chanute works at the corner of 15th and Clark Streets and started operations in May 1886. The following year, the two partners formed the Chicago Tie Preserving Company, a joint stock company, with Card as president and Chanute as secretary/treasurer.[25]

To publicize their tie treating business, Chanute placed his card in engineering publications and wrote similar letters to his many friends in the railroad community:

> Since publication of the preservation of timber report by the American Society of Civil Engineers, the railroads have manifested much interest in the subject. The Atchison, Topeka & Santa Fe contracted with Mr. Card of Saint Louis and myself to design and erect burnettizing works at Las Vegas, NM, which are now working to their entire satisfaction. We have contracted with the Union Pacific to erect similar works at Laramie, and with the Chicago, Rock Island & Pacific, to build works in Chicago. . . . Will you be kind enough to advise me whether the time has arrived when your line will be prepared to consider propositions of this kind?[26]

In 1886, with the nation's timber supply forecast to dwindle, the Department of Agriculture hired Bernard Fernow as chief of its Forestry Division. After studying the 1885 ASCE report, he frequently consulted with Chanute and incorporated his input into later Forestry Division publications.[27]

The Chicago & North-Western Railroad became the second paying customer of the Chicago works, but their one-time order was not large; they shipped 2,003 cedar and 5,934 hemlock ties for treatment in 1888. Chanute checked on their status in 1903 and again in 1906, but even though most ties remained solid in the tracks,[28] C&NW managers were not convinced of the economic advantage.

To stay close to the business and for easier commute to other consulting projects, the Chanute family moved from Kansas City to northern Chicago in October 1888, renting a row house at 5 Ritchie Place; some of Annie's relatives moved next door, to 7 Ritchie Place. Their oldest daughter Alice had separated from her husband after an unhappy marriage of five years, so she also moved to her parents' home in Chicago with her sickly son, Octave.

Except in the freezing winter months, the Chicago works were usually busy, using much water, which was released as steam into the air as part of the treating process. A new Chicago ordinance required companies to stop releasing smoke from chimneys; the chief inspector of the Chicago Health Department entered complaints in August 1890 against several businesses, including the Chicago Tie Preserving Company.[29] Chanute was pleased to explain that their "smoke" was not really a "nuisance of very smoky chimneys," it was just vaporized clean water.

To learn more about treating the various wood species, but also to entice new customers, Chanute offered to treat ties gratis as samples. Chief Engineer R. Angst of the Duluth & Iron Range Railway shipped eighty-five white pine, eighty-five tamarack, and eighty-six red Norway pine ties to the Chicago works in the fall of 1890. These trees were plentiful in the central and western part of the United States, and Chanute was curious how these soft and perishable woods would absorb the chemicals and how long they would last in the tracks. The road placed the treated ties into their mainline north of St. Louis in the spring of 1891. Upon inspection in 1904, Chanute counted almost 90 percent of the treated pine ties still in the tracks after more than thirteen years of heavy service,[30] but the Duluth & Iron Range management remained uninterested in erecting a plant.

While Octave kept busy with his various projects, Annie and her daughters usually spent the summer months at "Crest Haven," the country home of the Moores at Lake Minnetonka, Minnesota. Early in July 1890, Octave joined his wife and daughters, as did his oldest son Artie and his family, living in Denver, and Charley, living with his family in Kansas City. Having the complete family together inspired them to hire a photographer to document the occasion.

Back in Chicago, looking for a residence closer to town, the Chanute family moved in spring 1892 to 413 Huron Street, the first of the newly erected three-story duplex brownstone row houses in the Potter Palmer division of Streeterville. The house had a bay window to the street, a small fenced backyard, a basement, and space for a den on the top floor. Enjoying the backyard, Octave began to throw out crumbs for the sparrows, especially in the cold winter months.

The Rock Island extended their contract for another three years in 1891 and

The James Family at Crest Haven, Lake Minnetonka, Minnesota, July 7, 1890. Sitting in the front on the grass, left to right: Charles Chanute, Matie Chanute, Louis Chanute, W. Octave Chanute, Arthur Chanute. Seated in chairs, left to right: Annie Chanute, Elizabeth Chanute, Martha Cooke, Elizabeth Chadwick, Mary Chadwick. Standing in rear, left to right: Octave Chanute, Octavia Chanute, Alice Boyd, Latimus T. Moore, Virginia Moore, Charles James. Courtesy of the Chanute Family.

increased the quantity to 300,000 ties annually. To meet the demand, Chanute and Card added a third cylinder in 1892 and a fourth one in 1897. In the new contract, the road paid $0.16 for treating hemlock and $0.18 for red, black, or water oak ties, and Chanute received an annual railroad pass. In late 1893, the Chicago works moved to South Englewood, on land owned by the Rock Island.

Only a few months later, the Rock Island's president reported problems with previously treated ties, so Card and Chanute walked over the tracks of the Rock Island & Peoria Railway and counted ties in the roadbed and those discarded. Neither man had an idea why some ties had apparently decayed faster than others. Trying to understand the causes but also to document when a tie was treated, they began to stamp each tie with the year of treatment in 1895,[31] just as they had done at Las Vegas and Laramie years earlier.

The Chicago Tie Preserving Company provided steady income for the two families. Card operated the plant and Chanute solicited business, but also maintained his involvement with engineering societies and other consulting projects. This situation, however, changed in 1894. The Pullman strike of railroad men in the spring stopped all rail traffic from coming into the yard; plant operations resumed in mid-August, but there was no income for several weeks. Then the fifty-seven-year-old Joseph P. Card died on October 22, 1894.[32] To continue the business, Chanute took over as president, the widow of his former partner, Catharine Card, became part-owner, and their son, Joseph B. Card, became treasurer. Tragedy also struck the Chanute family. Octave's and Annie's oldest son Artie and his wife Mary, living in Denver, lost their ten-year-old boy Louis to scarlet fever in April 1894. Not even a year later, Artie died, leaving behind his wife, nine-year-old William Octave, and seven-year-old Matie.

The Kansas City Transfer Company finally sold in 1894; son Charley, who had married his brother's sister-in-law Emily Crocker Lockwood in 1889, moved with the two children from her previous marriage into the row house next door at Huron Street. Having Charley available as a trained chemist, Chanute wanted to find out why some ties decayed quicker than others even though they were treated the same way. Chanute then initiated systematic experiments in their small test cylinder, always treating three ties at the same time to confirm any result. In a step-by-step process, he evaluated each stage of the treatment by changing the strength of the chemicals and the treatment time and then extracting wood fibers from varying depths of the tie to analyze and confirm the amount of absorbed zinc chloride.[33]

Timber harvesters usually floated the logs down a river or lake. Checking the water-soaked wood, Chanute realized that the high moisture level prevented the antiseptic solution from penetrating, which could well have been the reason for the faster decaying of some ties. As a result, he established a seasoning procedure to lower the moisture content prior to the chemical treatment and urged others to do the same.[34]

Because the Wellhouse patent expired in 1896, Chanute introduced his three-step treating process in early 1896 and wrote Rock Island management, "I have not the least doubt that the ties treated in that manner will show a very long life. It considerably lengthens the entire process, and only the better results are a compensation for the higher cost involved."[35] Now timber was first piled for seasoning and the antiseptic solutions were then injected under vacuum, one after the other. As a final assurance, the crew weighed the retort truckload of forty ties before and after the treatment; if the ties absorbed less than expected, they were dried and retreated. Sharing his knowledge, Chanute submitted a well-researched article to the popular *Cassier's Magazine*.[36]

The new treating process ran smoothly, so Chanute decided to put the younger generation in charge of the plant operation. He had already hired two men to build his "soaring machine" and was anxious to watch the men fly (see "From Investigating to Trials" in chapter 7), thus he wanted to spend as little time as possible at the plant.

All operations ran smoothly in 1897, but then rumors spread among Rock Island men that ties treated by Chanute's works had to be replaced in less than seven years.[37] If this was true, the financial security of the Chicago Tie Preserving Company stood in grave danger. To verify the rumors, Chanute asked Sam Rowe,[38] now living in Chicago, to inspect the Rock Island tracks. Rowe could not confirm the rumor, so Chanute forwarded his report to the Rock Island's management.

Chanute still wondered how the rumor could have started. When the Rock Island began using treated ties in 1886, the foremen were instructed to record the movement of all treated ties, installed or removed, but record keeping became relaxed. Seeing Rowe's report, management wanted facts, so Hilton Parker, a vice president of the Rock Island, ordered a full accounting of every tie in every division. This count indicated a lifespan of 10⅔ years for treated ties in the tracks east of the Missouri River and a couple of years longer for ties laid west of the river,[39] verifying that Chanute's tie treating was done well. A few years later, Chanute wrote to Wilbur Wright, "There is a limit to the amount of neglect which a business will endure. I discovered in 1897, when after spending some $10,000 in experimenting [with full-size gliders] and assisting others, I found some danger of losing the business which gave bread and butter,"[40] recalling his tale of the responsibilities of being a business owner.

The Pennsylvania Railroad shipped their first batch of ties to Chicago for treatment in 1889. Three years later Chanute discussed with their chief engineer if the quality of the roadbed might affect the ties' lifespan. To find out, one-half of the next shipment of treated ties was placed in stone ballast and the other half on gravel. The chief engineer reported in 1903 that 38 percent of the hemlock ties in stone ballast were still in the tracks, while none of the ties in the gravel roadbed remained.[41]

European roads favored beech wood for ties, and Chanute was curious how they would stand up in America. Because beech ties were readily available along their line, the Pennsylvania road shipped a load for treatment. While they seasoned in the spring of 1900, these beech ties split more than usual, so Chanute wrote to the Pennsylvania's superintendent in Logansport, Indiana: "I found that French & German obviate the cracking by driving in the ends bands of hoop iron, rolled to a taper and bent to the shape of an 'S.' There are generally three sizes and these are driven in when the ties are first piled to season, in case of needs, also when new cracks appear during seasoning. This European remedy might be a good one to try." The Pennsylvania Railroad may have been the first American railroad to use these S-shaped thin steel clamps to prevent splitting. More than a century later, these steel clamps are still seen in ties embedded in roadbeds across the country.[42]

Checking on timber bridges he had built in the past, Chanute thought of retarding decay by applying the burnettizing solution between the opposing surfaces of the joints, after the structure's erection, to hermetically seal the cut

S-clamp for railroad ties, introduced in North America by Chanute
to prevent splitting, usually applied during seasoning.

ends. He submitted his patent claims "Preserving Timbered Bridge Structures"
in late 1886;[43] the office granted the patent in 1890. Applying the same thoughts
to ties, Chanute wondered if the antiseptic solution would penetrate differently
through the surfaces of hewn or sanded ends. The chief engineer of the Norfolk
& Southern Railway shared Chanute's curiosity and shipped five hundred pine
and green ties to Chicago in 1896 with their ends cut differently. Only time
would tell how long the ties would last in the roadbed.

Then came another blow to Chanute's business; the purchasing agent of
the Rock Island inquired how to cancel their contract early. Chanute initially
replied in anger, but then rewrote his reply. Exchanging many letters and telephone calls, the parties terminated their contract in late 1903. Four years later,
the president of the Rock Island reported to stockholders a heavy increase in
maintenance expense; among others, more ties needed to be renewed than in
the previous year. As the road encountered difficulties securing ties, they erected
a creosoting treatment plant in 1908 to again preserve ties to extend their life in
the roadbed. This confirmed Chanute's philosophy, but he no longer profited.
Between May 1886 and December 1903, the Chicago Tie Preserving Company
had treated more than 5.6 million ties for the Rock Island, with an average
lifespan of almost eleven years.

The Exposition Universelle de Paris, 1889

To celebrate the one hundredth anniversary of the French Revolution, the French wanted millions to visit their World's Fair in Paris. Having contemplated a European trip for some time, Chanute had several good reasons to go. He could visit his elderly mother and family members, see how the French operated their timber-treating works, and he could also attend the Aeronautical Congress (see "Aeronautical Research" in chapter 7) scheduled during the World's Fair. His wife Annie preferred to stay home, but his oldest daughter Alice, still emotional about her son's death in early January, and his middle daughter Lizzie, were anxious to come along on this, their first trip to Europe.

The Chanutes left Chicago on March 23 at 5:30 P.M., "after saying goodbye to my wife and Nina. Have been reading over my old journal and am profoundly thankful by the altered circumstances under which the present journey is undertaken. I am no longer anxious about the future of my family in case of accident to myself. I have amassed a competence and made my will so as to provide for the future," Chanute entered in his diary.[44] The *Lakeshore Limited* traveled at fifty miles per hour through the night, arriving in New York the next morning just before ten o'clock.

The Chanutes joined a small group of civil engineers and sailed for Europe on the *City of Chicago*. This steamship, owned by the Inman Line, had four masts rigged for sails, and a single-screw steam engine that propelled her at a speed of fourteen knots. "Daughters got sea-sick, but not I," were his diary notes. They arrived in Liverpool on April 6 and spent one week with members of the Institution of Civil Engineers (ICE); the group then traveled to the European mainland.

Spending spring in Paris was a dream of many young Americans; the thirty-year-old Alice and twenty-five-year-old Lizzie much enjoyed walking up and down the boulevards, where the *Parisiens* met the *Américains*. They visited the nearby Luxembourg Gardens and their father reminisced about playing here with his two younger brothers and their nurse. As in days long gone, street peddlers offered barley sugar candy, which Lizzie and Alice tasted for the first time. Chanute also bought some mirabelle plums in a local grocery store and remarked that they were just as good as he remembered from his childhood days.

Saint Sulpice was one church they visited more than once, because Alice and Lizzie were fascinated by the mystery surrounding this imposing structure. The giant clamshell near the entrance used as a baptismal font caught their attention, and their father confirmed that he, too, was baptized in it. The church interior was spectacular, especially the Chapelle des Mariages, with its beautiful stained glass windows, where Octave's parents exchanged marriage vows sixty-eight years earlier. Crossing the interior of the church, they stepped over the mysterious

Meridian Line, or "rose line." Impressive also were the rose windows with the letters P and S in their designs, honoring Pierre and Sulpice, the patron saints of the church.

They also walked through the extensive Père Lachaise cemetery and located the grave of Joseph Chanut. With great emotion, Octave retold the circumstances of his father's death twenty years earlier. Being interested in their father's family, Alice and Lizzie spent time with their grandmother, Elise de Bonnaire Chanut, who introduced them to other members of her family.

Walking through the city, Chanute could see the *Tour de Eiffel* from just about everywhere, it being taller than any other structure in the world. Some Parisians had initially protested against building the useless tower, but it soon became a focal point of the fair. Chanute agreed with fellow engineers that it was a thoroughly well designed piece of engineering.

The Paris World's Fair opened its doors on May 6, and the Chanutes were part of the throng. Six weeks later, a contingent of 275 American engineers of every branch of the profession arrived at Calais. The greeting committee, which included Chanute and his daughters, escorted them on a special train to Paris. Gustave Eiffel, the president of the *Société des Ingénieurs Civils*, introduced the American engineers to the mayor of Paris,[45] and a few days later, he provided an educational tour of his tower. Taking the lift to the upper platform, everyone enjoyed the view from 990 feet above ground and then stopped for lunch on the first platform.

This World's Fair was far more brilliant than any of its predecessors; beautiful as the buildings were by day, the scene became even more impressive at night, when rows of electric lights outlined the buildings. While some engineers and their families absorbed the sights of Paris and the Fair, Chanute and his friend Ernest Pontzen visited several tie-treating plants to learn how Europeans operated their works. Hearing that members of the American Institute of Mining Engineers (AIME) planned to tour Germany, visiting mines, steel mills, wine cellars, castles, churches, and gardens, Chanute joined them.

Selecting the best steamer from an engineering standpoint for the return trip to America was not an easy task. The major passenger lines usually scheduled their finest ships to sail on the same day of the week, racing to break the speed record for Atlantic crossings to the delight—or disappointment—of their passengers. The *City of New York* and *City of Paris*, launched in 1888 by the Inman Steamship Company, were the first liners with twin propellers that reduced the travel time to less than six days. The *City of New York* had just been overhauled and its owners hoped that she would win the coveted "Blue Riband of the Atlantic." The American engineers selected this steamer and left Liverpool on August 7. Much to the delight of her passengers, she crossed the Atlantic faster than the *Teutonic* from the White Star line.[46] Such an ocean race was electrifying and exciting, like watching a horse or greyhound race, except here one was part of the race by being on the boat!

Instead of continuing straight home, the Chanutes visited Toronto for the annual AAAS meeting. Chanute read his abstract on "Preserving Wood against Decay,"[47] but attendance at the mechanical engineering section was low, so little discussion resulted.

While Chanute was in Europe, the Chicago City Council appointed a committee of two hundred and fifty leading businessmen to bring the next World's Fair to Chicago, but New York, St. Louis, Washington, and Montreal were also interested in staging a fair to celebrate the upcoming four hundredth anniversary of Christopher Columbus's arrival in the New World. Naturally, each city proclaimed itself the most suitable venue to hold a prestigious fair.

The Chanutes returned to Chicago just in time for the next monthly WSE meeting, at which many wanted to hear news about the Paris Fair. Chanute recalled: "The President of the French Republic was also a civil engineer and he told the group that he received them, not as Americans, but as his comrades, and the mere fact that they belonged to that profession was a passport to all places of interest, but also to all social entertainment. They came away with the most gratifying impression of the high social position which our profession occupied abroad and with the hope that a similar recognition might be obtained here."[48]

Two members of the recently formed Chicago Fair Executive Committee, Edward Jeffery and John Carson, attended as guests. Realizing that there were just two years to create a fair grander than the one built by the French in four years, if Congress took action early in 1890, Chicago promoters wanted Chanute's input. To gather details about the exposition, organizers then asked Jeffery, the former manager of the Illinois Central, and Chanute, the civil engineer, to go again to Paris, offering an all-expense-paid trip plus $1,000 a month for up to three months. Even though he had just returned, he accepted, but added a note to the contract that the amount due him should go to his wife if something unforeseen should happen.[49]

Being the most experienced traveler, Chanute suggested buying tickets on the *City of Paris* that had just earned the "Blue Riband" for the fastest westbound travel. Jeffery and Chanute, with Charles Dawley and Charles Schlacks as stenographers and secretaries, left Chicago on September 28 and arrived in Paris on October 15. The fair closed its doors two weeks later, and during the teardown, workers took time to explain the various aspects of this successful event. Having learned from the past, the French had designed the buildings for the 1889 fair in a temporary fashion, quickly erected and easily taken apart, allowing the material to be sold after the fair closed. Watching the dismantling process proved educational, and the four Chicagoans carefully studied every aspect for their own planning back home.

Chanute saw his seventy-six-year-old mother a few times. On the last day, she gave him a letter to take home to her granddaughters, remembering them fondly from their visit half a year earlier. The elderly Elise then sold her house in Ivry and moved into a small apartment in Paris, where she died on October 6, 1893. Octave's uncle, Etienne Gérin, settled her estate.

The four engineers left Liverpool on December 7 on the *Umbria* of the Cunard Line, hoping for a fast trip home. The Atlantic crossing was slow and rough, taking nine days instead of the advertised seven, and Chanute recalled his first Atlantic crossing, fifty years earlier, when he and his father sailed for forty-four days. The four engineers used the travel time to compile their report on everything they had learned. Among others, they proposed all buildings and physical properties of the fair in Chicago to be of temporary construction, except for the Memorial Art Palace. Much to Chanute's surprise, Jeffery had their joint report printed in London without giving the coauthors' names.[50] This report was of considerable value in the ensuing political struggle to hold the next World's Fair in Chicago.

The Western Society of Engineers:
Preparing for the World's Columbian Exposition

Many easterners were convinced that Chicago, a "cow-town" situated eight hundred miles west of the Hudson River, could not stage a successful World's Fair, but Chicagoans put forward their claims with great nerve and bombast. The "Windy City" not only won the political struggle to host the World's Columbian Exposition but also profited from staging the prestigious affair.

The overall goal of the Columbian Exposition was to illustrate the progress of transportation in all its branches, on land, water, or in the air.[51] To achieve this, Chicago Mayor DeWitt Cregier asked WSE members to participate on the planning committee. Several engineers, including Chanute, subscribed to the fair's capital stock by "investing" $1,000, but their goal differed a bit from the mayor's. They wanted to stage an engineering congress, the first such congress to be held in the United States, and possibly in the world.[52] Chanute was convinced that the civil engineer would emerge from the lower plane of a trade into a true profession if this Congress would prove a success. To help this happen, he accepted the presidency of the General Committee of Engineering Societies, working closely with WSE President Elmer Corthell, but he also became first vice chairman of the World's Congress Auxiliary that would coordinate the various congresses, scheduled during the World's Fair, under the motto "Not things, but men."

To create an engineering triumph that dwarfed all its predecessors, American engineers had to trump Eiffel's tower. Charles Hastings from Kansas City proposed a tower that would shine in the daylight or by electricity during the night. George Morison suggested "The Chicago Sky-Piercer," a 1,120-foot-high steel structure, similar to the Eiffel tower but all-around bigger.[53] Then, George Washington Ferris of Pittsburgh proposed a huge rotating wheel, or a tower connected at each end to form a circle, from which visitors could view the entire fair and the city in the distance. The Ferris project was so grandiose that the committee dismissed it as unrealistic, but several WSE members, including Chanute, were intrigued. Organizers finally granted the concession to Ferris in December 1892, and workers erected the 264-foot-diameter Ferris Wheel in the Midway Plaisance, where it became an instant success with the visiting public.

With all the talk about engineering in the daily press, the Chicago commissioner of public works asked the WSE for the names of three engineers to look into a new waterworks tunnel project. This request stirred controversy within the WSE, as some members thought it unprofessional to accept such an appointment. Chanute strongly disagreed and explained that ASCE members had frequently made investigations of public works, and WSE members should comply because they were best qualified to perform the required tasks. Going back and forth, the WSE president determined that the investigation of public works was a function for members to exercise. Chanute's motion was adopted and the president presented several names to the commissioner.[54] This project set the stage for future civic involvement by WSE members.

The 1892 annual WSE banquet provided a good reason to celebrate, and more than 150 members attended, with most involved in the planning for the upcoming fair.[55] Thousands of visitors were expected to come, as Chicago's rail network was second to none, but the local transit system was inefficient and outdated.[56] In an effort to ease the local commute, two Chicagoans, Max Schmidt and Joseph Silsbee, had invented a multiple dispatch railway. With the help of several WSE members[57] they formed the Multiple Speed and Traction Company of Chicago, and Chanute became a director. Chicago city fathers briefly considered this moving sidewalk but then favored the construction of an electrically powered elevated railway. When the fair opened on May 1, 1893, the "brain-numbing noisemaker," later known as the "L," became a commuting success for Chicagoans, while the moving sidewalk with seats running between the Lake Michigan dock and the exhibition grounds became an attraction to fairgoers. The Traction Company dissolved in April 1894, paying each incorporator a profit of $1.50.[58]

American Society of Civil Engineers

Ever since joining the ASCE in 1868, Chanute wanted every competent civil engineer to be a member, wherever they lived. Claude Kinder, a British civil engineer working for the Imperial Railways of North China, applied for membership in spring 1886, but did not include the names of five ASCE members to

Annual meeting and banquet of the Western Society of Engineers, January 1893.
From WSE Archive.

verify his qualifications as a civil engineer, so the directors did not act upon his
application. Chanute heard of this accidentally, made some inquiries, signed
the application, and then forwarded it to the ASCE board for action. Curious
how this engineer handled railroad construction through virgin land in north
China, Chanute contacted him. Kinder then described the infancy of railways
in China and mentioned that the Chinese empress had been reportedly warned
in a dream to beware foreigners. Kinder had adopted the American principle of
building the line quickly and improving it later, and he did not think that American civil engineers were needed. The ASCE elected Kinder a member in 1890.

Late in 1890 Chanute received for a second time the nomination for the ASCE
presidency, and this time he accepted. His election in January 1891 went uncon-
tested; Chanute received 690 votes out of 694. Members knew that the society
had to emerge out of its organizational rut, but Chanute also knew that he needed
strong support from the membership for the upcoming engineering congress.

As president of the prestigious national society, Chanute received an invitation to speak at the American Patent System Centennial Celebration. He
looked forward to participating as one of the prominent speakers and to meet
the many visitors, but he came down with a high fever and had to ship his talk

Octave Chanute,
president of the American
Society of Civil Engineers
(1891). Chanute Papers,
Manuscript Division,
Library of Congress,
Washington, D.C.

to Washington. Howard Gore from Columbia University read Chanute's paper, "The Effect of Invention upon the Railroad and other Means of Intercommunication," in which he discussed the many inventions that created a better life for everyday people.

One month later, Chanute delivered his presidential address on the progress of engineering at the annual ASCE convention at Lookout Mountain, outside Chattanooga, Tennessee. In his lengthy address, paying splendid attention to detail but sometimes lacking "writerly finesse," he discussed the leading engineering works, illustrated with "stereopticon views." Chanute also stressed that members should submit quality articles to the various engineering publications, not only to the ASCE *Transactions.* Looking ahead, he recalled the hospitality extended by European engineers in 1889 and hoped to return the favors during the World's Fair in 1893. Chanute's final statement was reminiscent of what he had said in 1880: "If the present high standard of professional honor and integrity be maintained, there are good grounds for believing that there will be results within the next few years, and this will show a marked improvement in the independence, in the standing, and in the emoluments of the civil engineer."[59] An enthusiastic crowd of more than 250 engineers attended, and Chanute hoped that each of these competent engineers would attend the upcoming Engineering Congress in Chicago.[60]

During his term as ASCE president, Chanute made many trips to New York to preside over the twice-monthly meetings. When he handed the gavel to his successor, Mendes Cohen, Secretary Bogart announced that the society now had more than 1,400 members, the largest membership in its history, with 156 new members added during 1891. Chanute considered this his accomplishment and felt proud of it.

The Engineering Congress at the World's Fair in Chicago, 1893

Chicago was not just a "cow-town" and its citizens could indeed stage a prestigious fair. City planner Daniel Burnham laid out the Jackson Park area, and upon completion of the swamp draining, a team of architects and sculptors created the site for the World's Fair. As recommended by Jeffery and Chanute, all buildings were utilitarian and of temporary construction;[61] their facades were made of metal mesh covered with white stucco. Using electricity on a grand scale for lighting provided a stunning appearance to all structures in this "White City." The fair needed one permanent building within city limits, and organizers appropriated $200,000 to erect a "Memorial Art Palace" at Michigan Avenue. It was used by the World's Congress Auxiliary from early May through October, then the Art Institute of Chicago took over as the new owner.

In Chanute's opinion, every professional who called himself an engineer should participate in this first American Engineering Congress, and he encouraged related societies to hold a congress or conference at about the same time. Organizers established the headquarters of the Engineering Congress at 10 Van Buren Street, and Chanute was ever-present. His three daughters participated in the "Ladies Committee," showing the city from canal boats or from carriages. During July and August, informal gatherings took place every Monday to socialize and establish contacts with appropriate industrial firms. Because many engineers and their families extended their stay in Chicago, the committee scheduled excursions during the two weeks before and after the congress, offering guided tours of the Pullman works, Union Stock Yards, Armour meatpacking houses, the Standard Oil refinery, city waterworks, and, of course the Chicago Tie Preserving Company. Another engineering attraction was the Benz automobile imported from Germany, but Chanute considered "motocycles" noisy, cumbersome, and dangerous; he was sure that they would not become a success for mankind. Lizzie, Chanute's middle daughter, recalled that her father always hired a horse-drawn carriage, because he did not like riding in an automobile.

The Engineering Congress opened on Monday, July 31, 1893, in the Hall of Washington in the Memorial Art Palace. ASCE President William Metcalf and the chairmen of the seven divisions representing the major branches of engineering delivered short speeches. Then Chanute, president of the General Committee of Engineering Societies, welcomed everyone and declared the congress open. For the remainder of the week, each division worked in separate sessions. On August 5, Chanute chaired the final meeting of the Engineering Congress and noted proudly that "the number of valuable additions to the literature on the different subjects is so great that it is impossible in this summary to do them all justice, and it is thought best not to even attempt it. It may however be asserted that the results of the sessions of this Congress will be far-reaching and of great benefit to the profession all over the world."[62] A total of 3,132 engineers, including about seven hundred practical experts from seventeen countries, had registered for this pioneering congress; many more came for only one day and did not register.

Chanute spent much time and effort making the Engineering Congress an outstanding event; he succeeded, and accolades came from many groups. The ASCE resolved, "This Society tenders the warmest thanks to Messrs. Chanute, Whittemore, Strobel and their associates for the wisdom, skill and knowledge displayed in organizing the congress, which has resulted in bringing together the largest and most representative body of engineers the Society has ever witnessed and which will long live in the memory of all so fortunate as to be present."[63]

Two years later, the ICE elected Chanute an Honorary Member. The circular stated: "Because as former Chief Engineer of the Erie Railway; as President of the American Society of Civil Engineers; as Engineer of many important works, Mr. Chanute has held a prominent position amongst American Engineers for the past 30 years; on the occasion of the great Engineering Congress at Chicago he presided at the Inaugural Meeting and delivered the Address of Welcome on the part of his countrymen to the Delegates of the Institution of Civil Engineers, and of the corresponding scientific bodies throughout the world, officially invited to the Congress." This was the highest honor the ICE could bestow on an American civil engineer.

The Columbian World's Fair had a profound impact on urban planning, architecture, and transportation development, not only in Chicago but also in the rest of the country. Once again, Chanute had used his talents to bring together his vast number of acquaintances, his knowledge, and his experience to synthesize success in a new endeavor. In so doing, he extended his personal network even further, creating a resource that would aid him throughout life's next challenges.

O. Chanute & Company, Mount Vernon, Illinois

Early in 1895 oil prices soared,[64] and lumber prices soon doubled on Chicago markets. To increase awareness of the necessity to preserve timber, Walter W. Curtis researched and reviewed progress since the publication of Chanute's 1885

ASCE report. Naturally, Chanute shared his knowledge acquired since,[65] just as he did when Curtis worked with him on the Fort Madison Bridge a decade earlier. Curtis read his paper in May 1899.

The Chicago & Eastern Illinois Railroad signed a contract with Chanute in the fall of 1898 to erect a plant to treat black, red, and water oak timber, which absorbed the chemicals better than hemlock. Having read of the method using a zinc chloride and creosote mixture, developed by Julius Rütgers, Chanute wondered if this would prove economical, and he explained to a chemist in Germany: "My promotion of tie preserving for fourteen years has resulted in so little business (because wood is still cheap) that I naturally wish to spend as little money as possible. I now have closed a five-year contract for 100,000 ties annually and have the option of treating them either by the zinc-tannin or the zinc-creosote process. I want to use the best process, even if I make less profit, for the circumstances are such that I can make a first class record."[66]

William Curtis of the Southern Pacific and John Isaacs had patented a portable tie treating plant in 1896, so Chanute purchased from them a license for $2,000. "I hope to secure some business in the South to treat in the winter, so I am doing missionary work for wood preserving and portable plants,"[67] Chanute explained to Isaacs. It would cost less to haul the tie-preserving plant instead of a trains loaded with ties over miles of tracks. The Chicago Tie Preserving Company then erected a portable plant at the corner of South Fourth and Southern Streets in Mount Vernon, Illinois, about 270 miles south of Chicago, ideally located at the intersection of several railroads. Chanute designed the lay-out and installed the equipment on flatbed cars, so that the company could easily move the gear to another location for another customer, especially during the cold Illinois winters. Operations began on July 17, 1899, and Chanute submitted a detailed news release to the press, providing information to anyone interested in repeating the process.[68]

Charley Chanute became the superintendent of the Mount Vernon works, while his father acted as consultant and promoted their service. The works used Chanute's three-step zinc-tannin process up to 1905. But in the mean time, they also experimented with the zinc-creosote mixture, and Chanute explained in a news release: "Upon sawing a specimen tie across, they were fairly well impregnated to the very center, showing that the tar oil found its way into the wood quite as well as the zinc-chloride. . . . We spent about $1,000 in tar oil, and about the same in remodeling our plant to do the work; we hope to get our money back in time, although we shall not know how well we did the work for 10 or 15 years."[69]

To promote timber preservation, Chanute assisted John B. Johnson with his treatise on "Materials of Construction." Johnson incorporated the information into the text but he also included Chanute's chapter on wood preservation as appendix.[70]

In March 1900, the American Railway Engineering and Maintenance-of-Way Association (AREA) held its first convention in Chicago. In his inaugural address, President John Wallace told the audience that the society's goal was to combine technical and practical elements, "and it was hoped that the intercourse between the technical engineer and the practical railroad man would be of mutual benefit, that the technical man would become more practical and the practical man more technical." Every committee was to serve as a clearinghouse for methods relating to railroad construction, economics, location, operation, and maintenance. Chanute, a charter member, served on the tie committee.[71] Wallace served as president for one year, long enough to start the society on the right track.

The November 1900 AREA *Bulletin* carried Chanute's advertisement that the Chicago Tie Preserving Company was prepared to erect works or treat ties by the zinc-tannin or zinc-creosote process in either their Chicago or Mount Vernon plants. He also introduced his new motto: "Thoroughly Good Work Guaranteed." This ad brought at least one customer; Chanute designed works for the Hawaiian Commercial & Sugar Company, purchased the equipment in Chicago, and shipped it to Maui. Sam Rowe assembled it and made it operational.

Even though the Mount Vernon plant site was ideally located, the works needed to shut down in August 1901, in the middle of the treating season. The area experienced a drought and the local water company could not supply a sufficient

Advertisement from *AREA Bulletin*, November 1900.

quantity. Business was steady during the following years; six lumber companies shipped ties for treatment, bringing in more than $33,000 in 1908 alone.

Date Nails and the Recording of Treated Ties

Most railroaders considered ties big costly sticks, but management wanted the expense monitored. To keep a record, railroad men had applied notches or stamped the year of treatment with a four-pound hammer,[72] but these markings usually lost their sharpness in the very tie of which it was important to know the year of treatment. Looking for a more reliable method, Chanute checked with Ernest Pontzen in France. "What is the best way of marking railroad sleepers for future identification? In one case we nailed on pieces of tin, with date, but they got knocked off. In another case, we have stamped each end of the sleeper with a hammer, with its face cut to mark the year, but I am told that the German RR prefer to drive in a nail with the date on its head."[73] Pontzen mailed a few *clou millésimé*, as the French called them, or *Jahresnägel*, as the Germans called them. These "date nails" displayed the last two digits of the year on the flat head and looked like perfect record keepers.

The Rensselaer Polytechnic Institute in Troy, New York, one of the oldest engineering schools in the country, started a guest lecture series in fall 1890; Palmer Ricketts, a professor of mechanics, asked Chanute to present the first lecture. He selected a little-known subject, "which in my judgment is destined to have a considerable application—I mean the artificial preparation of wood to prevent its decay. . . . We are consuming at present at least twice the quantity of timber, which is produced in the ordinary growth of our forests, and the cost of timber is sure to rise. It is certain that economy must be practiced to lengthen its life." Chanute then described the various treating processes and explained the difficulties in determining if the work was done well, but he also highlighted the economics of the business. In closing, he stated that ties "can be marked in one or two ways, either by stamping an indentation at the end, or, as the Germans do, by driving in a nail with the year of its preparation in raised letters on its head. In that way, the year in which the tie was prepared is at once seen, and the decay and wear is accurately known."[74]

The lecture was published in the school paper, *The Polytechnic*, with the editorial comment: "The large attendance at the lecture delivered by Mr. Octave Chanute displayed the lively interest taken by the students in the subject. The lecture, delivered by the nominee for President of the American Society of Civil Engineers, was a most valuable one." The *Railroad Gazette* and the *Scientific American Supplement* republished the lecture almost in full, while other publications carried abstracts.

When rumors circulated in 1897 that ties from Chanute's plant decayed quickly, he had no way to confirm or deny the rumors, but this episode convinced him that he had to protect his business and mark all ties permanently. He purchased the first batch of date nails in February 1899 from the American Steel & Wire Company in Chicago. "We hereby accept your proposal to make for us 100,000 to 200,000 nails with the figures 99 raised on the head, 2 ½ inches long of No. 1 wire, galvanized, at $3.60 pr c. wt. Kindly make 105,000 to 110,000 of these nails, and ship to our works here. We are especially desirous that the heads shall not come off, and that the figures can be distinctly read fifteen years hence."[75] In a second letter, Chanute listed seven railroads that used treated ties, so that American Steel & Wire could sell the same nails to them as well. In the second batch, ordered in April, the numbers were indented and wire makers used a smaller wire size.[76] The galvanized nails, costing one penny, were furnished gratis to the Chicago & Eastern Illinois Railroad and the crew drove them into the ties when laying them into the tracks.

As part of the news release about the Mount Vernon plant, Chanute also included a galvanized date nail. The write-up in *Engineering News* stated in closing: "In preparing statistics and records of life and renewals of ties, great difficulty has always been experienced in getting certain and accurate dates of the placing of the ties in the track. To obviate this difficulty in the future, Mr. Chanute has each treated tie marked by means of a stamp hammer and also by a 2 ½-in galvanized iron nail in whose head are stamped the last two figures of the year of treatment."[77] The editorial staff added a drawing of the date nail to the write-up.

The general manager of the Great Northern Railway had requested a proposal in 1892 to erect a plant operating under the Wellhouse process. Six years later, in 1898, Chanute erected a temporary plant in Minnesota that began treating ties in 1899 and the road bought the same nails with the "99" marking from American Steel to identify their treated ties. Needing a permanent plant, they again contacted Chanute, who passed the request along to Sam Rowe. He designed and built new works near Kalispell, Montana,[78] using Chanute's three-step modification of the Wellhouse process.

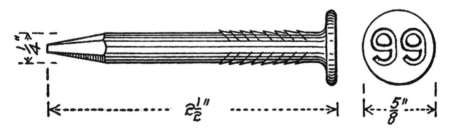

Galvanized nail, with date stamped on head, for marking treated ties.
From "Portable Plant for the Preservative Treatment of Railway Ties,"
Engineering News, August 17, 1899.

The purchasing agent of the Mexican Central also inquired about date nails. This was not surprising, because Albert Robinson, the road's president, knew the benefits of using treated ties and had discussed with Chanute the possibility of moving the Mount Vernon works south for the winter. This did not materialize; Robinson built a plant in Mexico a few years later.

In the late 1880s the "Wire Trust" consolidated most steel and wire manufacturers. An editorial in the *New York Times* described this as "one of the most gigantic trusts ever organized in this country. It is the complete monopoly by a few firms of the entire trade in barbed and ordinary wire."[79] Eventually, wire price increased from $1 to $5 a ton and the price of nails rose accordingly. To keep the cost down, Chanute ordered the next batch with the numbers "oo" for $0.06 a pound in cash from Crerar, Adams & Company in Chicago. They then raised their price, prompting Chanute to order the next batch from the Orr & Lockett Hardware store at Randolph Street in Chicago, paying $4.50 for a keg. Early in 1901, Chanute instructed his son: "I would order the nails, tinned, from Orr & Lockett as having served us well in the past. The A T & S Fe has ordered 1,000,000 nails tinned, with a big 1 on the head. We may either adapt the same or change to 01."[80]

The Chicago Tie Preserving Company continued stamping treated ties and supplying date nails to each customer.[81] The following table lists the nails, ordered by Chanute in the first three years after introducing them, where they were used, and how the numbers were applied to the heads.

In the next few years, date nails were used by the Chicago, Burlington & Quincy, Chicago & Eastern Illinois, Cotton Belt Route, El Paso & Southwestern, Long Island, Great Northern, Pittsburgh & Lake Erie, St. Louis & Belleville

Galvanized Date Nails, Ordered and Used between 1899 and 1901
by the Chicago Tie Preserving Company

Date bought	Supplier	Amount ordered	Where used	Pattern of head
Feb. 2, 1899	American Steel & Wire Comp.	~110,000 nails	Chicago Plant. Misc. customers	99 raised figures
Apr. 22, 1899	American Steel & Wire Comp.	~100,000 nails	Mt. Vernon Plant. Chicago & Eastern Illinois. Misc. customers	99 indented figures
Mar. 22, 1900	Crerar & Adams	~200,000 nails	Mt. Vernon Plant. Chicago & Eastern Illinois. Misc. customers	00 raised figures
Aug. 16, 1900	Orr & Lockett	10 kegs (~28,000 nails)	Mt. Vernon Plant. Chicago & Eastern Illinois	00
Feb. 2, 1901	Orr & Lockett	~100,000 nails	Mt. Vernon Plant. Chicago & Eastern Illinois. Misc. customers	01

Electric, Southern Pacific and Wabash Railroads.[82] Chanute advocated that a date nail should be driven into every treated tie, but starting in 1909 Frank Angier of the Chicago, Burlington & Quincy emphatically promoted the use of date nails in test sections only. This somewhat divided railroads, because they leaned toward Angier's proposal in order to cut costs. In the 1920s railroads realized the advantage of having a nail in every treated tie, and the practice of using nails in ties, poles, and bridge timber spread slowly throughout the community. The style, size of head, length, and thickness of shaft and metal used for the date nail changed little over the next century; they eventually became collectors' items.

Back to Europe in 1899

Timber prices rose as predicted in 1880. Chanute's treating process increased the life of ties to ten or twelve years; however, Europeans guaranteed a lifespan of fifteen to thirty years. Why was there a difference? Did European chemicals have a slightly different composition or trace elements? Was the actual treating process different? Another trip to Europe might provide answers. Having decided to visit Europe, Chanute needed to renew his passport. So he mailed $1 to the State Department, together with his old passport, on which he had updated in pencil his age and the color of his hair.

The contract with the Rock Island was renewed as hoped, and the Chicago and Mount Vernon works readied for their annual shutdown. On October 31, 1899, Chanute and his youngest daughter, twenty-eight-year-old Nina, sailed for Europe on the *Kaiser Wilhelm der Grosse*. This steamship was the newest, finest, and fastest ship of the North German Lloyd, and she was the first German steamer to win the "Blue Riband of the Atlantic," traveling at 22.4 knots.

Father and daughter stayed for two weeks in Hamburg and then traveled to Berlin, spending three weeks with Julius Rütgers, a seventy-year-old contractor who had treated ties for almost fifty years. Most ties used by German and Austro-Hungarian railroads were treated in one of his sixty plants across Europe. Rütgers spoke some English and gave Chanute an extended tour through several of his plants. He also explained how his men mixed the zinc chloride with creosote prior to injecting the solution into the wood. This process was significantly less expensive than straight creosoting and seemed more efficient than Chanute's zinc-tannin process. Having brought along two samples of American creosote, Chanute asked Rütgers to have them analyzed in his lab at the Theerproduktion Fabrik in Erkner, just east of Berlin. Both chemists declared the American creosote unsuitable for tie preserving, so Rütgers gave Chanute samples of his tar oil to experiment with back home.

Chanute and Nina continued their travel to France and visited the tie treating plant of the Western Railway. The French requirements for creosote were specific, but apparently there was little adherence to the standards. Their final stop was in England to meet Samuel Boulton, who had established guidelines for the British tie treating community. However, Chanute could not confirm if British treaters actually followed Boulton's guidelines.

The Chanutes returned to Chicago on December 24, 1899. The trip was a success. Nina had thoroughly enjoyed her first journey to Europe and her father had acquired useful information on European practices of timber preservation, but he now wondered how much to adopt in his own business. In the next few months, he shared what he had seen and learned in Europe with those who were interested.[83]

The Beginnings of the Creosote Controversy

Visiting European works had provided information on the treating processes then in vogue, but Chanute had also noticed that European railroad beds had better drainage and ballasting than their American counterparts. Using creosote as an additive was a good idea, but tar oil was expensive and not readily available, so Chanute looked for a less costly oily substance to carry the antiseptic into the wood.

At the end of the nineteenth century, an estimated 100,000,000 ties were laid annually across the country, either for renewals or new construction. This represented about 500,000,000 cubic feet of forest grown material. However, only about 12 percent of these ties were chemically treated either by burnetizing or creosoting in sixteen major treating plants. Chanute treated about 700,000 ties annually in his two plants, or about 6 percent of all ties treated in the United States.[84]

To determine the merits of the various processes, Bernard Fernow of the Forestry Division had suggested comparison testing in 1893, but Congress did not approve the expenditure. Now Hermann von Schrenk of the Shaw School of Botany at Washington University in St. Louis discussed such a test with Chanute, who mentioned the subject at the next WSE meeting. To speed the decaying process, James Dun of the Santa Fe suggested placing the ties on one of their lines in southeastern Texas, where temperature and humidity were high year-round and untreated ties usually decayed within one to two years. Later in 1901 von Schrenk had everything organized. Eight railroad companies donated 4,900 hemlock ties, which were treated gratis in three plants across the country. To document each of the six different processes, zinc coated nails were applied to each treated tie, giving the date, type of wood, and treatment process. Workers then placed treated and untreated ties into the Santa Fe's tracks as recommended by its chief engineer. Eighteen months later, the untreated ties had seriously decayed; of the treated ties, Chanute's ties were in the best condition.[85] In 1909, crews again inspected the tracks, and a large number of the ties treated by Chanute remained solid in the roadbed.

Naturally, every businessman looked for the least expensive procedure, modifying the mode of injecting, the strength of the chemicals, or the duration of the treatment to save time and money. Unfortunately, not all changes were for the better, because it usually took half a business life to verify the results of any small-scale experiment. This search for the most effective and least expensive wood preserving process, combined with the secretive business tactics of the promoters of the just-patented "empty cell process," is often called the "creosote controversy."

Two engineers introduced a different method to treat ties with creosote that became known as the "empty cell process." Max Rüping from Berlin, Germany, patented his process in the United States in 1902, and Cuthbert B. Lowry patented a similar process in 1906. In the traditional "full cell" process (Chanute's favorite), the injected creosote remained inside the cavities of the wood cell structure. In the "empty cell" process, the tar oil was injected into the wood under pressure to fill the cell cavities but was then withdrawn under vacuum, leaving the cell walls coated and the cell cavities "empty." Many traditional treaters, including Chanute, believed that removing the creosote was simply fraudulent and bad craftsmanship. They "knew" that the lifespan of a treated tie decreased when not enough creosote remained in the wood, referring to the experience of the Western Railway in France as solid proof.[86] Chanute also did not understand why the promoters of the empty cell process did not want to explain their method and why they believed theirs superior to the old-and-tried procedures. After all, it would take almost a decade to verify any assumptions.

Like any other businessman, Chanute also looked for ways to cut cost. One of the steps used by the Southern Pacific looked promising and was worth repeating, so Chanute instructed his son: "Work a short vacuum after the blow back of the zinc and the 2nd solution, to get deeper penetration of the leatheroid and in connection with this, to inject tannin next to the zinc and the glue last. Try this at once. We will then compare the results with Joe [Card] who is doing the same here in Chicago. If successful, we shall want to keep the process to ourselves until I can patent it."[87] Chanute submitted his claims for an improved process to the Patent Office on December 3, 1900. They granted it a year later,[88] but he requested the patent placed into the Patent Office's secret archives until November 1931, the last allowable date.[89]

As chairman of the AREA subcommittee on tie preservation, Chanute submitted his report prior to leaving for his annual winter sojourn in southern California. He discussed several processes introduced by other treaters but did not mention his findings. Instead, he wrote in his 1903 report: "While it is not improbable that some new methods may prove to have value, and while it is important that they should be tested experimentally, it is assumed that our railroads will chiefly hold to the processes which have stood the test of time. The most successful of these are (1) Creosoting, (2) Zinc-creosote, (3) Burnettizing, and (4) Zinc-tannin or Wellhouse process."[90] And he continued urging railroad men not to jump to conclusions just because his competitors claimed to have a better process.[91]

A potential investor forwarded confidential papers about the empty cell process to Chanute, who responded: "I would not advise you and your friends to put much money into this process. The parties who treat at 18 cents a tie will lose money. And there is not a particle of evidence to substantiate that the 18 cents treated ties will last in the track 20 years. Experience hitherto has been that ties treated by current creosoting methods last very nearly in proportion to the amount of tar oil injected. Numerous efforts have been made with small quantities of oil, but all such attempts have proved failures."[92]

With an increased demand for treated ties and a higher cost of chemicals, new plants opened and Chanute's business slowed. This "empty cell controversy" sharply divided the wood preservation community. Treaters suspicious of the new method adopted the zinc-creosote process, and others continued burnettizing, while the "newcomers" used their patented creosote process. This division among treaters had reached its peak when Chanute left the business in 1910; it only subsided years later when records proved that empty cell treating did work as projected by its promoters.

The Western Society of Engineers in 1901

There were many reasons to celebrate the 30th Annual WSE Meeting in January 1900; the membership had reached five hundred and included most prominent engineers from the United States but also additional members from Canada and Europe. Seven of the original charter members, or "sixty-niners," were still active and shared stories of the past. Former Secretary Louis P. Morehouse read sections from old WSE minutes, and everyone chuckled hearing that Chanute was the first who paid money to the society when he attended his second meeting, years after joining as a charter member. Toastmaster Theodore Condron then introduced the next speaker "as one who had soared to great heights" but who had generally been kept on earth, bridging rivers. Chanute talked about the "Kansas City Bridge of '69" and its Pier No. 4, which kept fighting him and other engineers. It was a lively gathering with entertaining tales.

In December 1900, Chanute accepted the presidency of the WSE, even though he thought that a younger man should take the office. Throughout his working life, Chanute had encouraged budding engineers to gain confidence among equally minded professionals, to grow in their career and then give back to the profession. Warren Roberts was one of these younger engineers, who recalled years later: "Many of the senior engineers were too occupied by business to take interest in the younger generation, but Chanute was an exception. He never seemed to be in a hurry to leave at the close of a meeting, and he was so approachable that a younger engineer readily made his acquaintance. He always appeared to be happy to take the time to explain clearly any problem we presented to him or to discuss the article presented at that meeting, answering any question, which we had been too timid to ask in public. Chanute seemed like a walking library to a young engineer. He was surely the most versatile engineer I have known, and on most lines of engineering he was an authority."[93]

One of the jobs of the WSE president was to solicit speakers, who would then submit well-researched papers for the *Journal*, making the society better known in engineering circles. Knowing so many prominent engineers, Chanute invited good speakers throughout 1901.

Hermann von Schrenk, the newly appointed agent for the Department of Agriculture,[94] agreed to present a paper on "Factors, which Cause the Decay of Wood." This became a pivotal talk in von Schrenk's career, as he reached across the barrier separating the scientists from the engineers.[95] During the next decade, von Schrenk, the scientist, and Chanute, the pioneer wood preserver, collaborated on various approaches to retard the decay of timber.

The following month, Frank Lloyd Wright presented a paper on "The Art and Craft of the Machine." Wright had opened his architectural firm a decade earlier, defining his "Prairie School," in which he combined the "soul-less" machine-made process with the native surrounding. Wright believed that the artist, or the architect, the engineer, and the machine would need to work together to accomplish art and craft. In the ensuing years, Wright's Prairie Style became a signature building design, still imitated a century later. Next, Chanute invited Victor C. Alderson, another non-WSE member, to present a paper on the "Ethics of the Engineering Profession," discussing the relation between the engineer and the laws of nature, the principles underlying the profession, and their effect on the mind and character of professional men.

In mid-June, Bernard Fernow, the former chief of the USDA's Division of Forestry and now the director of the New York State College of Forestry at Cornell University in Ithaca, presented his paper, "The Forester, an Engineer." He explained that the forester was not interested in preserving the forest but wanted to cut it down and then reproduce it. With many members interested in timber, a lively discussion followed Fernow's presentation.

Needing to attend an engineering function at the Pan-American Exposition in Buffalo, Chanute thought of a more exciting mode of travel. The Northern Steamship Company had opened their passenger service from Chicago to Buffalo in the middle of June, using the finest steamships on the Great Lakes. Chanute's oldest daughter Alice and his widowed daughter-in-law Mary, with her two children, joined him on the steamer *Northwest*. They left Chicago on Wednesday, July 10, at 2:30 P.M. and stopped at Milwaukee, Harbor Springs, and Mackinaw to tour the island. The trip continued down Lake Huron, stopping in Detroit and Cleveland before arriving in Buffalo on Saturday morning. It was a fun mini-vacation, especially for the two grandchildren, who had never been on an "ocean voyage" before. After enjoying the fair for a few days, the Chanutes returned to Chicago by train.

In September, Chanute invited Wilbur Wright to talk about his and his brother Orville's aeronautical experiments (see "1901: An Eventful Year for Research" in chapter 8). This talk was a turning point in the careers of the Wright brothers, just as von Schrenk's talk half a year earlier had helped shape his future development.

More than one hundred members and guests attended the thirty-second annual meeting in January 1902. In his address as the retiring president,[96] Chanute reflected on the society since its founding thirty-three years earlier. He was proud of all the papers presented during the past year, but especially the five talks by nonmembers, "men who were experts and who could elucidate subjects new to us and who had done novel things." These talks added interest and helped the society grow.

The feature of the after-dinner talks was the announcement by President Finley that Chanute, the retiring president, had presented the society with a check for $1,000 to be used as an endowment for three engraved medals to be given annually for the best papers on civil, mechanical, and electrical engineering presented by members to the Society during the past year. These awards were an important milestone for the engineering profession, which still lacked the breadth of published material needed to grow the profession and the development of its members. The criteria for the "Chanute Medal" changed over the years, but the WSE still presents it more than a century later.

Chanute's leadership as president or chairman of societies, including the American Society of Civil Engineers in 1891, the General Committee of Engineering Societies in 1892 and 1893, the Association of Engineering Societies between 1893 and 1895, the Western Society of Engineers in 1901, and vice chairman of the World's Congress Auxiliary Committee in 1892 and 1893, were major accomplishments. Here Chanute met, encouraged, and influenced the careers of many engineers, and their thinking influenced him.

1902: A Year of Change and Another Trip to Europe

Charles James sold his home in San Diego in 1901 and moved to 91 South Grand Avenue in Pasadena, California. As in previous years, Annie and her daughters joined her brother and other family members in early January 1902 and rented a house at Maylin Avenue. There was a large garden, filled with blooming flowers and orange trees, a perfect surrounding to spend the winter. Later in January Octave also escaped the Chicago winter and celebrated his seventieth birthday among family members with a bottle of red wine from the Napa Valley and a fresh box of his favorite cigars. After years of limited mobility, Annie's lameness improved a little and she started to walk again, but her overall health did not recover. On April 3, 1902, Octave's wife of forty-five years died of pneumonia and other complications. She was brought east and interred in the family plot at Springdale cemetery in Peoria, Illinois.

It was not easy to accept the loss of Annie, the wife and the mother, because each family member needed her so much. Her death left a large vacant spot in Octave's heart, but life continued. The seventy-year-old engineer had always been a workaholic, so he now delved deeper into the various businesses and lived by the clock. Emily Chanute, Charley's widow, recalled that the family sat down for dinner at 6:30, after which her father-in-law would go back to his study to retire at the stroke of 10:30.[97]

Nina, the youngest daughter, took charge of the household and reportedly tried to save money from the food budget so that she and her sisters could attend an extra opera performance. They hired an additional servant, but with the new help came a different kind of problem. "My den was invaded by a house cleaning cyclone during my absence, and I may need a little time to find things,"[98] Chanute wrote to von Schrenk.

Social gatherings during the fall of 1902 included discussion on where to spend the winter. An advertisement for a tour to Egypt, arranged by Thomas Cook & Company, sounded exciting, so the Chanutes reserved space, but the daughters also wanted to spend springtime in Paris. The aging engineer felt reasonably healthy and accepted the appointment to promote the aeronautical activities of the upcoming St. Louis World's Fair while in central Europe (see "The Aeronautical Mission in Europe" in chapter 8).

Chanute and his three daughters left Chicago on December 29, 1902, taking the *20th Century Limited* to Boston. They boarded for a few days at the Hotel Touraine[99] and celebrated the arrival of the New Year with friends. On January 3 they embarked on the *SS Commonwealth* of the Dominion Line. This steamer, launched just two years earlier, sailed with a service speed of sixteen knots, much to Chanute's disappointment, and offered little luxury, not even in the first-class section. After brief stops at the Azores, Naples, and Genoa, they arrived on January 19 in Alexandria, Egypt, and boarded at the luxurious Shepheard's Hotel, where every guest received a booklet describing the land of the pharaohs. Following this guide, the Chanutes went sightseeing, like other tourists.

On February 1, they ferried to Italy, staying for a week at the Grand Hotel in Naples. Chanute noted in his diary that this hotel "was close to the sea, had an open and healthy situation, with a splendid view." They spent a week in Rome, where Chanute celebrated his seventy-first birthday, then visited Florence, Milan, and Venice before departing for Nice in France. On March 13, the daughters took the train to Paris while Chanute went to Austria and Germany to promote the 1904 St. Louis aeronautical contests (see "The Aeronautical

Mission in Europe" in chapter 8) but also to meet Julius Rütgers in Berlin again. Rejoining his daughters, Chanute and family spent a few weeks in Paris and London and then returned on April 28 on the recently launched *Kronprinz Wilhelm*, the most luxurious steamer crossing the Atlantic, arriving home in Chicago on May 9, 1903.

Shortly after returning home, Octave's brother-in-law, Charles James, died in Pasadena and was buried in Peoria. Chanute's friend Thomas Clarke had died a year earlier, George Morison died in July,[100] and Robert Thurston in October. The loss of his wife, family members, and friends unquestionably dealt a blow to the aging Chanute, but he tried to recover some of his own mental balance, though finding it at times difficult to stay focused on what was important.

Chicago Timber Treating Plants:
Paris, Illinois and Terre Haute, Indiana

Locating new business after losing the contract with the Chicago, Rock Island & Pacific in late 1903 required some serious salesmanship. Because the Chicago works had treated ties gratis for many railroads in the past, Chanute now contacted their chief engineers to check on previously treated ties. "Should the result of your investigation be such as to indicate that the chemical treatment of ties of inferior woods promises economical results to your company, I shall be glad to make you a proposal for the erection of a plant and the treatment of ties somewhere along your line."[101]

Even though he had watched the men at the Theerfabrik, owned by Julius Rütgers in Germany, perform the zinc-creosote process to treat their ties, Chanute encountered many problems duplicating it. The American creosote had a different chemical composition, so he had to experiment extensively to make the chemicals penetrate deep into the wood. Chanute finally succeeded and offered this choice to any interested customer: "I imported different grades from Europe and by mixing them, I could produce the required raw material and I treated 15,000 ties last year by the zinc-creosote process with good results. If a road will adopt this process, concerning the efficiency of which there is no doubt, I can arrange to procure the necessary supply of oil. The cost of zinc tannin process may be estimated at 14 cents a tie, and of zinc creosote at 22 cents."[102]

At the next AREA meeting, Chanute mentioned to George Kittredge, president of AREA, that he was looking for commercial contracts because he had the zinc-creosote treating process all figured out. In his professional life, Kittredge was the chief engineer of the Cleveland, Cincinnati, Chicago & St. Louis Railway, commonly called the "Big Four." The networking worked, and a few months later Chanute signed an annual contract with the Big Four to erect a plant at the Midland Yard, just outside Paris, Illinois, about 180 miles south of Chicago. Operation began in April 1904. "It [the moveable plant] will travel over the various railroads of the United States and multiply by three the life of every tie with which it comes in contact," was the comment in the Paris paper.

Trying to satisfy the needs of his newest customer for zinc-creosoted ties, Chanute encountered more problems, because no dealer or manufacturer could supply the needed tar oil. The Rütgers Company in Germany then agreed to ship creosote in five hundred wooden barrels. However, they had never exported oil across the Atlantic and Chanute had never imported casks with oil, which supplied another learning experience. Nevertheless, Chanute's company treated almost 55,000 ties in the first year.

Late in 1903 Chanute succeeded in signing a contract with the Southern Indiana Railroad to treat ties for their Chicago extension. Two treating cylinders of the former Chicago works had already been transferred to the Paris works, and now the other two were shipped to Terre Haute, Indiana. Joe Card, the former manager of the Chicago works, took over the operation in Terre Haute and treated almost 60,000 ties with the zinc-tannin process in the first year.[103] With all the talk about the various treating processes, Card improved Chanute's three-step process and submitted his claims for patent, granted in March 1906. One year later, in April 1907, Card joined a competing timber preserving company to operate a plant in Waukegan, Illinois, using his patented zinc-creosote process. Charley then agreed to manage the Terre Haute works as well, which had only four customers in 1908, bringing income of just over $9,000.

In spring of 1905, Chanute's competitor, Cuthbert Lowry, opened a plant at Shirley, Indiana, and began treating ties for the Big Four using his patented empty cell process, while the Paris works noticed a major decrease in the ties it received according to its contract. The concerned Chanute wrote to the purchasing agent: "Not only is the lack of ties causing us expense in keeping our force idle, but we are suffering loss from leakage of creosote which we import in barrels. Please take steps to allow us to resume steady work." Unfortunately for Chanute, the Big Four did not renew their contract in October 1907.

Feeling his declining health, Chanute simply wanted to exit the earthy tie business. "I am 75 years old, my health is failing and I want to go out of business. My partner, Mrs. Card wants to keep on. Would you, as you once intimated, like to buy my half interest in our three plants, or would you buy a divided interest in one? If you entertain the idea, make a low price," he wrote to a potential buyer. Writing to another person, "we still desire of selling our Midland tie treating plant, will accept $25,000 when our contract with the Big Four terminates."[104] Chanute enclosed a blueprint and invited an inspection, but no sale occurred.

Fellow WSE member Bion Arnold inquired on the best treating process for ties to be used by the Chicago Traction Company. At their November 1907 board meeting, Chanute explained the merits of the various wood preservatives, and the board resolved to have their ties treated at Chanute's Paris works; they were the only customer in 1909, and the plant was dismantled later that year.

Preparing for his own mortality, Chanute and his partner Catharine Card divided the Chicago Tie Preserving Company early in 1910. Charley Chanute inherited the Mount Vernon works, and Catharine Card and her son Joe became owners of the Terre Haute plant, changing its name in May 1910 to Indiana Zinc Creosoting Company.

Business Owner and Family Man

The holdings of Octave Chanute were varied and, with taxation becoming more sophisticated, Chanute occasionally became the target of tax authorities. Assessors from the city of Chicago found several properties in late summer 1901 that were either assessed low or had been overlooked in the past. Chanute's personal property was on the top of the listing, as published in the *Tribune*, and reviewers estimated the taxable property at $70,000.[105] Chanute had to pay back taxes and put his estate in shape.

Meanwhile, his consulting business occasionally put him in the role of establishing tax rates. In 1899, the state of Michigan requested an evaluation of how to tax its railroads. Mortimer Cooley was in charge of this project, and Chanute was one of seventy-five engineers who in 1901 trudged through the wilds of Michigan, collecting all available data on each railroad company. To avoid personal error, Cooley asked four engineers to serve on a board of review and appointed Chanute the chairman.[106] After much deliberation, the board concluded that gross earnings should form the basis for taxation, and tax officials should have the right to investigate the railroad's corporate books. During the next decade, several states adopted similar valuation methods for their roads.

In spring of 1904, President Theodore Roosevelt appointed John Wallace the first chief engineer of the Isthmian Canal Commission to build the Panama Canal. The WSE sponsored a farewell dinner in May, presided over by Chanute, who felt proud of his friend having been selected for this notable position. As part of the evening conversation, several engineers discussed the sickly living conditions at the Isthmus; Wallace then stood up and said: "I am not going to the Isthmus of Panama to make a reputation as an Engineer. I am going because my country calls me. The United States has decided to build that canal and some one has got to go down there and do it."[107] He stayed for two years in Panama and needed several months to recuperate after returning to the United States.

Only a few months later, ASCE members, traveling to the Engineering Congress at the 1904 World's Fair in St. Louis stopped briefly in Chicago. Chanute and two of his daughters were part of the Chicago contingent, who then joined the group on the train to St. Louis. The Engineering Congress took place in the first week of October, and Chanute chaired the ASCE's Section H, "Miscellaneous" session. He had invited almost one hundred papers, but owing to very poor attendance, they were read by title only.[108]

On one of the last days of the fair, managers of several timber treating plants met at the Missouri Botanical Garden. Von Schrenk was the driving force to form an association where members could discuss the merits of the different treatment processes. The newly formed Wood Preservers' Association (WPA) held its first convention in New Orleans in January 1905. Octave and Charley wanted to attend, so the family combined the trip with a visit of family members.

Soon after returning to Chicago, the lonesome widower came down with a severe case of influenza and felt his energies failing. Planning for the future, he contacted Henry Gardner, the son of his former mentor, who had offered his legal assistance. Chanute owned several rental properties, which he then transferred to his children. The building on Broadway in Kansas City was deeded to his daughter Alice; Charley received sundry property in Fort Madison, Iowa, while Nina and Lizzie received rental property in Chicago and land in Kansas City. Gardner's law firm managed their money wisely for many years.

In May 1905 the 7th International Railway Congress, the first to be held in the United States, convened in Washington. Feeling better, Chanute attended and agreed to act as interpreter and secretary/reporter. The goal of this two-day congress was to increase communication between railway men. During the discussion on ballasting roadbeds, one European speaker mentioned that his road placed creosoted wooden shims between the rail and tie for cushioning.[109] Chanute shared this information with fellow WSE members, and James Dun of the Santa Fe introduced such shims on his road, with other roads quickly adopting them as well. The social highlight of the Congress was a garden party, hosted by Edith Roosevelt on the south side of the White House, and Chanute was pleased to socialize with President Theodore Roosevelt, whom he had met more than a decade earlier in New York.

Even though Chanute did not have an academic degree, he always enjoyed lecturing to university students, hoping to provide another "tool of knowl-edge."[110] "You will note when I began my career, knowledge had to be acquired from private schools, tutors and personal study. I hope that the members of your fraternity will realize that while they are favored over the preceding generation by many opportunities to learn the State of Knowledge and the use of mental tools, they will have to combine study with practice in order to become experts."[111]

On October 18, 1905, Chanute's secret dream became reality. The University of Illinois installed a new president and several WSE members attended the inauguration at Urbana. After Dr. Edmund James delivered his presidential ad-dress,[112] Professor Ira Baker announced that the "Honorary Doctor of Engineer-ing" degree was awarded to Chanute for his contributions in timber preservation. Chanute was pleasantly surprised, but told his daughters that the tassels on his mortarboard got in his way; and he did start occasionally adding the academic title to his name, Dr. Octave Chanute.

Octave Chanute (circa 1905). Chanute Papers, Manuscript Division,
Library of Congress, Washington, D.C.

By the early 1900s, the Potter Palmer subdivision had slowly deteriorated.
Charley's home had been broken into a few years earlier and his family had since
moved to Jefferson Avenue. The situation worsened when squatters moved in
and several houses were burglarized in early 1905. Chanute's daughters insisted
on not renewing their lease.

To make the move easier, Chanute needed to clean up. Reading of the San
Francisco earthquake and the fire that burned everything owned by the Technical Society of the Pacific Coast, he donated $75 and shipped many of his
duplicate books and journals to start a new library. He gave other technical
books to the John Crerar Library in Chicago and the Universities of Illinois and
Wisconsin. Packing up his nicely bound *Harper's* magazines was emotional, as

he had started buying each issue in the early 1850s; they were shipped to the Chanute, Kansas, library. The family donated books from his deceased brother-in-law to the Peoria Public Library.

In May 1906 the Chanute family moved to a large mansion on 61 Cedar Street that had an enormous room on the top floor for Octave's large, flat-top desk and his many bookshelves. The monthly rent was a little higher, but everyone felt much safer. Soon a terrier named Tacks (because he was sharp as tacks) joined the family, and everyone enjoyed walking the dog through the neighborhood and along Lake Michigan. The big house was also perfect for entertaining, and Alice recalled a dinner party, arranged by Charley and his wife Emily at the new home.[113] Everyone wore elegant evening clothes and looked very fashionable. Octave was working in his library and had promised to "dress up a little," but time slipped by. When Charley called his father for dinner, he just put on his dress suit coat, still wearing his comfortable, baggy trousers, stating that no one would see his trousers once he was sitting down. Charley was a bit perturbed with his elderly father, but the guests reportedly had a good time.

In September 1906, Alice and Lizzie joined their father on the summer excursion of the Canadian Society of Civil Engineers traveling for three weeks from Montreal to Vancouver on the Canadian Pacific Railroad. A few months later, the society elected Chanute an honorary member, which greatly pleased him.

Even though he was slowing down, Chanute maintained his interests in timber preservation and enjoyed the fellowship of meetings and conventions. For the third WPA Convention in Memphis in January 1907, Chanute submitted a paper on the steaming of timber. He believed that the necessity to steam (or not) depended on the actual condition of the timber. Wood, he asserted, should be thoroughly seasoned, but customers usually required treated wood for immediate use, thus the European practice of drying wood in kilns could be introduced in America. In closing Chanute hoped that "members of the association would contribute their experiences to the discussion of the paper and that the results would produce still better work in wood preserving."[114]

At the annual WSE meeting in January 1909, the group named Chanute an honorary member, the second in the history of the society. Reflecting on his chosen profession, he stated: "The profession has grown in standing and in prestige. But above all, the one thing for which I am most grateful is that thus far, instances of dishonesty and graft have been very rare amongst engineers. The profession has maintained a high standard and I feel sure that all of you gentlemen will do your utmost to maintain and increase that standing."[115]

A couple of weeks later, Chanute presented a paper on the history of wood preserving at the fifth WPA convention, held in Chicago, discussing the growth of the industry but warning that some contractors "have produced inadequate treatment through ignorance, as it takes several years to find out positively whether good work or bad work has been done. You can tell something by chemi-

Chanute and his two older daughters on the transcontinental railroad excursion across Canada (September 1906). Courtesy of the Chanute Family.

cal analysis, but it takes a number of years to try to work it out practically. . . . So
that I, who is about to retire from that business, would urge on you gentlemen
to do your work as thoroughly and as well as you can."[116] Chanute was elected
an honorary member later that evening.

A few weeks later, after all required appearances were completed and new
ones declined, the family went south. Octave spent his seventy-seventh birthday
among his extended family at the rice plantation of his niece Amelia Gueydan,
who prepared a cake with seventy-seven candles.

Back in Chicago, Chanute attended WSE meetings regularly. In October 1909
Charles Barnum of the Forest Service presented a paper on wood preservation
from an engineering standpoint. When the discussion turned to using creosote,
Chanute commented, "I will say that if the process can inject an average of
twelve pounds of solution to the cubic foot, we do not need anything better."[117]
As expected, the debate became heated and Barnum politely stated that the

Forest Service was only interested in helping engineers solve problems with the nature of timber.

The 1910 WPA convention was again held in Chicago, with Chanute, son Charley, and Joe Card in attendance. As expected, the discussion soon turned to how much creosote to inject and how much should remain in the wood.[118] After the meeting, Chanute wrote to a friend, "the agent for the New York Central had investigated the works at Shirley, Indiana. And Curtis had inspected the Ayer & Lord Company using the Rüping process. In both instances creosote was subsequently withdrawn to save money. Several members hotly disputed this and the discussion assumed such violence that the Chairman stopped it abruptly. But some of us had lots of fun."[119] The elderly Chanute seemed unwilling to accept anything related to the empty cell process.

One More Trip to Europe in 1910

Even though the Chanutes lived in a large mansion, in a nice neighborhood, and his three daughters attended concerts, operas, and theaters regularly and had opened their home for social gatherings, Alice, Lizzie, and Nina still did not usually receive invitations to major social affairs. Perhaps thinking it more fitting for their social status in Chicago society, they wanted to own a large residence instead of just renting one. These things did not bother Chanute, but he went along with his daughters' desire to move again.

The large, four-story, brick and stone residence at 400 Dearborn Avenue, on the corner of West Elm Street, built by the veteran banker and early Chicago settler James M. Adsit, was offered for sale. After Adsit's death, the residence with its spacious tiled reception hall became neglected. It required extensive renovating and painting, but Chanute bought the residence for his daughters for an undisclosed price. In March 1909 the family moved to this fashionable mansion, which even included a stable in the back of the property, in a well-to-do neighborhood. Chicago streets were renumbered late in 1909, which changed their address to 1138 North Dearborn. Coincidentally, the next issue of *Who is Who in Chicago Society* listed Octave Chanute and the names of his three daughters.

Early in 1910 every family member was ready for another trip to Europe. Chanute's daughters wanted to enjoy the baths at Carlsbad, a famous health spa in the Austro-Hungarian Empire, while Chanute was interested in attending the 8th International Railway Congress in Berne, Switzerland, early in July. He had already prepared a paper on record keeping of ties and was excited on being elected an honorary member of the French Society of Civil Engineers.

Dr. Allport refilled Chanute's heart medicine prescription and instructed him not to do any extraneous walking. Feeling good, Chanute wrote to his many correspondents, thanking each for all they had done, perhaps anticipating health problems while in Europe.

Chanute and his three daughters left Chicago on May 14, 1910 and crossed
the Atlantic on the *Kaiser Wilhelm der Grosse*.[120] After spending a few weeks in
Paris, they continued their trip to Carlsbad, where Chanute became ill. Not
speaking English or French, the local doctors suggested moving their patient
to Paris. When the train pulled into the Paris train station, a waiting ambu-
lance rushed him to the American Hospital, where doctors diagnosed him with
pneumonia. For the next two months the daughters boarded with an English
lady and visited their father regularly in the hospital. By the end of September
Chanute felt strong enough to travel and the daughters hired one of his nurses,
Mary Willingale, for the trip back home. Charley and Emily waited in New
York and everyone arrived safely in Chicago. Though weak, Chanute continued
a subdued life at home.

A Look Back

Department of Forestry statistics forecast in 1880 that the nation's supply of
white pine would be exhausted in eleven years and hardwood within twenty-five.
Chanute was arguably the first American to show his concern for preserving
natural resources; he demonstrated the commercial feasibility of preserving
wood to slow deforestation. Gradually, railroaders recognized the need to make
timber last as long as possible and they realized that increasing the life of ties
decreased the cost of track repair.

The need for an industry was real; were it not for Chanute's early efforts and
his insistence on collaboration, the explosion in wood preservation between 1897
and 1905 would not have proceeded as rapidly. Chanute and his partner Joseph
Card designed and erected the first large-scale American commercial wood
preservation works for the Santa Fe in Las Vegas, New Mexico, in 1885, for the
Union Pacific in Laramie, Wyoming, in July 1886 and for the Rock Island and
other customers in Chicago in May 1886. These were the first and only works
treating ties on a large scale until other American railroads followed more than
a decade later.

For a quarter century Chanute brought the economic advantages of timber preservation to the attention of the railroad community and the American
public. He attended most meetings of the American Railway Engineering and
Maintenance of Way Association and later the Wood Preservers Association,
willing to listen and to share his knowledge.[121] Chanute knew that a perfect
preservative must be an antiseptic, insoluble in water, nonvolatile under normal
temperatures—and inexpensive. Competitors started to promote their patented,
lower cost, empty cell processes and obtained long-lasting, lucrative contracts,
but they kept their procedures secret. In Chanute's opinion, these promoters
possessed no empirical data to verify their results and furthermore, he thought,
they should be "noble enough" to report their difficulties and explain their

mistakes. A decade later, records proved that ties treated with the empty cell process did indeed last as long as their promoters had speculated.

The wood preservation business provided income for the Chanute and Card families, but their earthy treatment plants and competitive business environment were not really part of his character, especially as he aged and hoped to retire from the daily effort of running the business. Chanute's newest "side issue," mechanical flight, provided more exciting challenges, as he believed that the problem of flight could only be solved through an interchange of thoughts. Even though work in aviation would clearly cost him money rather than make him money, Chanute delved deeper into the fascinating topic of flying machines. The collaborative opportunity and the technological excitement seemed to be waiting in his final foray of life.

CHAPTER 7

From the Locomotive
to the Aeromotive

STILL YEARNING FOR CHALLENGES, excitement, and new discoveries in the engineering field, the fifty-one-year-old Chanute left the Erie in 1883, to embark on a self-directed career as a civil engineer. Three years later, he explained:

> The busy men who are developing this country need to keep up with new discoveries and progress even before they are reduced to practical account, and to look into the future as well as in the past; they especially need that personal contact, which nothing can replace, with men of science, to learn of what is being done and hoped for, and to make known what new information is needed to remove obstacles to their own progress. Engineers particularly owe it to themselves . . . to make the other members acquainted with whatever new facts or ideas they may have acquired outside of the routine of their profession. . . . Here they can indulge in pure science without regard to the practical use or bearing of the facts which they have discovered, and, provided always that they stick accurately to the facts, the resulting discussion cannot fail in being of value to them as well as to the listeners.[1]

Early in his working life, Chanute had become interested in the unconventional topic of manned flight, but in the interest of his career and social standing, he did not discuss it publicly. Now approaching what he considered the end of his professional career, this seemed an opportune time to investigate mechanical flight,[2] but only as a "side issue." Believing that the well-informed professional should "think out" the unresolved problems of aeronautics and not just "think of" them, Chanute tackled this subject with an unbiased engineering mind and with the approval of his family. He firmly believed it possible that the century that had seen the development of the ocean steamer, the submarine, and the railroad might witness success in aerial navigating. But for the flying machine to join the other modes of transportation, visionary engineers needed to become involved, so that "the subject could be cleared of much rubbish, and placed upon a scientific and firm basis,"[3] Chanute wrote in 1882 to Fred Brearey, the secretary of the Aeronautical Society of Great Britain.

The American Association for the Advancement of Science (AAAS) reorganized its Mechanical Science Section (or Section D) in 1884, and Chanute's longtime friend and frequent collaborator Robert Thurston accepted its chairmanship for the 1885 meeting. To broaden the scope of Section D, he wanted members of every branch of engineering to participate. Chanute, who had been elected an AAAS Fellow in 1877, attended the 1885 meeting in Ann Arbor, Michigan, and Thurston convinced him to chair Section D in 1886. Chanute was at first reluctant, as he handled a heavy consulting workload, but he also knew that being the vice president of Section D would allow him more opportunities to entice others to help advance the sciences and possibly share his curiosity, aeronautics.

The 1886 Buffalo Meeting

Having read a communication in the Aeronautical Society report on the flying habits of soaring birds, submitted by Israel Lancaster from Chicago, Chanute wondered if these birds could provide insight into the potential for humans to fly. Traveling back to Kansas City after signing the contract for his next tie treating plant, Chanute stopped in Chicago, met Lancaster, and listened to his observations that soaring birds used air currents to remain airborne. This topic would undoubtedly interest engineers and scientists, so he invited Lancaster to speak at the next meeting of Section D in August 1886.[4] Lancaster agreed, so Chanute mailed guidelines on how to prepare a talk for a scientific audience and submitted his name for membership in the AAAS.

Vice President Chanute had invited twenty papers on a wide range of mechanical engineering topics to be read at the 1886 AAAS meeting in Buffalo, New York. In his keynote address, he pointed out that men depended on each other to develop abstract scientific knowledge into useful applications, because "information is so widely scattered, and covers so many different fields of science, that it is only by patient effort and much searching that the needed knowledge is gained."[5] Because flying machines were what imaginative writers discussed when talking about future mechanical inventions, he thought a paper on bird flight would interest many listeners.

Lancaster, the last speaker that afternoon, described his models that flew against the wind, and soaring birds that traveled on motionless wings to gain altitude in upward moving air.[6] The next day, on Friday August 20, instead of demonstrating his model of soaring birds or "effigy" as listed in the program, the Illinois farmer tried to explain the lifting action of the wind mathematically and caused a row. Chanute was not present, and therefore could not help. Making some quick calculations, De Volson Wood proclaimed Lancaster wrong on the spot. Thurston remained unconvinced that the data was erroneous, but that a mechanical engineer should perform more careful research. Another attendee, the astronomer Samuel Langley, was intrigued and decided to research the power needed to sustain a given weight in the air. "It seemed to him that this inquiry had to precede any attempt at mechanical flight, which was the very remote aim in his efforts,"[7] Chanute recalled years later.

The press gave Lancaster's talk mixed reviews. Buffalo newspapers described him as a "practical joker" and the *Science* editor reported that Section D probably needed recreation.[8] The *Scientific American* discussed Lancaster's talk at length. "A rather fanciful and highly wrought, yet interesting and suggestive paper was read by Mr. Lancaster of Chicago who has for many years made a special study of the flight of birds." The reporter then described soaring flight but also thought that Lancaster should have launched at least one little model.[9]

In one of the last meetings, the AAAS board accepted the recommendations of Thurston and Chanute to change the name of Section D to "Mechanical Science and Engineering Section." This meeting of Section D in 1886 had stimulated interest and Chanute had achieved his goal. "How often is it that the imagined things of today become the accomplished things of tomorrow?" Several intellectuals became curious about aerodynamics and lift, thus creating a new branch of scientific inquiry.

Collaborating to Learn

To proceed in his learning process about aeronautics, Chanute took the heuristic approach[10] and defined a series of questions: What are the basic aerodynamic requirements? How does air flow over or under a wing? What are the forces acting on wing surfaces as they cut through the air? Creative thinking, observation, and experimenting would perhaps yield answers.

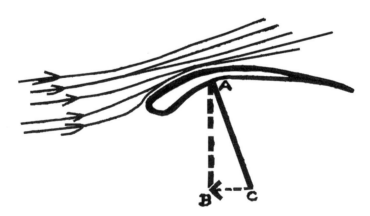

Shape of airflow over curved wing. Chanute's letter to A. M. Herring, April 26, 1895. The solid line A-C indicates lift; the dotted line A-B supports the weight (or gravity); the horizontal line B-C pulls the machine forward, thus producing "aspiration" (thrust).

The January 6, 1888 issue of the *Railroad Gazette* printed a letter by Charles Latimer, who had received primitive flying machine plans and wondered if some profit could be made from them. Chanute responded to Latimer directly: "This is an old hobby of mine, but I say very little of it, partly because the scoffers would laugh, and partly because I came to the conclusion that aerial navigation was only practicable by the use of a motive power a good deal lighter than any motor known . . . I have sometimes thought of devoting some leisure time to making an inquiry."[11]

After moving back to Chicago in October 1888, Chanute began actual experiments with flying models, including a *parachute dirigeable*, or a gliding model with two cells, that was described in a French magazine. Chanute built one and launched it with a brick as "passenger" from the top floor of his three-story home "in the early morning when only the milkman was about." The model glided steadily in all sorts of breezes, but its angle of descent was much steeper than that of the birds.[12] This did not provide answers, so he determined that he should first acquire knowledge of the laws that govern flight.

Keeping his ears and eyes open for anything aeronautical, he read of World's Fair organizers planning to hold an international aeronautical congress in 1889 in Paris. Participating would provide a good opportunity to meet like-minded professionals and become known in the aeronautical community, so Chanute volunteered to serve as the United States delegate and to deliver a paper. To avoid confusion at the conference, the Aeronautical Commission informed all presenters to use the word *aéronautique* for the art of mechanical flight, *aéroplane* to describe craft flying horizontally like the birds, and *aérostat* for balloons.

Atmospheric resistance had interested him for a long time, so Chanute selected the reactions of air flowing against various-shaped surfaces as the subject for his Paris paper. He ceremoniously removed the red tape from his aeronautical clipping collection that he had rolled up fourteen years earlier and started to study in seriousness. On December 2, 1888 Chanute put down his ideas on manned flight in a letter to Charles Hastings from Kansas City, who was also interested in aerial navigation. In closing, he asked his friend to sign a statement that he read the letter. "This may prove useful in establishing a date, should I ever folly this subject further."[13] The untimely death of Hastings in 1892 ended the relationship quicker than anticipated.

Another letter went to De Volson Wood: "When a current of air impinges upon a plane, producing a certain pressure, what is the measure of the normal component which presses on the plane? Or in other words, what is the 'lifting force'?"[14] Looking for input on the wing shapes, Chanute included tables, in which the retarding and supporting forces on inclined plane surfaces were first calculated, then measured in natural wind and in artificial airflow provided by a fan.

When next in New York to attend an ASCE meeting, Chanute discussed the forces of the wind with Wood, but he also picked up the just published novel *The Clipper of the Clouds,* in which the author Jules Verne postulated that a successful aeroplane should be heavier than the air. Through his brilliant but fictional engineer Robur, Verne postulated there was no need to copy nature, even though she never made mistakes. "Locomotives are not copied from the hare, nor are ships copied from the fish. To the first we have put wheels, which are not legs; to the second we have put screws, which are not fins. Besides, what is this mechanical movement in the flight of birds, whose action is so complex?"[15] Chanute underlined these words in his book and, as recommended by Robur, he proceeded to move "from the locomotive to the aeromotive!"

Seeing a report on the recently patented airship of Peter Campbell in the *Chicago Tribune,* Chanute noticed several errors and wrote to the editor that Mr. Campbell was fortunate to have experimented on a perfectly still day to display his airship. "He will not accomplish a real success until he develops a motive power with much greater energy in proportion to its weight than any with which we are now acquainted." In his cover letter, Chanute asked not to include his name if this communication was printed, because he did not want neighbors to point fingers at members of his family in the grocery store! His comments were published on December 28, 1888, signed with "Engineer."[16]

Continuing his research, Chanute explained to Thurston: "I trust you will not think me a lunatic if I say that I have had in mind for years to devote part of my leisure time to the opening up of an inquiry, whether man can ever hope to fly through the air!! A single person can not carry out such an investigation, and I am looking for a number of other visionary students of science who will give thought to the subject and correspond with me. . . . Can you refer me to tables of experiments on air pressures on inclined planes at various angles?"[17]

With thoughtful comments coming from his friends, but also discovering data on moving bodies in fluids and information on wind pressures and velocities in sails, his paper made good progress. Was artificial flight really as absurd as some people thought or could an intellectual resolve it? While Chanute studied the theoretical aspects of aerial forces, Samuel Langley, the new secretary of the Smithsonian, began his research using flat wings. He soon discovered that the only useful publications came from the French and British aeronautical societies, "but in these, as in everything then accessible, fact had not yet always been discriminated from fancy."[18]

Matthias Forney had sold his interest in the *Railroad Gazette* a few months before Chanute resigned from the Erie. Three years later he bought the *American Railroad Journal* and *Van Nostrand's Engineering Magazine* and went back into the publishing business. Then, on January 1, 1887, the fifty-one-year-old Forney introduced the *Railroad and Engineering Journal.* Naturally, Chanute offered his help as a friend to make this publishing enterprise a financial success. Forney had an idea, to which Chanute replied cautiously: "Why have you put temptation in my way, by proposing such an excellent subject for a book, and such a good plan? I will not promise to undertake it, for I am a slow literary workman, and I might not be able to get it done as soon as you wish, but I will talk it over with you when next in New York."[19]

Early in 1889, Chanute submitted his first aeronautical contribution to Forney, an unsigned commentary describing a just granted patent in Forney's magazine. Chanute's article "The Latest Rapid Transit Scheme" discussed Reuben Jasper Spalding's patented "Machine for Navigating the Air," a purportedly easy-to-operate apparatus for an imaginary way of travel by the most direct route. The aeronaut, wearing a leather jacket with two attached feathered wings, was to be suspended from a balloon. Chanute pointed out that one of the patent engravings showed "the passenger just ready for flight with his broad pinions spread, and attention may be called to the magnificent and stately appearance presented by these pinions and by the—well, steering apparatus, to speak politely." Spalding also claimed that the balloon was not an absolute necessity, because the wings were to operate with the same effect as the wings of an eagle. To this Chanute added his lighthearted comment, "which might be a dangerous admission were the eagle in a position to file objections in the Patent Office."[20]

This article appeared in print while Chanute and his two older daughters traveled to Europe (see "The Exposition Universelle de Paris, 1889" in chapter 6), arriving in mid-April 1889; Alice and Lizzie soon enjoyed spring in Paris. Meanwhile, Chanute became acquainted with Gustave Eiffel, who had studied gravity using his 990-foot tower and was interested in Chanute's interpretation of air resistance of various shaped objects.

The *Congrès International d'Aéronautique* began on July 31, with more than one hundred researchers attending. The president of the French Commission introduced the five foreign delegates, including Chanute, and members of his commission and declared the congress open. Samuel Langley was the only other American registered; most likely, neither man realized how much they both would shape the development of aeronautics during the next decade.

The first speaker at the morning session on August 1 was Étienne-Jules Marey, presenting the current state of aerodynamics based on his twenty-year study of bird flight. He tried to explain the much-debated question of how the bird soars, but he did not discuss the sustaining reaction derived from the air, or the power required for flight. Chanute was the next speaker; being unfamiliar with the French technical language, he asked to read his paper in English.[21] A few thoughtful questions and comments from attendees followed the reading. Participating in this four-day conference allowed researchers to "disseminate information on the scientific principles involved, the mechanical difficulties to be surmounted and the practical details of constructing flying machines."[22] Listening to the other speakers and talking with attendees, Chanute also detected "utter confusion" about mechanical flight.

Traveling back to the United States, Chanute incorporated his newly acquired knowledge into his next paper for the AAAS meeting in Toronto later in August.[23]

Because of poor attendance, organizers read it by title only. However, in talking with attendees, the majority seemed to agree that a civil and a mechanical engineer, a mathematician, and a practical mechanic should work together to solve the problem of aerial navigation, but a syndicate of capitalists would also help.

Robert Thurston had moved to Ithaca, New York, in 1885 to become the director of the College of Mechanical Engineering at Sibley College, Cornell University. To broaden the horizon of his students and increase the status of the school, Thurston introduced lecture series by nonresident lecturers. Interested in aeronautics himself, he invited the well-known researchers Walter Le Conte Stevens, Samuel Langley, and Octave Chanute to discuss the results of their aeronautical work during the 1890 winter term. The fifty-eight-year-old Chanute went to Ithaca on May 2, 1890, and Thurston announced, "This lecture will be especially valuable, treating as it does a subject, which has usually been considered visionary in the extreme."

Lecturing on the unconventional topic of aeronautics, Chanute selected his words carefully. He explained that the topic of aerial navigation should be divided into two categories, each again being subdivided to deal with powered and motorless craft. Members of the first group, the aeronauts, believed that success would follow with a lighter-than-air apparatus, while the second group,

Octave Chanute, reading (early 1890s). Courtesy of the Chanute Family.

the aviators, hoped to achieve flight with a heavier-than-air machine. "A mea-surable success had been attained with navigable balloons, but much greater speeds can perhaps be attained with aeroplanes. . . . Once a partial success is attained, it is not impossible that improvements will follow each other so rapidly that the present generation will yet see men safely traveling through the air at speeds of 50 or 60 miles per hour."[24] By explaining both schools of thought, he hoped to encourage his listeners to become involved.

Going public was a breakthrough for Chanute, but he looked for inventive young minds to supply fresh ideas to help him solve the centuries-old problem. One of the attending students, twenty-eight-year-old Albert Zahm, had studied flight in the past. He recalled the lecture three years later and described Chanute as "a silver-haired gentleman, full of faith in the art, but apologetic for identifying himself with a pursuit so generally condemned. Mr. Chanute began with a hesi-tancy amounting almost to reluctance, seeming to entreat the young men not to believe that the study of such a subject was a more than probable indication of failing mental vigor."[25] Zahm subsequently designed America's first significant wind tunnel and developed instrumentation for accurate measurement of air velocity and pressure, making valuable contributions to aeronautics.

In the next few months, Chanute expanded his Sibley lecture with drawings and tables and offered the manuscript to Matthias Forney, hoping that his write-up was suitable for his readers. Forney published it in five consecutive issues in his *Railroad Journal.*[26] Separately, the *Scientific American* printed Chanute's comments on motors for future flying machines.[27]

Having prestigious journals and a highly regarded engineer tackle a hereto-fore-questionable topic gave credibility to the subject of aeronautics; the public began to show interest in the possibility and success of manned flight, not only in its many failures. To capture the imagination of those readers who longed to imitate the birds, the editor of the new *Engineering Magazine* asked Chanute for a popular write-up,[28] which the *Chicago Tribune* reviewed with the pro-phetic title, "Progress in Flying Machines. Inventors of such structures no longer thought to be insane."[29] When the editor of *Appletons' Cyclopedia* considered a reference book on the most useful engineering advances of the past decade, he thought that "Aerial Navigation" should be included, so he, too, asked Chanute for an entry.[30]

But in the real world, good people usually left the room when the conversation turned to flying machines, or they tapped their foreheads while exchanging sig-nificant glances. Years later, Robert Woods recalled a social gathering in Kansas City at which the talk turned to hobbies. When Chanute was asked about his, he smiled and said, "Wait until your children are not present, for they would laugh at me." After the children had gone to bed, the question was repeated and Chanute replied: "My hobby is flying machines, and I guess I would spend twenty-four hours working on them if my family would let me."[31]

Chanute's writing indicated his belief that the future flying machine would employ rigid wings and a powerful, light motor with a screw to move the craft forward without flapping wings. Because there was no motor available in the early 1890s, he considered the ascending trend of the wind as a possible energy source, but he also prophetically warned that "no amount of motive power would avail unless the apparatus to which it is applied is stable in the air."[32]

Looking for aeronautical news, Chanute subscribed to two services offering one hundred newspaper clippings on flying machines from around the country for $5. He also ordered clippings from the Durrant's Press Cutting Agency in London and the Courrier de la Presse in Paris late in 1890. Unfortunately, he could not read the German language, remarking, "I fancy that I am missing a good deal of valuable material." He saved these clippings in wooden file boxes or pasted them into scrapbooks. Even though they provided what he called "mostly gush and slush" information, he uncovered names of investigators and began a worldwide correspondence that soon intensified.

Seeing a request for information on air resistance at very high velocities in the *Railroad Gazette*, Chanute explained to its editor Henry Prout that no accurate information was available.[33] He noted that the field needed much more research to reach a clearer understanding of the resistance, which varied with the shape of the object and the angle of attack. Having studied this fascinating topic in the past, Chanute offered to share his sundry notes with other interested researchers.

To make his *Railroad Journal* interesting for his readers, Forney now wanted to publish aeronautical material every month. An article in *Nature* on the "Soaring of Birds" looked promising, so Chanute abstracted it and added his comment, "The problem of soaring is one of the oldest problems in mathematics, but unfortunately the formulae given by mathematicians do not agree with the facts,"[34] admitting that he could not yet provide a mathematical solution.

Seeing in a French magazine a description of Clément Ader's steam-powered aeroplane *Éole* that flew for 150 feet on October 9, 1890, outside Paris, Chanute translated this for the September 1891 issue.[35] According to historian Charles Gibbs-Smith, this was the first tentative flight of a powered, man-carrying aeroplane in history.[36] But Ader did not want the public to know details, so Chanute shared the opinion as expressed by the French reporter: "Until we have witnessed a convincing experiment, at which we shall have seen with our own eyes the generator of the power employed, we shall remain skeptics."

Aerial Engineering: Looking for Solutions

In the company of open-minded engineers, Chanute felt comfortable talking about his eccentric hobby. "I feel sure that if any professional men are entitled to wear wings, they are the civil engineers. In making preliminary surveys, I can see that an engineer soaring over the country with a 'Kodak' would have great advantage over the present tedious method of plodding along the ground. . . . The principal difficulty hitherto has been the lack of adequate motive power to accomplish what birds daily perform." He ended his mini-presentation at the WSE Annual Meeting optimistically: "It is not impossible that to future Annual Meetings our members will come, not only by rail or water, but perhaps in flying machines through the air."[37]

Elected the ASCE president in 1891, Chanute had several opportunities to speak at prestigious meetings. Preparing a paper for the American Patent System Centennial Celebration in April, Chanute found it remarkable how improvements had benefited regular people:

> Improvement in transportation has followed upon improvement, because invention has been more active and successful than at any period in the world's history. The sea, the land, the air are experimented on to gain higher speeds or more economical modes of transit. . . . I know personally of eight or ten perfectly sane men throughout the world who are experimenting with real flying machines, depending like the birds upon the reactions of the air for their support. If one inventor hits upon the right combination, there seems to be no reason why man may not emulate the flight of the swallow, whose speed is computed at 150 miles per hour or that of the swifter marlin, which is said to flash through the air at 200 miles per hour.[38]

One of these "perfectly sane men" was the next speaker, Samuel Langley, who ventured to predict, "that the air may probably be made to support engine-driven flying machines before the expiration of the present century." Langley had experimented with flat plates on a "whirling arm" that created an artificial wind of forty miles per hour and published his research later in 1891. In a letter to the assistant secretary of the Smithsonian, and probably to the delight of Langley, Chanute commented in April 1896: "It is significant that prior to the publication of Langley's work, it was the rare exception to find engineers and scientists who would fully admit the possibility of man being able to solve the century old problem of aviation. Since the publication of 'Experiments in Aerodynamics' however, it is the exception to find an intelligent engineer who disputes the probability of the eventual solution to the problem of man flight."[39]

Chanute's ASCE presidency was winding down, and having collected much material on flying machines, he was ready to begin writing about the progress made worldwide. Because Forney had asked for a monthly series on aeronautics five years earlier, he now offered payment, but Chanute enjoyed writing and did not wish to be paid. His first article appeared in October 1891, and Forney gave this new series special editorial mention: "The navigation of the air has heretofore been hardly considered a practical question, chiefly from the reason that those who have attempted it have generally been without sufficient theoretical or practical knowledge to meet the conditions involved. It has now been undertaken by another class of men, who are thoroughly equipped for dealing with the problems involved."[40] In his first article, Chanute discussed the principles of air resistance, but he also thought that someone needed "to bring order out of chaos in aerodynamics and reduce its many anomalies to the rule of harmonious law."[41] He felt reasonably sure that all discrepancies could be explained by some simple explanation that had been overlooked so far.

Throughout his engineering career Chanute had insisted on using proper terminology, so he now formalized words that would become standard nomenclature: "aeroplane" (as recommended by the French in 1889) even though he thought that "aerocurve" would be more suitable; "lift" (the force perpendicular to the direction of the airflow); and "drift." Chanute used "drift" to describe the resistance of the wing, but this word was later replaced with drag or hull resistance.

Chanute also studied Francis Wenham's paper "On Aerial Locomotion and the Laws by which heavy bodies impelled through air are sustained," read at the first meeting of the Aeronautical Society of Great Britain in 1866. Here Wenham discussed the efficiency of propeller blades after having observed that the swiftest-flying birds possessed long, narrow wings, while the slow, heavy flyers had short, wide wings, recognizing the importance of the ratio of the length of the wing to its width.[42] Wenham also theorized that the same lifting power of a large-area wing could be obtained by staggering several smaller "aeroplanes." Chanute wanted to learn more from this civil engineer, so he wrote: "I am writing a series of articles upon 'Progress in Flying Machines' and have been endeavoring to get your address, so that I might communicate with the pioneer who first advanced a rational engineering view of the subject. . . . My general idea is to pass in review what has hitherto been experimented, with a view to accounting for the failure, clearing away the rubbish, and pointing out some of the elements of success, if I can."[43] He described Wenham's work in his September 1892 article.

Another Englishman, Horatio Phillips, exposed wings with differing curvature to artificial air currents of varying strength in a wooden trunk. He had mounted the wings on two wires attached to the leading edge and measured the "drift" once the wings assumed a lifting attitude. Phillips patented the most efficient wing shapes in October 1884 and published his findings in *Engineering*. Reading this report, Chanute wondered about the angle of the wing toward the airflow and considered repeating the experiments. He sketched a wind tunnel with a lift balance and mailed it to some correspondents for comment. In lieu of experimenting, he determined the lifting angle as 15 degrees and discussed the work of Phillips in his March 1893 article.

Late in 1892 the Smithsonian Institution received several issues of the *Prometheus* magazine with articles by a German experimenter, Otto Lilienthal. George Curtis translated them for Langley and then mailed the material to Chanute, who found Lilienthal's progress in practical flight inspiring. Lilienthal

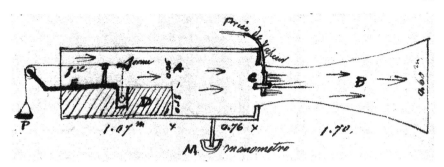

Chanute's wind tunnel idea, adopted from Horatio Phillips design. Letter to
A. Goupil, December 17, 1892. Chanute Papers, Manuscript Division, Library
of Congress, Washington, D.C.

clearly explained how the operator should handle the machine in turbulent
air, but Chanute also thought that "learning to ride the machine would probably be quite as difficult as designing it." The Smithsonian published extracted
translations in their report for 1893, and Chanute discussed Lilienthal in his
July 1893 article.

During the winter, the press reported more frequently on Lilienthal's flying;
an article in the *Chicago Tribune* described how the "human angel" accomplished the aerial movements of the bird. If the tales were true, Lilienthal was
very near to success, so Chanute wrote him: "I need hardly say that I will be
pleased to correspond with you, especially if you write in English, as I do not
read German"[44] and added, possibly as a character reference, that he was the
past president of the ASCE and the president of the Associated Engineering
Societies. Chanute mailed the letter to a Berlin book dealer, requesting that they
forward it. Not receiving a reply, Chanute wrote a second letter after receiving
Lilienthal's correct address from another correspondent. This time Gustav Lilienthal, who spoke English, replied and mailed several recent articles. A WSE
member translated them, and Chanute submitted one for Forney's December
1894 issue as, "Why is artificial flight so difficult an invention?"[45] These translated
articles linked the "Flying Man" in Germany with aeronautical development
in the United States, with Chanute in the middle.

On the opposite side of the globe, Lawrence Hargrave of Sydney designed
flying machines of increased size and published full accounts of his experiments.
Chanute contacted Hargrave in 1891 and was surprised to read in one of Hargrave's early letters about his philosophy: "Workers must root out the idea that
by keeping the results of their labors to themselves a fortune will be assured to
them. . . . The flying machine of the future will not be born fully-fledged and
capable of a flight for 1,000 miles or so. Like everything else it must be evolved
gradually. The first difficulty is to get a thing that will fly. When this is made, a
full description should be published as an aid to others. Excellence of design

and workmanship will always defy competition." Chanute discussed Hargrave's early work in September 1893, his cellular kites in October 1893, and his later work in April 1895.

Thanking Hargrave for sending photos and publications, Chanute wrote, "I had already reached the conclusion that longitudinal stability was best to be obtained by placing aeroplanes behind each other, but I must say that the cellular idea is new to me. Do you find that the vertical planes between the horizontal surfaces add very much to the stability?"[46] This was an interesting question, because Chanute had experimented in 1888 with a two-cell model similar to a Hargrave box kite, but it did not perform to his satisfaction. At that time he did consider flying his model as a kite; not having done so, he may have missed the distinction that attaches to Hargrave's name.[47] The following year, Chanute arranged for Hargrave's No. 14, the best of six models driven by compressed air, to be donated to the Field Museum at Chicago; it arrived in November 1894. The model was transferred to the National Museum in Washington in 1906.

Investigating the progress made with flying machines worldwide was more intense than Chanute had first envisioned. Because past failures had resulted from so many different causes, he listed in his concluding article in January 1894 what he considered the ten most critical problems. Each problem, he believed, needed to be resolved separately and all the separate solutions should then be combined to achieve a final result:

1. The resistance and supporting power of air.
2. The motor, its character and its energy.
3. The instrument for obtaining propulsion.
4. The form and kind of the apparatus.
5. The extent of the sustaining surfaces.
6. The material and texture of the apparatus.
7. The maintenance of the equilibrium.
8. The guidance in any desired direction.
9. The starting up under all conditions.
10. The alighting safely anywhere.

In his personal learning process, just as he had done when researching timber preservation a decade earlier, Chanute acted as a clearinghouse and introduced similarly minded researchers to each other. More engineers became involved, fresh ideas surfaced, and the readership of Forney's magazine grew, to the satisfaction of both men.

Continuing Development: Louis-Pierre Mouillard

Several attendees at the 1889 Aeronautical Congress had mentioned Louis-Pierre Mouillard and his book *L'Empire de l'Air*, published almost a decade earlier. Chanute located a copy while still in Paris and read about Mouillard's observations on bird flight with fascination. To share the information with English-speaking investigators, he translated the book, but regretted that his English was not as vivid as Mouillard's French. Langley included an abbreviated translation in the *Annual Report of the Smithsonian Institution for 1892.*

In March 1890 Chanute wrote his first letter to Mouillard,[48] who lived in Egypt, and mailed him a reprint of his paper, presented in Paris six months earlier. Half a year later, Chanute explained his ideas on manned flight: "I should not even dream of imitating the bird, but I believe it is possible to apply the same principles to obtain a stable equilibrium. The question now is to limit as much as possible any movement of the center of pressure, without decreasing the lifting surface."[49] In his response, Mouillard explained his thoughts: "When the bird is about to turn, the wing tip twists. . . . The feathers obstruct the motion, they stop it, forming a long lever on this point. This causes a variation of motion and a rapid change of direction."[50] Chanute now wondered if the same principles of a locomotive banking (see chapter 4) could apply to a flying machine. Could extra drag be applied to one wing to turn the machine in flight?

The following year, Mouillard mailed Chanute the manuscript of his second book, *Le Vol sans Battements* (Flight without Flapping), in which he described his flying experiences years earlier. Chanute translated Mouillard's story for his January 1893 article:

> It was on my farm in Algeria that I experimented with my apparatus. . . . Nearby there was a wagon road, raised some 5 ft. above the plain. I thought that I might try it armed with my aeroplane; so I took a good run across the road, and jumped at the ditch as usual. But, oh horrors! Once across the ditch my feet did not come down to earth; I was gliding on the air and making vain efforts to land, for my aeroplane had set out on a cruise. I dangled only one foot from the soil, skimming along without the power to stop. At last my feet touched the earth. I fell forward on my hands, broke one wing, and all was over; but goodness, how frightened I had been! I then measured the distance between my toe marks, and found it to be 138 ft. . . . I cannot say that on this occasion I appreciated the delights of traveling in the air, and yet never will I forget the strange sensations produced by this gliding.[51]

Continuing their correspondence, Mouillard wrote in early summer that he wanted to build an aeroplane to do maneuvers, not only glides.[52] Chanute thought that he could sacrifice 50,000 or even 100,000 francs in a promising enterprise like a flying machine, so he suggested applying for a patent, and mailed Mouillard 500 francs for expenses. The two men compiled their claims and Chanute submitted the papers and fee to the U.S. Patent Office in September 1892.

Much to Chanute's dismay, his patent attorney, George Whittlesey, reported that the patent examiner was quick in declaring "Mouillard's device incapable of ascension," because there was no gas field. This initial objection was shocking news, and Chanute quickly responded that everyone must have seen a kite rise, and kites did not require gas fields, and neither did their machine. "Such an objection indicates that we shall have more serious trouble when we come to the important part of getting the claims allowed."[53] Chanute included $100 for expenses so that Whittlesey could discuss their claims with the examiner, or even the examiner in chief. After much debate between the applicants, the patent lawyer, and the examiners, the Patent Office granted the patent "Means for Aerial Flight"[54] five years later.

In this patent, the steering of the machine was described as: "A portion J' of the fabric at the rear of each wing is free from the frame at its outer edge and at the sides. A pull upon one of these handles causes the portion J' to curve downward, and thus catch the air, increasing the resistance upon that side of the apparatus and causing it to turn in that direction. . . . For turning in the horizontal plane, the resistance on one wing is increased, so that it tends to travel at a less velocity than the other. This is done by warping the apparatus provided for the purpose." Creating drag worked as an effective brake in a two-dimensional scheme, but banking a flying machine in three dimensions was different. Neither Chanute nor Mouillard considered the lifting effect, which would turn the machine in the opposite direction of what they intended.

Chanute wanted Mouillard's invention to succeed and his aeroplane built. Even though the 1893 financial panic had also affected his financial holdings, he mailed Mouillard a draft for 2,500 francs and promised an equal amount, if needed. In an exchange, he asked for a description of the machine and periodic progress reports. Over the next few months, Chanute provided constant encouragement. Sending the July 1893 *Railroad Journal*, he commented that Lilienthal

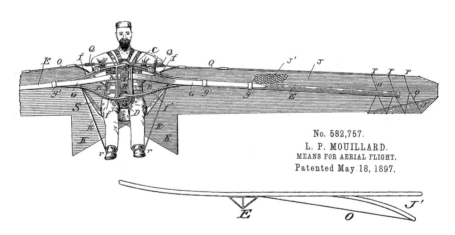

No. 582,757.
L. P. MOUILLARD.
MEANS FOR AERIAL FLIGHT.
Patented May 18, 1897.

Means for Aerial Flight. United States Patent No. 582,757, by Louis-Pierre Mouillard, assignor of one-half to Octave Chanute. Top: front view of the glider, with the operator in position. Bottom: the horizontal steering as envisioned by the two inventors.

could only glide, and he hoped "that yours will make us perceive ascension, and perhaps aspiration."[55] In Chanute's vocabulary, "ascension" meant gaining altitude after take-off, while "aspiration" was the performance of a bird using air currents to be drawn forward to achieve soaring flight.

Early in December 1895, Mouillard mailed photos of his aeroplane. Much to Chanute's disappointment, the machine looked cumbersome, and he wondered if Mouillard copied the bird too closely. "Still, that is not so important, the essential thing is to know if your invention will give you the two controls of direction and equilibrium. Don't get discouraged too soon, but be careful that you don't get hurt."[56] Mouillard did not respond for several months.

Finally, Mouillard wrote in September 1896 that his machine was too heavy and needed to be rebuilt before he could fly it. This was disheartening, even though Chanute had secretly shared the same opinion. "I regret that you have dismounted it without having made a serious experiment, for, be it ever so heavy and imperfect, it would have indicated if one could expect gliding flight from it. If it does not permit it, it would be of little use to make another lighter and finer machine."[57] Two months later, Mouillard wrote that he was partially paralyzed. Feeling sorry, Chanute sent money and hoped that his correspondent's health would return, but his collaborator died in September 1897.

Fifteen years later, a number of Frenchmen wanted to show Mouillard as a true pioneer and that Chanute stole his patent. Wilbur Wright was asked to clarify "What Mouillard Did," and he stated: "As a missionary, Mouillard stood at the top along with Lilienthal and Chanute. As a scientific student of the laws and principles of aerodynamics he is not to be mentioned in the same class with such men as Cayley, Wenham, Pénaud, Langley, Lilienthal, Chanute and Maxim. He was a careful observer of birds, and possessed a genius for expressing his thoughts and feelings in words, but beyond that he was mediocre. He made a few feeble attempts to construct soaring machines, but their design and construction were so crude that he failed to surpass the futile attempts at gliding made by Cayley and Wenham who long antedated him."[58]

The 1893 International Conference on Aerial Navigation in Chicago

No one really knew whether human ingenuity would ever lead to aerial transportation, but the subject was fascinating. Having studied aerodynamics for several years, Albert Zahm, who taught mechanical engineering at the University of Notre Dame, suggested to Chanute in early September 1892 to stage an aeronautical conference during the upcoming World's Fair, similar to what the French had done in 1889. At first Chanute was concerned about an invasion of cranks, but he agreed that such a gathering would "enlighten a number of worthy, but ill-equipped, inventors who are re-trying old experiments with no proper understanding of the enormous mechanical difficulties involved."

Several WSE members, including Bion Arnold, Ira Baker, Elmer Corthell, and William Karner, offered their support and World's Fair organizers accepted the proposal. Chanute became the chairman and Zahm the secretary of the first American conference on aerial navigation. In December 1892 they prepared a circular letter asking for papers and mailed it to aeronautical societies and experimenters around the world.[59]

Writing to Carl Myers, owner of a "Balloon Farm" in Frankfort, New York, Chanute explained that the object of the conference was to discuss the scientific aspects of aeronautics. "We shall desire to elicit facts, positive knowledge and records of experiments, rather than theories or novel proposals."[60] Myers, whom Chanute later introduced as an "aeronautical engineer," giving him the same title as the military had assigned to balloonists during the Civil War, agreed to solicit speakers for the session on ballooning. To prevent uninvited spectators, or "cranks," from attending and disrupting the conference, organizers decided that every person should purchase a "personal card of admission" for $3.

The Conference on Aerial Navigation ran in conjunction with the first American Engineering Congress. Chanute wanted both events to achieve full success, so he became deeply involved with their planning. Rather unexpectedly, the aeronautical conference ran into several problems just prior to the opening. Papers were slow arriving, and then speakers canceled, including Langley, who had agreed to chair the first session. At the last moment, Chanute, Zahm, Thurston, and Myers assigned time slots for the speakers who were present.

Chanute, president of the General Committee of Engineering Societies, opened the engineering congress on Monday morning, July 31. The next day, the aeronautical conference opened in the afternoon in Hall 7 of the Memorial Art Palace. In his address, Chairman Chanute summarized the state of affairs: "The truth seems to indicate that it is not unreasonable for us as engineers, as mechanics and as investigators to meet and discuss the scientific principles involved and to interchange our knowledge and ideas."[61]

Chanute chaired the first session on scientific principles, covering the properties of the air, propellers, motors, and materials of construction, and he read Langley's paper "The Internal Works of the Wind" first, because it held promise as an important breakthrough. Langley wrote that "this internal work of the wind might conceivably be so utilized as to furnish a power which should not only keep an inert body from falling, but cause it to rise. . . . They are paradoxical at first sight, since they imply that, under certain specified conditions, heavy bodies, entirely detached from the earth can be sustained indefinitely without any expenditure of energy from within. This is, without misuse of language, to be called a physical miracle." In the discussion, De Volson Wood agreed that certain birds could soar against the wind, but that there was clearly some principle not yet understood, "and the possibility of reproducing it artificially does not yet appear to be within the grasp of man."[62] It took many more years to understand how a heavier-than-air craft could maintain altitude and fly against the wind, using the lift provided by the upward vector of the wind as its energy source.

The recently deceased Charles Hastings had prepared two papers on the theoretical aspect of aeronautics and William Breithaupt from Kansas City, a mutual friend, read them both. Carl Vogt from Denmark discussed in his paper the differences of a screw pushing a ship in the water and a propeller driving a balloon or aeroplane through the air, describing the peculiarities of either supporting medium. Thurston read his paper "Materials of Aeronautic Engineering" last, presenting facts on the properties of light but strong metal alloys, some of which were new even to the attending scientists. His metal of choice for the future aircraft was steel, but his research had shown that adding small quantities of magnesium to aluminum strengthened it considerably, thus he recommended this metal as his second choice.

The next day, Wednesday, Thurston chaired the session on aviation. Twelve papers discussed heavier-than-air flying machines that had to propel themselves forward and overcome gravity. About one-half of the papers detailed the sailing flight of birds, while others discussed aeroplane designs and learning to fly. Zahm read a paper on "Soaring Flight" by an up to that time unknown experimenter from Tennessee, Edward Huffaker, and then his own paper, in which he described his models that flew steadily in wind and calm.

The final session on ballooning had the advantage that it was already successfully practiced. Looking for someone to chair the session just prior to the opening, Thurston suggested Lieutenant Colonel W. R. King of the Army Corps of Engineers, who had recently read a paper to Sibley students on the "Military Engineer and his Work." King agreed, even though his knowledge of ballooning was limited. Discussing natural gas balloon ascensions, Myers described a flight by his wife Carlotta, in which she had reached an altitude of 21,000 feet and landed safely after traveling 90 miles. Because the gasbag had no precedent in the ornithological world, De Volson Wood suggested in his paper, "Flotation vs Aviation," a departure from the bird method in flying. "Discarding the flapping wings as a motor, the continuously rotating propeller is the only form of driver that commends itself." In closing, he stated: "A speed of 30 miles per hour produces relatively a brisk breeze, and to talk of carrying a man 100 miles per hour in an airship we consider rather 'flighty'."

About one hundred persons attended each session, and a good discussion followed the reading of every paper. Zahm noted in his diary: "Large attendance and papers pronounced very valuable and interesting. Three hours all too short for presentation of papers. Reporters and private stenographer in constant attendance."[63] Even though attendees made advances in understanding the principles, they agreed that without motive power and an aerial screw, combining maximum efficiency with minimum weight, manned flight would take years to become reality.

In his Sibley lecture in 1890, Chanute had expressed hopes of forming an American aeronautical society, so now he wondered if twenty persons would be interested in starting a national aeronautical society and advancing money to build and experiment with flying machines.[64] Even though he found little interest in either proposal, Chanute was encouraged that several prominent engineers, creative giants in their respective fields, attended the conference and participated in the discussions. These men included Thomas Edison, inventor of the light bulb and interested in helicopters; John P. Holland, the inventor of the modern submarine, who considered building a flapping wing machine; Alfred Mayer, professor at Stevens Institute, who had experimented with flying models in the 1870s; Francis A. Pratt, cofounder of the Pratt & Whitney company with whom Chanute had worked twenty years earlier; Charles Steinmetz, who later invented the system of distributing alternating current; Nikola Tesla, who had long puzzled over a flying machine; and George Ferris, the originator of the World's Fair's main attraction.

The amateur bird watcher Samuel Cabot, a wood preserver, became interested in manned flight but also in establishing an aeronautical society in Boston. He began corresponding with Chanute soon after returning home. Spending a few months in Europe in 1896, he visited with Maxim and Lilienthal and obtained drawings for a Lilienthal glider, but decided against building one. He kept up his friendship with Chanute and eagerly promoted aeronautical activities.

Because the country still suffered from the aftereffects of the 1893 financial panic, money was tight, but Chanute really wanted the conference papers with their discussions published. A Chicago printer showed some interest, but Forney had a better idea. He proposed to publish the papers in a new monthly, *Aeronautics*, charging a subscription price of $1 for twelve issues for new subscribers and $0.50 for current subscribers of the *Railroad Journal* and then publish the *Proceedings* in book format.[65]

The aeronautical conference was a step forward in engineering science and attracted legislative attention. Senator Cockrell introduced a bill in December 1893 to secure aerial navigation, but politicians ignored it. Chanute and James Means rewrote the bill and resubmitted it. "The trackless ether presents such limitless highways that the regulations for travel would be comparatively simple and travel in a given direction would follow laws both as to altitude and latitude. Rapidity of flight would be by no means the chief advantage. As the railroad across the plains, in comparison with the ox cart, has made travel delightful, so the little aerial car fitted up for the family party would excel the costliest parlor car of Mr. Pullman."[66] Even though the Committee on Interstate Commerce recognized the value of aerial navigation, they did not recommend passage of the bill.

The aerial age was clearly on its way, and Chicago politicians wanted the railroad center to take the lead in developing aviation as well. Over the next decades, vast sums of money were spent on promoting aeronautics in Chicago, partially owing to Chanute's early influence.[67]

Progress in Flying Machines: An Aviation Classic

More than seven years after Forney suggested this project, the article series *Progress in Flying Machines* was ready to be published as a book.[68] In writing the monthly articles, Chanute hoped to provide an understanding of the principles and the known experiments, so that any investigator could distinguish between a proposal that was sure to fail and a design that might succeed.

Defining the requirements for a practical flying machine, Chanute thought that two types might eventually evolve:

> The soaring type may or may not be provided with a motor of its own. This must be a very simple machine, as this type will have to rely upon the power of the wind, just as the soaring birds do, and whoever has observed such birds will appreciate how continuously they can remain in the air. Such an apparatus for one man need not weigh more than 40 or 50 lbs., nor cost more than twice as much as a first-class bicycle. . . . The journeying type of flying machine must invariably be provided with a powerful and light motor, but they will also utilize the wind at times. . . . It seems quite certain that flying machines can never carry even light and valuable freights at anything like the present rates of water or land transportation, so that those who may apprehend that such machines will abolish frontiers and tariffs are probably mistaken. . . . So may it be; let us hope that the advent of a successful flying machine will bring nothing but good into the world; that it shall abridge distance, make all parts of the globe accessible, bring men into closer relation with each other, advance civilization, and hasten the promised era in which there shall be nothing but peace and good-will among all men.[69]

Forney shipped the typeset pages to Chanute to proofread and to create an index, which turned out to be a "devil of a project." The author found a number of typographical errors, but he also wanted his name on the title page to show "O. Chanute." Questioning why there was no publication date, Forney explained that the text had been reprinted from his copyrighted magazine, thus no date or copyright statement was needed. The family discussed the color of the cloth binding, and one daughter suggested gray-green—gray for the author and green for the purchaser, but Forney selected a light-blue material instead. The artwork on the cover engendered more discussion; Forney wanted to use a balloon engraving, but Chanute did not think this appropriate for a book on dynamic flying machines. Seeing a full-page drawing of Lilienthal's glider in flight in *La Science Illustrée*, he suggested using this for the cover.

Forney corrected the text, printed the pages, and everything was ready to go to the binder when Chanute received Lilienthal's latest article, in which he questioned why no one had previously discussed the benefits of curved wing surfaces. In Chanute's opinion, Lilienthal's article "so fully sustains the views set forth in this book, and holds out such promise of success in the near future,"[70] that including this article would help sell the book. Forney reluctantly agreed, but only as an appendix.

In late April the 308-page, nicely bound book *Progress in Flying Machines* arrived in Chicago. Forney had printed about five hundred sets of pages and had two hundred books bound in the spring of 1894; Chanute received twenty-five unbound page sets for distribution as gifts and bought forty books for $2.50, minus 45 percent discount, to send to his friends. To reach a broader audience, Forney mailed books for review to influential people, magazines, and larger libraries, resulting in favorable reviews. "As an engineer Mr. Chanute is able to prick many a bubble and fallacy, while his genuine enthusiasm for the subject leads him to examine carefully into the most unlikely project," was the reviewer's comment in the *Electrical Engineer*.

Effective January 1, 1896, Forney sold the *American Engineer and Railroad Journal* but agreed to stay on as editor and inserted book ads periodically. Up to the end of 1898, Forney sold ninety-nine books and Chanute bought a similar number, receiving a royalty payment of $.14 for each book sold. Needing more books in early 1899, Chanute inquired about a soft-cover binding, but Forney did not favor it. The title page of the resulting binding shows "M. N. Forney" as the publisher and the date 1899. The cover also differed slightly; the word "Progress" on the spine was positioned below the embossed border on the front. In April 1902 Forney had the remaining page sets bound into books.

Even though *Progress in Flying Machines* was never a bestseller, it quickly became the basic handbook for would-be aviators. Its editing by an engineer of Chanute's status gave the subject a prestige not afforded to most aeronautical literature of the day.

Manned Flight is Possible

The 1893 Conference on Aerial Navigation had attracted several experimenters whose work had not been previously known. One of them was John J. Montgomery of San Diego, who was visiting the World's Fair when he heard of "scientific men discussing flight."[71] He introduced himself to Zahm and Chanute and told them that he had experimented with flight ten years earlier. Participating in the discussions, he explained that he took his glider to the top of a sloping hill in 1884 and "placed himself within the central framework; ready to sit down, he faced a sea breeze steadily blowing from 8 to 12 miles an hour, and gave a jump into the air without previous running" and landed 100 feet downhill; a second attempt resulted in a smashed glider. Chanute discussed Montgomery's flying activities in his December 1893 article.

With Mouillard progressing only slowly, Chanute mentioned to his family that he really would like to see a glider in flight and collaborate with Montgomery. Spending the winter of 1894 in San Diego, Annie Chanute and her daughters decided to meet Montgomery. They took a ride to his farm in Otay Mesa, but Montgomery's mother told Annie that her son was not home. A few days later, Montgomery wrote Chanute that he could not resume his studies.[72]

Arriving in San Diego a few weeks later, Chanute did not try to meet Montgomery. Instead, he "idled" his time studying birds. Chanute had been interested in photography for many years, so he bought the recently patented Kodak box camera that was advertised to make hundreds of photos instantly, using a rolled negative film instead of glass plates that always broke at inappropriate times. Could this new Kodak capture birds in flight just by "pressing the button?"

This was not an easy task, because the birds were either too far away or moved too rapidly, or there were no objects to judge dimensions or positions. Back in Chicago, Lizzie succeeded, and her snapshot of a seagull won second place in a photo contest sponsored by Sam Cabot[73] and the Boston Camera Club.

Spending a few late winter weeks in 1895 with Annie's relatives in Thomasville, Georgia, provided more opportunities for bird watching, and Chanute described his observations to Wilhelm Kress: "The soaring vultures are abundant here and are aloft all day long. When the wind is light, they sweep in circles, but when the breeze is brisk, they progress in straight lines in every direction, even dead against the wind, without losing height, giving a practical demonstration of what the French call 'aspiration.' I am convinced that there is a phenomenon which we do not yet understand, and which, when thoroughly mastered, may enable us to join the birds in the sky."[74] Listening to the peculiar buzzing noise of birds in high-speed flight, there had to be some kind of driving force. "The bird is generally sailing at a speed of 20 to 25 miles

Octave Chanute in southern California. Courtesy of the Chanute Family.

per hour, with a wind blowing from 12 to 18 miles per hour and the impulse is received by the bird when sailing with the wind. My own opinion is that it is received on the quarter, when the action is similar to that of a ship sailing 'close hauled,' but I am not yet prepared to publish my views."[75]

Assuming that this pursuit in aeronautics would eventually succeed, engineers wondered to which branch of engineering it should be assigned. After listening to Zahm's lecture on "Aerial Navigation" in January 1894, John Trautwine, a fellow ASCE member, questioned if this was part of civil, mechanical, or electrical engineering or even military engineering. Preparing his monthly column for *Engineering Magazine*, Trautwine mailed the manuscript to Chanute, who commented, tongue-in-cheek: "There would be time enough later to discuss to what branch an engineer belongs when he has succeeded in flying in the blue."[76]

Otto Lilienthal from Berlin had discovered the secrets of flight by using soaring birds as models; he regularly flew his glider, without an engine or airscrew. Chanute did not meet Lilienthal, but curious Berliners, reporters, engineers, and scientists from around the world were frequent visitors. One American engineer described his visit: "Dr. Lilienthal began by putting on his fair-weather wings, and, ascending to the top of the hill, he jumped off. Something however was wrong and in a moment he landed prostrate on the ground. Fortunately, he was not injured, and taking off the unfaithful wings, he put on a second pair, similar in construction but of smaller spread, being destined for windy days. With this apparatus he essayed a second flight, with success, passing over the heads of the spectators, and landing safely on his feet a little beyond the foot of the hill."[77]

The American inventor Sir Hiram S. Maxim, living in a London suburb, took a different approach to discover the secrets of flight. After spending about $100,000, he launched his steam-powered aeroplane from an inclined railroad track on July 31, 1894. On the third attempt, the white bird, with four wings and two seventeen-foot, two-bladed propellers, was airborne for a few seconds. This was the second successful attempt of powered flight carrying a man, but Maxim's eight-thousand-pound machine did not fly at ninety miles an hour or more than one thousand miles, as predicted by its inventor. Chanute discussed Maxim's aeroplane in his September 1894 article.[78]

Chanute's article series on flying machines, followed by the publication of the monthly *Aeronautics* magazine, had generated much interest among Forney's paying subscribers, and he wanted to continue publishing "all matter relating to the interesting subject of Aerial Navigation, a branch of engineering, which is rapidly increasing in general interest."[79] Starting with the October 1894 issue, the *American Engineer* featured the subsection "AERONAUTICS," with Chanute acting as the associate editor.

Discussing the latest aeronautical news and reviewing articles and books every month in the *American Engineer* encouraged authors to mail their publication to Chanute. Early in June 1895, he received the *Taschenbuch* from Captain

Hermann Walter L. Moedebeck of the German Artillery.[80] Reviewing the book's contents, Chanute noted, "nothing marks better the recent rapid advance in aeronautics and the growing faith that practical success is not afar off, than the present publication, in German, of a pocket-book for practitioners and students of this inchoate art." He liked the advertisements in the back that even included one for his book *Progress in Flying Machines.* But the article by Lilienthal on "Kunstflug" (artificial flight), with its graphs, table, and photos, really piqued Chanute's curiosity, however someone knowledgeable about aeronautics had to translate it for him.

The world was, at times, small. In the early 1890s, looking for an explanation for the constantly alternating sea and land breezes in San Diego, Chanute had introduced himself to the chief of the local Weather Bureau, Willis Moore. Later, Moore transferred to Chicago and in 1895 to Washington to become chief of the U.S. Weather Bureau. To provide a better weather forecast, Moore liked Chanute's idea of raising kites to higher altitudes with instrumentation to better understand the weather patterns. The bureau introduced kites late in 1895 and Charles Marvin, working under Willis Moore, won the "Chanute Prize" in 1897 for his published research with kites.[81] Both Moore and Marvin continued their collaboration with Chanute over the next decade.

Hearing of a privately published thirty-page pamphlet titled *Manflight*, Chanute ordered several copies from the author James Means and the two men quickly became friends. Early in 1895 Means published a more ambitious literary project, the *Aeronautical Annual*,[82] in which he reprinted several classic articles, and Chanute was more than pleased to review the book.

Hiring a Helping Hand: Augustus M. Herring

While studying mechanical engineering at Stevens Institute of Technology, Augustus Herring had built several rubber-powered models and a glider.[83] Seeing Lilienthal's German patent, he built three such gliders in 1894, but none performed particularly well. "The first machine cracked up due to my lack of knowledge in sailing it. The longest flight was ~80 ft., with a descent of 17 ft. The second machine was built stronger to withstand any heavy shocks, which might occur from my unskillful handling. I was able on one occasion to fly 150 ft or more."[84] Sailing through the air was exciting, so Herring wrote about his experiences and mailed the manuscript to the *American Engineer*. This was the beginning of an at times fractious relationship with Chanute.

With Mouillard progressing very slowly in building his glider and Montgomery not interested, Chanute wondered if Herring might like to collaborate. In mid-December 1894 Chanute wrote Herring: "Are you so circumstanced that you could undertake to build an experimental machine for me, at your leisure?" Herring, who at the time worked as chainman on a railroad, replied immediately that he would be pleased to undertake its construction. A week later, Chanute mailed his sketches. "This present design is for a soaring apparatus for one man, but it is intended to add eventually a motor and two screws, revolving in opposite directions. . . . The primary object is to provide for automatic stability, by causing the center of gravity to move, so as to coincide with the center of pressure in horizontal flight, so that the operator need only intervene when he wants to change his direction, either up or down or sideways."[85]

Having accepted the appointment as consulting engineer for the Rapid Transit Commission, Chanute needed to travel to New York in early January 1895. This also gave him the opportunity to meet his newest correspondent to discuss their mutual field of interest. The twenty-seven-year-old Herring liked the Chicagoan, who was about the same age as his father who had died a few years earlier, and the almost sixty-three-year-old Chanute liked this young engineer, who seemed to have the qualifications he wanted to foster. Without hesitation Herring accepted Chanute's job offer for $5 a day of work. At that time a carpenter made about $2.80 and a brick mason about $4 a day, working eight hours.

Chanute's first assignment was a twelve-wing "ladder kite" on which the wings could be adjusted to test the optimum position and possible interferences. He also asked Herring to build three different wings to determine the most efficient wing curvature and test each design by mounting it on a bicycle, driven at a steady speed to collect reproducible data. When the model was ready to be flown, Herring asked Chanute to come to New York, but Chanute wrote back that "Business must go before pleasure, I can not come right now to join in the flying." Herring reported back that the model performed better than the ones he had built previously.

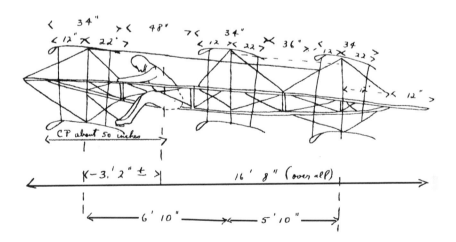

Chanute's proposed design for a multiwing glider. Letter to A. M. Herring, December 31, 1894.

The "Ladder Kite," built by A. M. Herring for Chanute, flown early in 1895 in New York and later in Chicago. Chanute photo album, Manuscript Division, Library of Congress, Washington, D.C. WB039P-018-67.

Late in April 1895, the secretary of the Smithsonian paid Herring an unexpected visit. Assuming that Langley was looking for someone to hire, Herring showed his latest Lilienthal-type glider and the models he had built for Chanute. Langley liked what he saw and wrote Chanute about his intent to hire Herring, whereupon Chanute told Langley that he wanted Herring to build him a full-size soaring machine, "but you are so much better equipped than myself, that I now believe that greater progress will be made toward a solution of the problem by me foregoing my purpose, and Mr. Herring joining you. If you can make an arrangement with him, I propose to let you avail of whatever novelty and value there may be in my own models or ideas. I should expect in return a like frank access to your results, when the time arrives."[86]

Herring accepted Langley's offer to work for the Smithsonian, but wrote to Chanute a week later: "Started work for Langley on Monday, 20 May, for $150/ month. Even in a days work under him I could foresee a possibility that he and I might not be able to agree."[87] Langley had several steam-powered models built by his two recently hired men, Herring and Edward Huffaker from Tennessee; however, none performed as envisioned. Langley wanted no input, not even from a well-meaning friend. So Chanute commented to Herring: "I am rather sorry that he did not accept my offer to communicate my knowledge to him as frankly as it was made; and yet it is perhaps best that each of us should remain in his portion of the field. Professor Langley working out dynamic flight, and I experimenting on soaring flight."[88]

Spending the summer of 1895 in Europe, Langley also stopped in Berlin and reported to Herring: "Although I was only part of two days in Berlin, I managed to go out to Lilienthal's place and see him take several flights. They were interesting, but I did not feel that I learned much. His flying machine looks unnecessary heavy. Could not describe the curves, as I had no opportunity of measuring, but the aspect of the whole was heavy and clumsy—however, handsome is as handsome does."[89] Needless to say, Herring shared this letter with Chanute.

After Langley returned to Washington, working conditions became challenging for Herring, who wrote Chanute in November: "One of the disagreeable features is Mr. Langley's inability to distinguish between the ideas of other people & his own, whether this is intentional or not I can't say—the effect is the same." A couple of weeks later, he wired: "Mr. Chanute: Expect resign December first. Have you any work for me?" Even though Langley had confided that Herring's reliability was questionable, Chanute ignored Herring's perhaps ominous statement about ownership of ideas and wired back, "I want you here to build a soaring machine."

Early in 1896 Chanute worked several consulting jobs, and his wife and daughters had gone south in late December, so he offered Herring the use of their house. The Herring family arrived with their two rat terriers in late January, and the servants relayed Chanute's instructions: the plans for the soaring machine were on the library table and Herring should use Bill Avery's shop to start building at once. "Make the main arms of the soaring machine from spruce, the front edge of the wings of ½ bamboo, and the ribs of willow. Make one wing first, then proceed with the other."

The new year had just begun; 1896 would prove a landmark in the history of aeronautics.

From Theory to Investigating:
The Lilienthal-type Glider

After spending more than a decade researching, lecturing, and writing, even experimenting with a variety of flying models, Chanute was highly aware of the many unknowns in aerial navigation. Impressed that the equilibrium of the bird was basically automatic, he studied the bird's complicated flying maneuvers but could not translate flight into a mathematical formula. Looking for input from fellow researchers, he wrote an article on "Sailing Flight"[90] and sent it to Means for his second *Annual.*

Chanute had highlighted the ten major issues in aerial navigation in *Progress in Flying Machines* and considered the maintenance of equilibrium of utmost importance. First, experimenters needed to solve the problem of stability in the wind, Chanute firmly believed that without built-in stability, "it will be exceedingly dangerous to proceed to apply a motor and a propeller. Birds preserve their balance by instinct, by skill acquired through long evolution and tentative practice. Man will have to work out this problem thoroughly, even to the temporary disregard of the others, if he is ever to make his way safely upon the air."[91]

In the mid-1890s Otto Lilienthal was the only person to achieve limited lateral and longitudinal control by energetically swinging his body to move the glider's center of gravity, thus affecting its pitch (nose up and down movement) and roll (sideways movement).[92] Chanute hoped to combine automatic stability with some form of weight shifting to provide a safer design. Agreeing with Lilienthal, Chanute advocated that flying a full-size glider was probably the quickest, cheapest, and surest way to learn the exact conditions in practical flight. "Herr Lilienthal says that his experiments have taught him that there is no mystery about sailing flight. Inventors need not look for some new mysterious force, nor need they be afraid that if they propose to experiment with soaring machines they will be considered lunatics."[93]

To maintain his status in aeronautical matters and move closer to manned flight, Chanute had to begin experimenting with practical flying. He knew that he had to validate the work done by others to separate fact from error; only then could he venture into the unknown to discover the unanticipated. Taking a closer look at photos of Lilienthal's glider in flight and studying his patent claims, Chanute was concerned that this apparatus "requires constant personal interference to maintain the equilibrium and that when a motor is added, it will be more than a single man can handle, run the motor and to persevere the balance at the same time." To make the aircraft easier to operate, he determined that, "Instead of the man moving about, to bring the center of gravity under the center of pressure, it was intended that the wings should move automatically so as to bring the movable center of pressure back over the center of gravity, which latter should remain fixed. That is to say, the wings should move instead of the man."[94]

Chanute submitted his claims for an improved "Soaring Machine,"[95] based on Lilienthal's patent, to the U.S. Patent Office in December 1895. He proposed to include "a stationary seat and to pivot each wing on an upright pintle, so that it can be moved bodily forward and backward, as may be required to preserve the balance of the machine and the aviator." Wire braces would keep the wings in position, "but permit them to be turned forward and backward. Strong springs between the hoop and the leading edge of the wing were to exert a constant forward pull on the wings. In operation, the springs are arranged to yield sufficiently to allow the machine to be properly balanced when soaring in wind of ten or twelve miles per hour. If the wind strengthens, the wings are forced farther backward. If the wind lessens, the springs pull the wings farther apart. In each case the shifting of the wings compensates for the change in the center

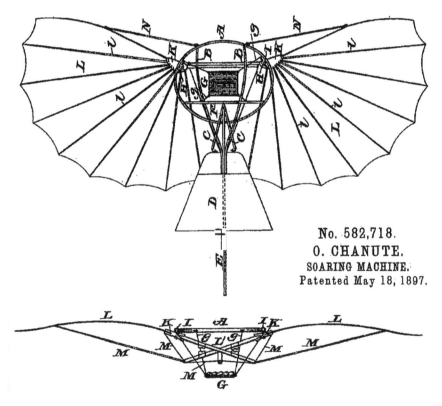

No. 582,718.
O. CHANUTE.
SOARING MACHINE.
Patented May 18, 1897.

Top and front view of Octave Chanute's "Soaring Machine," an improved Lilienthal-type glider. United States Patent No. 582,718, granted May 18, 1897.

of the wind-pressure due to the change in the velocity and keeps the center of pressure coincident with the center of gravity, thereby automatically preserving the equilibrium." The patent protected his ideas, so now he could take the next step and build a machine that would fly!

Shortly after moving to the Palmer subdivision in 1892, Chanute had met William "Bill" Avery, whose father owned a jobbing shop across the alley at 404 Superior Street. Avery was an avid sailor with good knowledge of wind and weather and knew how to fix everything. Now Chanute humorously announced: "Have made contact with an intelligent carpenter to build my own machine and we will see how much of a failure we can make."[96] But seeing Chanute's sketches for an improved Lilienthal-type made Avery realize that this was probably beyond his capabilities. Fortunately, Chanute then hired Herring, who brought his latest Lilienthal-type along. The men rebuilt this glider in Avery's shop, to incorporate the patented improvements and to provide a sturdier construction with a larger wingspan.

Initially, Chanute had considered experimenting with gliders in San Diego, but the Weather Bureau's newsletter listed Chicago and the southern Lake Michigan shore with a higher average wind velocity than other locales. So Chanute and Herring took the Lake Shore & Michigan Southern Railway to Michigan City, Indiana, and inspected the sixty-mile stretch of "wild waste regions of sand hills" along the lake. Millers, later an eastern suburb of Gary, Indiana, with its almost treeless, sloping dunes rising to ninety feet, seemed like the perfect spot; the men could launch the gliders there into several wind directions and land on wide, soft, and sandy beaches.

Early in June 1896, a visitor came to the Chanute house and introduced himself as William Paul Butusov, but wanted to be called Mr. Paul. He explained that he was born in the early 1850s, raised in St. Petersburg, Russia, and became a sailor. In 1882, he settled in Chicago and had built and flown a soaring machine in 1889. Chanute doubted his story but hired him after quickly checking his references.

Six weeks after Langley's successful launching of his unmanned steam-driven aerodrome No. 5, the excitement began for Chanute's team. On Monday morning, June 22, 1896, they met at the Chicago Tie Preserving Works and carried gear and supplies for the next two weeks to the nearby Englewood Station. The Lake Shore train left at 7:24 A.M. and arrived at Millers an hour later, causing a hum of excitement among the locals. It was a colorful group stepping off the morning express train: Chanute, a distinguished-looking man with a gray mustache and imperial, his son Charley, Avery, Butusov, and Herring with his two dogs. Each person carried odd-looking luggage over the mile-long path to the dunes. After pitching their tents, they assembled the Lilienthal glider. Herring was the first to launch into the wind, looking like a huge butterfly with concave wings, slowly settling to the ground. Butusov tried next but had an upset. After watching the others, Avery asked to fly, and wrote a decade later: "My heart was in my throat, but in spite of my fears, I succeeded in making a very creditable flight."[97]

Much to Chanute's surprise, a *Chicago Tribune* reporter located their camp the next day, resulting in front-page headlines: "Men Fly In Midair. Chicago Experts Make Experiments on Indiana Soil."[98] The *Kansas City Journal* picked up the story, adding three laudatory paragraphs about their former resident O. Chanute, who was now so deeply interested in aerial navigation.[99] A reporter from nearby Westchester, Indiana, was curious about what was going on in his neighborhood, and his report appeared on June 27.[100]

Chanute had corresponded with Lieutenant William Glassford of the U.S. Signal Corps since the early 1890s, and the two men had met during the World's Fair in 1893. Glassford subsequently published an article in the *Journal of the Military Service Institution*, in which he lobbied for the Signal Corps to develop an expanded air arm, stating "it is the tactics of this new engine of the art of war which we must study."[101] Chanute agreed that the first nation to provide a flying machine would indeed possess a decided advantage in future wars. As part of their correspondence, Chanute mentioned his upcoming flying experiments, and Lieutenant Joseph Maxfield, stationed in Chicago, visited the camp in an unofficial capacity on June 24. Watching the Lilienthal glider in flight, Maxfield reported that experimenting with these machines would be worth the military's interest.

The last visitor to arrive in camp with his sketchbook was Frank Manley from the *Chicago Record*, who had interviewed Chanute a few months earlier. His eye-catching report included drawings of the Lilienthal glider and the kites in flight[102] and was "less inaccurate than the others," so Chanute mailed this two-column clipping, with his added penciled corrections, to his many worldwide correspondents.

Watching the Lilienthal-type glider in flight confirmed Chanute's concerns about stability, as the pilot had to be an acrobat to keep it under control. After making about one hundred jumps, the glider turned over on landing without hurting its pilot and was broken past mending. "Glad to be rid of it," Chanute wrote in his diary.[103] To make this a true sport, flying had to become much easier.

From Investigating to Trials: The *Katydid* Multiplane

While the eagles and the gulls flew effortlessly overhead, demonstrating how flying ought to be done, the men assembled Chanute's "soaring machine" next. This glider had twelve wings, each six feet long and three feet wide. Closely woven cloth was stretched tightly over the wing surfaces and coated with pyroxelene. Upon drying, the wings became airtight and rang like a drum when tapped with the knuckles. Lilienthal had used collodion, but Chanute preferred using pyroxelene,[104] a mixture he had used in the past on windmill blades in Kansas and Nebraska.

STEAL THE BIRDS' ART

Investigators Whose Camp Is on the Lake Shore Study the Way of the Eagle in the Air and Learn to Fall Slowly from Great Heights.

AERONAUTS MAKE PROGRESS.

MR. HERRING'S FLIGHT WITH THE IMPROVED LILLIENTHAL SOARING MACHINE.

The improved Lilienthal-type glider, as depicted by Frank Manley in the *Chicago Record*, June 29, 1896.

Chanute expected his machine to glide at angles of three to seven degrees in light breezes and eventually soar in stronger winds, either in spirals in a ten-mile breeze or in aspiration in a wind blowing at twenty miles an hour. But after only a few flights, it became apparent that the wings greatly interfered with each other. Methodically, Chanute altered the glider's architecture in a step-by-step process to ascertain the most efficient grouping of the wings to achieve maximum lift and the greatest stability. "The paths of the wind currents in each arrangement of the wings were indicated by liberating bits of down in front of the machine, and, under their guidance, six permutations were made, each of which was found to produce an improvement in actual gliding flight over its predecessors."[105] Analyzing each flight, he recorded the pertinent data in his little brown notebook:

June 24, 1896 Tried the 12-winged plane. Found it steady but lifting too much in front. Tied on 12-lb. bag of sand in front.

June 25 Reset machine as 8-winged. Located c. of pressure just under 1st pair. It proves steadier than Lilienthal's and promises to soar. Found it better balanced and easier to handle than with 12 wings. Made tests of wind currents with down and found it was deflected downward after leaving wings.

June 27 Rigged up machine with 4 wings in front & 8 behind, the latter superposed to 0.4 their breadth. Found this more stable than any arrangement yet tried.

June 29 Found 4 wings front & 8 behind better balanced when standing still in wind, but the front was insufficiently sustained in flight. Next tested with 8 wings front & 4 behind. Found machine nearly as easy to control when standing still and sustaining much better in flight. Both Herring & Avery made glides of 76 ft against a north wind and pronounced machine more stable in its present shape than the Lilienthal.

June 30 Raised top wings to test whether the increased leverage of wind would make machine more difficult to control, and whether the increased distance between wings would make the lift greater. Added a 5-lb. bag of shot in front.

July 1 Got flights of 30 ft. and found it as easy to handle as with the top wings lower down. After waiting till 5 P.M. for more wind, took off the upper back wing and inserted it in the wide space under front top surface thus having 5 sets wings in front (148 sq. ft.) and 1 set at bottom in rear (29 sq. ft.). Called this the "*Katydid.*"

July 4 Made a number of excellent jumps. Best one, Avery, 78 ft.; Herring, 82' 6".

Slowly, the machine evolved from a ladder kite to an ordinary multiplane glider. The resulting *Katydid*, named after the midwestern grasshopper, was easy to handle in twenty-mile winds and the operator needed to move only two or three inches and not the acrobatic fifteen to eighteen inches, as the Lilienthal-type had required. In a thirteen-mile breeze, the *Katydid* would glide with its pilot in an angle a little better than one in four.

Chicago papers still discussed Chanute's glider flying when the Democratic National Convention began on July 11. The thirty-six-year-old William Jennings Bryan, who favored Free Silver, campaigned hard for the presidency of the United States. Cartoons, poking fun at political events, were at that time as important as headlines on the front pages of newspapers, and several papers showed cartoons of Bryan flying to "cold height" in imaginary flying machines adapted from Chanute's gliders. Chanute owned stock in several silver mines, but it is not known if he supported the Democrat Bryan, who lost the election. Republican William McKinley won the final election as a champion of the gold standard.

The evolution of
the multiwing glider
Katydid, June 1896.
Chanute photo album,
Manuscript Division,
Library of Congress,
Washington, D.C.
WB039P-004-13;
WB039P-004-11;
WB039P-005-15.

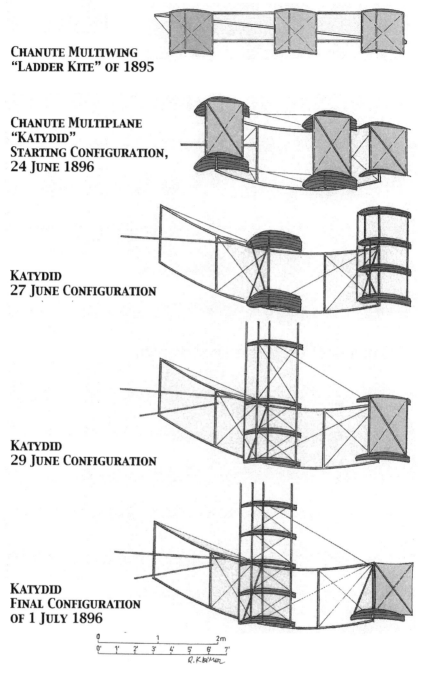

CHANUTE MULTIWING "LADDER KITE" OF 1895

CHANUTE MULTIPLANE "KATYDID" STARTING CONFIGURATION, 24 JUNE 1896

KATYDID 27 JUNE CONFIGURATION

KATYDID 29 JUNE CONFIGURATION

KATYDID FINAL CONFIGURATION OF 1 JULY 1896

Engineering development of the multiwing glider design (1896). Drawings by Reinhard Keimel.

During their two weeks of practical flying, everyone in Chanute's team had learned about the constantly changing wind and the updrafts along the dunes, but they also learned how to launch and control an aircraft in flight and then land safely. With the knowledge gained, Chanute had more ambitious plans: "I shall now proceed with the construction of two or three machines, to pass from the toboggan stage of air jumping to some attempt at soaring. I find that there is a vast difference between experimenting with models and with full sized machines with a man on them, as the wind is constantly changing in trend, in direction and in form, and gliding becomes an acrobatic exercise."[106] Writing to Wenham, Chanute explained that he now endeavored "to glide at right angles to a wind blowing up along the hillside,"[107] which Zahm called "naval soaring." In more modern language, Chanute hoped to "ridge soar" along the front of the dunes, but he also stated, "I may have something interesting to report in a week or two—or I may have a failure."

The men rebuilt the *Katydid* in early August. The wings now pivoted on ball bearings, placed at the top and bottom of wooden uprights to automatically adjust the center of lift with the wind. Chanute explained: "When the forward motion of the machine in flight is slackened, the springs pull the wings forward, thus advancing the center of pressure. The center of gravity, which has remained stationary, causes the supporting surfaces to oscillate towards the rear and to increase the angle of incidence, thus restoring the needed support. When the speed increases, the reverse effect takes place."[108] However, saving the expense of using the old wings proved a mistake, because they were distorted from prior use and the *Katydid* did not perform as hoped. But the basic design was a step toward a stable flying machine that Chanute wanted secured by patent. After some discussion with his patent attorney in Washington, he submitted his claims to the British Patent Office.[109]

The improvements over previously patented machines were mainly an attempt to achieve automatic equilibrium. "The rear portion of each wing is left free to flex upward by reason of the elasticity of the transverse ribs. This action compensates for abnormal changes in wind pressure when the machine undulates in flight. A vertical rudder is hinged to the stern post between the tail wings and may be governed by tiller ropes." Suspended below the frame was a seat for the operator, who would operate the wing-swinging system with foot-operated levers. A cord with a spring "interposed" ran from the lever around a pulley on the frame to the front edge of the multiple wings and another cord acted as a return device. "By shoving on both pedals, the wings will be pulled back, and the machine will tilt over slightly to the front and glide at a flatter angle. On letting both the levers swing back, the springs pull the wings forward, increasing the angle of incidence, tilting the machine up in front, and slowing its headway. By operating one pedal alone, the machine may be steered to the right or left." Unfortunately, this method of turning did not work as envisioned.

From Triplane to Biplane Glider

On August 12, one headline in the *Chicago Tribune* caught Chanute's eyes: "His invention cost him his life." The much-admired forty-eight-year-old Otto Lilienthal had lost his life in a flying accident and the telegraphic news stated: "The apparatus worked well for a few minutes, when suddenly the apparatus got out of order and man and machine fell to the ground."[110] The sixty-year-old Wilhelm Kress was just one correspondent who wrote in the old German handwriting about the accident and his earlier visit with Lilienthal. Chanute recalled this hard-to-decipher letter years later: "Herr Kress, the distinguished and veteran aviator of Vienna, witnessed a number of glides by Lilienthal with his double-decked apparatus. He noticed that it was much wracked and wobbly and wrote to me after the accident: 'the connection of the wings and the steering arrangement were very bad and unreliable. I warned Herr Lilienthal seriously, and he promised me that he would soon put it in order, but I fear that he did not attend to it immediately.' He unwisely made one too many and, like Pilcher, was the victim of a distorted apparatus."[111] Later reports confirmed that Lilienthal made his last flight in a well-designed monoplane, but the information that he had flown a distorted biplane perpetuated through the literature.

Lilienthal's spectacular successes, followed by his abrupt death, sad as it was, were an attention-getting catalyst. "The death of Lilienthal, the aviator, and the fatal accidents which have been so common of late among balloonists, are likely to check somewhat the work of experimentation in aerial navigation; but it is not probable that it will put a stop to research in this seductive though dangerous field," Thurston wrote in his news release for *Science*.[112]

Chanute's next flying machine showed ingenuity; its boxy structure mirrored his bridges and provided similar three-dimensional strength and rigidity. Trying to reproduce the wing of the soaring birds, he had consulted with John Johnson of Washington University about the characteristics of rattan. Using this wood allowed shaping the ribs in a circular-arc curvature with a height of about 1/12th of the width. Silk cloth was stretched tightly over the wing surfaces and made airtight using pyroxelene. Having learned how to efficiently position the wings in the *Katydid*, Chanute sketched a triplane and handed the paper to Herring. The glider, built in Avery's shop, had three superposed surfaces with a sixteen-foot wingspan and a four-foot, three-inch chord. The uprights were made from spruce and steel wire braced the wings in a Pratt truss pattern. Chanute later explained to Sam Cabot how he determined the wire size: "Having decided upon the weight to be supported, we calculated the amount coming on each post at the panel point, and its diagonal resultant along the wire. This gives us diameters varying from two to five hundredths of an inch, with a factor of safety of five, and we took the nearest commercial size to that."[113]

No one wanted Lilienthal's accident repeated, so Chanute asked his physician, Dr. Walter Allport, in his usual methodical manner, for a young colleague who would join him in camp, to serve as doctor for medical emergencies and as cook.[114] A few days later, Dr. Howard T. "Jim" Ricketts joined for $2 a day. Fortunately, his medical skills were not required and opinions differed on his cooking abilities. Ricketts reported that his breakfast consisted of coffee, roasted potatoes, and eggs scrambled with small chips of bacon, which everyone ate ravenously. Chanute later told his doctor: "Ricketts may have been a good surgeon, but as a cook he was certainly the rottenest I ever knew." A decade later, Ricketts achieved fame when he discovered that wood ticks transmitted the "tick fever" and lice spread the epidemic "typhus fever."

To avoid the world of meddling curiosity and sensationalism on their second expedition to the dunes, Chanute had all equipment loaded on the sailing vessel *Scorpion*, owned by the Chicago Tie Preserving Company. They then traveled on Thursday, August 20, over Lake Michigan, well out of sight of curious reporters, to the 71st Street pier to load Butusov's machine, the fourth glider built during the summer. Because it was getting late, they anchored the boat overnight at South Chicago. The next morning, the *Scorpion* headed toward "Experiment Hill," but ran aground near Windsor Beach. The men freed the boat from the sandbank after two hours of hard work, and then beached five miles east of Millers, north of Dune Park Station. "Pitched the tent and got settled. A fearful storm came up from S.W. at 3 AM. Blew the tent down and rained very hard. All the party got very wet and provisions were damaged."[115] At daybreak, Chanute walked to the station to wire for a new tent; it arrived later that afternoon, disclosing the group's presence.

The storm had also damaged the triplane; it was repaired and first flown on August 29. Because the bottom wing usually got caught in the sand, Avery suggested removing, it and Herring repositioned the Pénaud-type tail with his "regulator" (an elastic cord with a spring and a hinge assembly between the main frame and the tail). The resulting machine was a Pratt-trussed biplane glider. "This machine proved a success, it being safe and manageable,"[116] and Avery soon flew the glider nearly as well as Herring.

This time, the *Daily Inter Ocean* was the first newspaper to bring news of Chanute's flying activity. Hunters had "caught distant glimpses of a white-looking something that sailed slowly from the summit of a tall hill, landed a couple hundred feet away, and crawled slowly back to repeat its performance. Sometimes there were two of them, looking like baby specimens of a lizard-bird taking first lessons in the art of navigating the upper regions of the air."[117] Reading stories such as this strengthened Chanute's aversion to reporters; he did not like them to embellish facts and "put words in his mouth," his oldest daughter recalled years later.

Not wanting to miss eye-catching news, the city editor of the *Chicago Tribune* assigned a reporter to find Chanute, hidden in the dunes, and get an exclusive story. The twenty-seven-year-old Henry Bunting located the camp after some initial difficulties and stayed for almost three weeks. His first report

Chanute holding biplane, with side panels as first envisioned (late August 1896).
Chanute photo album, Manuscript Division, Library of Congress, Washington, D.C.
WB039P-014-52.

made front-page news in the *Tribune* a week after he arrived in camp.[118] This
enticed other reporters to come; Harry Macbeth of the *Times-Herald* was the
next visitor, followed by Frank Hemingway of the *Chronicle*, traveling on horse-
back. Bob Armstrong of the *Morning Record* was the last reporter to join the
campers. As a gracious host, Chanute offered food, cigars, and a bed in one of
the smaller tents, if the uninvited guest wished to stay. Bunting later reported
that the evenings around the campfire took on the banter and jollity of a Press
Club stag night. "The boys of the press took to camp life and gliding naturally,
with some usefulness to the boss, Mr. Chanute."[119] They kept as busy as hired
men, tugging gliders back up the slope and running errands to camp for tools
and equipment, and "some were bold enough to take a try at gliding." Other
reporters, including William Jackman of the *Chicago Journal*, only stayed for
a few hours. Whether Chanute liked it or not, his gliding activities were daily
news and distributed worldwide, courtesy of the news wire service.

On September 11, steady northerly winds of twenty-five to thirty miles per
hour blew against the dunes, allowing several long flights with a much shal-
lower angle of descent than the glides with no wind.[120] Bunting reported in the
Tribune: "With the high wind the practice was full of excitement. One wholly

Camp life in late August and early September 1896. From left to right: Davy Crockett, Augustus Herring, Octave Chanute, Henry Bunting (*Chicago Tribune*), Paul Butusov, Harry Macbeth (*Chicago Times-Herald*), and Frank Hemingway (*Chicago Chronicle*). Chanute photo album, Manuscript Division, Library of Congress, Washington, D.C. WB039P-025-94.

new freak of the air was experienced by Mr. Herring when his machine rose with a sudden gust forty feet higher than the starting point, then coming to a sudden poise, balancing like a bird, swooping at a right angle, traveled a long journey, and alighted gracefully upon a hillside. It was seen that Mr. Herring's flight with the wind alone caught and held the machine, then let it descend gradually and alight safely every time."[121]

Just for fun, Chanute suggested staging a gliding competition, with Avery flying the *Katydid* and Herring the biplane. Watching the two, Charley Chanute also wanted to try this sport, but tore his pants on landing, the only reported incident. The reporters and Jim Ricketts also took turns in "doing some cruises" in the biplane. Herring was absolutely thrilled about the performance of "his" biplane but somewhat annoyed that he did not outperform the others even though he had the most flying experience. The following day, he went off alone to fly "his machine"; returning later that afternoon, he reported his best glide as 14 seconds in duration and 359 feet in distance. He was ready to install an engine, but Chanute thought this premature. Maybe Herring was irritated by the older engineer's caution or annoyed about not being the best performer, but whatever his reasons, he packed his belongings and left camp.

Chanute also went home for the weekend. Early in the afternoon on Monday, Herring stopped at the house and said that he disbelieved Butusov's story and considered his machine dangerous; he preferred to withdraw. Knowing of the conflicts between Herring and his other team members, Chanute let him go.

The biplane glider in flight (September 1896). Chanute photo album, Manuscript
Division, Library of Congress, Washington, D.C. WB039P-005-16.

Statistics for the Best Glides in the Chanute-Type Biplane in September 1896

Operator	Length in feet	Time in seconds	Angle of descent	Height fallen, ft.	Speed, feet per second	Descent ratio
September 11						
Avery	199	8.	10°	34.6	24.9	1 in 5.75
Herring	234	8.7	7½°	30.4	26.9	1 in 7.69
Avery	253	—	10½°	46.	—	1 in 5.50
Herring	239	—	11°	46.3	—	1 in 5.24
Herring	220	9.	—	—	24.4	
Herring	235	10.3	—	—	22.8	
Avery	256	10.2	8°	25.5	25.1	1 in 7.18
September 12 (reported by Herring, without observer)						
Herring	359	14.	10°	62.1	25.6	1 in 5.75

Arriving back in camp in early evening, Chanute explained the situation to the still-present reporters, and Avery and Ricketts told of various incidents and remarks showing unfairness on Herring's part in the tests of the machines, except the one he called "his." A few days later, Chanute wrote to Edward Huffaker that his former coworker Herring had left "to go on his own hook and endeavor to build a power machine. I am afraid that Professor Langley was right about him."[122]

Just before breaking camp, Manley came again and sketched Avery flying high above the dunes in the "aerocurve," as Manley called the biplane glider. The *Chicago Record* published his report on September 28,[123] and Chanute mailed this clipping to his many correspondents.

Given the depth of Chanute's talents as an engineer and his encyclopedic awareness of aeronautical activities around the world, it is not surprising that his 1896 glider represented the state of the art. During their seven weeks of flying, Avery and Herring increased the contemporary knowledge of stability and the interplay of aerial forces on a flying machine's wings, amassing flight time second by second. Several hundred glides were made with the biplane, repeating and improving on Lilienthal's efforts, although no real comparison could be made, because Lilienthal did not measure his flights or fly in winds stronger than about ten miles per hour.

Chanute was pleased about the overall progress but disappointed that the trials had not achieved sustained flight over longer distances. "Not even the birds could have operated more safely than we; but they would have made longer

OCTAVE CHANUTE'S AEROCURVE BEING TESTED AT DUNE PARK. IND.

William Avery flying the Chanute biplane in its final configuration, as depicted by Frank Manley in the *Chicago Record*, September 28, 1896.

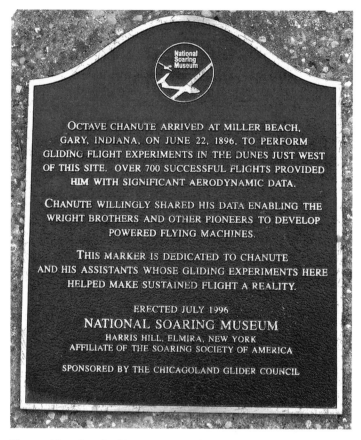

National Landmark of Soaring No. 8, honoring Octave Chanute.
Sponsored by the National Soaring Museum and the Chicagoland
Glider Council (July 1996).

and flatter glides, and they would have soared up into the blue."[124] He noted
in his diary that neither machine could perform soaring flight, and explained
to Means: "The bird easily glides [in an angle of] 1 in 10, and I have not been
able to get an artificial machine to do better than 1 in 6"[125] or six feet forward
for every foot in vertical descent.

All these aeronautical experiments were made in a genuinely scientific spirit,
without thought of pecuniary benefit. Forty years later, the Western Society of
Engineers dedicated a glacial boulder in Millers, Indiana, with a brass plaque:
"Octave Chanute / An eminent engineer / Father of Aviation / Made the first
successful flights in heavier-than-air craft from these dunes in 1896."[126] On the
one hundredth anniversary of Chanute's gliding experiments, the dunes around
Millers became a National Landmark of Soaring,[127] dedicated by the National
Soaring Museum and the Chicagoland Glider Council.[128]

William Paul Butusov and his *Albatross*

The Russian sailor, Mr. Paul, claimed to have soared for forty-five minutes in 1889 near Mammoth Cave, Kentucky, "imitating the maneuvers of the soaring birds, gliding about in zigzags and spirals, raising upon the wind and going in any direction."[129] He believed that he had the soaring machine idea worked out and looked for an investor. Reading of Chanute, he introduced himself in early June 1896. Even though Butusov's story sounded incredible, Chanute hired him to first help in camp and then build another "aerial boat"[130] capable of soaring. Prior to building his *Albatross*, similar to Le Bris's 1857 design, Butusov submitted the patent claims for his "Soaring Machine" with Chanute's input and money to a Chicago attorney. The patent, with its five engravings, was granted two years later,[131] but gave only Butusov's name, so Chanute requested a correction in the patent book but not a full reissue of the patent.

The 190-pound *Albatross*, with its 40-foot wingspan and a boatlike fuselage, could not be launched like the other gliders; it needed a ramp to assist the take-off. Charley Chanute and Joe Card delivered the timber for the trestle ramp and Chanute hired several locals from the sparsely populated area to erect it; one of the men was Davy Crockett, a former slave who lived with the Carr family in the nearby dunes around Millers.[132] The ramp's fixed position required a north wind to launch the *Albatross*, but during this September, the winds mostly came from the south. Bunting reported on one attempt and highlighted it with a sketch of the "toboggan slide for flying machine."[133] Chanute described the final launch on September 26 in his diary: "The wind from the north finally set in at 8 AM at 3 miles per hour. By 3 PM wind 18 miles an hour. Paul in machine and all was ready. Cut the rope at 3 PM. Wind turned to N.E. Machine slid down slowly and stopped on level portion. Fastened a rope at once to haul machine to top of slide. Substituted 90 lbs ballast. Fastened ropes to front and rear of machine & manned them so as to give initial velocity. Sent machine off again at 3:30 PM. As soon as the front of machine had fairly left the chute, the side wind blew the head around. The left wing struck the trees west of chute and was broken."[134] Fortunately, no one was hurt.

Front-page news in the *Westchester Tribune* the next day summarized it all: "Ship fails to fly." The flock of mechanical birds was packed up, but Butusov's machine was left behind. "The white gull, circling high in the air over the sand dunes along the lonely shore of the lake, looked with pitying contempt at the wreck of the flying machine far below. 'In my opinion,' said the bird, 'no inventor will ever hatch a real flying machine out of his head. The human skull is too thick,'" read the editorial comment in the *Chicago Tribune*.[135]

Receiving the *Kansas City Journal* regularly, Chanute probably smiled a few days later upon reading that a replica of his flying machine would appear in Kansas City's annual parade.[136] The body of the float was an artistic cloud and

The *Albatross* on the launching ramp, with Chanute and Davy Crockett standing in the back (September 1896). Chanute photo album, Manuscript Division, Library of Congress, Washington, D.C. WB039P-025-93.

soaring high was the airship, surrounded by a flock of storks and wild geese. The newspaper write-up continued: "High hopes were entertained through out the scientific world for the machine of Octave Chanute, but only the other day it failed as ignominious as did that of Darius Green." But Kansas Citians were convinced that their former citizen, who had bridged the unbridgeable river, would sooner or later succeed.

Sometime during 1896, Butusov had compiled a "Memorandum of Agreement."[137] Chanute was to pay $500 for the construction of the *Albatross* plus material, $20/week for living expense while building the machine, and $200 for experimenting with the *Albatross*. They would submit the design to the U.S. Patent Office, with one-half of the patent assigned to Chanute, who paid all expenses for securing it. If the *Albatross* flew successfully, Chanute would furnish $15,000 to develop a flyer with motor and propeller. In January 1897 Butusov mailed this document to Chanute, who signed it but returned it with an accounting of money already drawn.

Because the *Albatross* did not perform as claimed, Butusov now became interested in entering the show business with a biplane glider, again asking Chanute for funding. Carl Myers, who was interested in expanding his business, had suggested establishing an airship farm just outside Chicago. This sounded interesting, so Chanute proposed, "A man could glide, in public, distances of ¼ to ½ mile, and perhaps circle in the air. This would prove an attractive exhibition,"[138] and Butusov could launch his newly designed glider from Myers's balloon to reach the required altitude. A contemporary newspaper report gave details: "Then he will proceed to give an imitation of a large and lively turkey buzzard and soar and soar and soar until he gets tired or the wind gives out and his machine settles gentle down to earth."[139] Sometime later, Butusov experienced a bad fall and was paralyzed for several years; he lost contact with Chanute. Seeing William Jackman's book *Flying Machines, Construction & Operation*, published in June 1910, and reading Chanute's description of the 1896 events, Butusov felt perturbed that his name went unmentioned. He approached the Chanute family a few months after Chanute's death and explained that their father had been his partner, and requested Chanute's half of the patent to be signed over to him. Charley, always good hearted, gave him $100.

The Flying Continues

The powered aeroplane seemed so close and Herring wanted to be the one to invent it. Shortly after leaving Chanute's team, he reported to have built a new triplane glider and flew it in October 1896, sometimes using a sandbag in lieu of an engine. He then reported glides of up to 927 feet in length, "all while 'quartering' on the wind. In a few of the flights it was found quite safe to turn the apparatus and it would have been possible to land on a higher point than the starting one."[140] Chanute was not invited to witness these glides, nor was anyone else, and many historians are skeptical, but Herring could have been "quartering" for forty-eight seconds and could have made partial turns with a strong wind blowing in the right angle to the sloping dunes.[141] As he described how he prolonged his flights, he understood the principles of soaring.[142]

Anxious to protect his ideas, especially his regulator and tail assembly, Herring contacted Chanute's patent attorney in Washington in May 1896; in December, he submitted his claims for a triplane.[143] The application was rejected. Herring then revised some claims but was rejected again. To prove his claims, he wanted to build his aeroplane, so he contacted several business entrepreneurs[144] for financial support, but one negative reply followed the other. Not giving up, he approached Chanute, who reluctantly rehired him in early June 1897. However, Herring was not interested in building, but only wanted his aeroplane patented. To keep peace, the two men combined their claims into one patent, for Herring's powered triplane and Chanute's biplane glider with a seat, and mailed it to Thomas Moy in London, who submitted it as a communication to the British Patent Office.[145] With his claims secured, Herring quit after one month employment, hoping to join Maxim in England. Chanute shared the news with Means: "I have given him a good letter to the latter, who, if there is an opening, may control him better for a while than I can. I do not think there is bad intent,

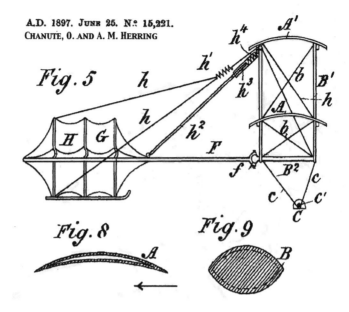

A.D. 1897. June 25. N.º 15,221.
CHANUTE, O. AND A. M. HERRING

Biplane glider, from British Patent No. 15,221. Communication submitted by Octave Chanute and Augustus Herring. Accepted on July
25, 1898. Figure 8 shows wing cross-section and figure 9 cross-section
of the aerodynamically shaped struts of the framework.

but his mind revolts at following other man's ideas. As I said before, I regret his
leaving very much."[146]

In late fall 1896 a mysterious airship was reportedly sighted over California,
and news reports about the "flying machine" appeared almost daily as it seemingly moved slowly across the country. Looking for an explanation, reporters
contacted Chanute, who simply stated that "he could not command patience
to read the full account in the newspaper because of its absurdities"[147] and that
he was convinced it was a "prodigious fake." The airship, with its multicolored
lights, then disappeared as mysteriously as it had appeared half a year earlier.

Staying in touch with his friends, Chanute mailed Thurston a brief report on
his glider flying. Thurston then suggested presenting a lecture at Sibley College,
following up on what he had said seven years earlier. Chanute gave his talk on
February 26, 1897.[148] Having watched his gliders in flight, he knew that much
more research was needed to fly a powered aeroplane, so he offered the use
of either the biplane glider or the *Katydid* to anyone interested in continuing
the experiments. The frontispiece of the *Sibley Journal* June 1898 issue shows
Chanute's "Two Modern Flying Machines."

During the winter of 1896–97, Chanute advertised his gliders for $300 each,[149] so that members of his team would continue gaining experience in building and flying. Matthias Arnot, a banker from Elmira, New York, wrote to Chanute in January 1897 that he had become interested in gliding after graduating from the Sheffield Scientific School in 1891. Chanute passed the request along to Herring, and in late June, Arnot and Herring signed an agreement for a biplane glider, which was cheaper to build and easier for an amateur to fly than the *Katydid* or the triplane. The "Herring-Arnot" Chanute-type glider was built in Avery's shop and delivered in early September 1897. Arnot paid the expenses for one week of flying camp and enjoyed this new experience. Fond of publicity, Herring wired directions for how to reach the camp to Chicago newspapers, and several reporters came to try "tobogganing in the air." When Chanute came for a brief visit, Herring asked him to fund a second week of flying. Even though no new knowledge would likely result, Chanute agreed, because he thought others might enjoy the experience. Back home, he wrote a quick note to James Means in Boston and extended an invitation to fellow WSE members. Bion Arnold, William Karner, Warren Roberts, and Leland Summers accepted, and on September 16, the group took the train to Dune Park Station. The facilities were rather primitive, but the weather was perfect. Seeing the many contemporary reports and photos, participants apparently enjoyed this exciting new sport,

Tobogganing on the air from the dunes of Lake Michigan (September 1897). Chanute photo album, Manuscript Division, Library of Congress, Washington, D.C. WB039P-008-29. Inset shows line of flight, from *Chicago Sunday Times-Herald*, September 12, 1897.

because the glider was fun to fly.[150] Chanute took a photo of the forty-four-year-old Means holding the glider, but he declined to fly.

A few years later Chanute recalled a "curious thing" that happened in camp:

> We had taken the machine to the top of the hill. A gull came strolling inland and flapped full-winged to inspect. He swept several circles above the machine, stretched his neck, gave a squawk and went off. Presently he returned with eleven other gulls, and they seemed to hold a conclave, about 100 feet above the big white bird, which they had discovered in the sand. They circled round after round, and once in a while there was a series of loud peeps, as if a terrifying suggestion had been made. The bolder birds occasionally swooped downward to inspect the monster more closely. After some seven or eight minutes of this performance, they evidently concluded that the stranger was too formidable to tackle, if alive, or that he was not good to eat, if dead. They flew off to resume fishing, for the weak point in a bird is the stomach.[151]

Chanute agreed to talk about his "Gliding Experiments" at the October 1897 WSE meeting. He first described the technical aspect of flight with its ten problems, followed by a narrative of the flying activity in 1896 and 1897. Herring attended as Chanute's guest and described how it felt to fly along the sloping dune that furnished support and propulsion. Chanute described why the glides in calm air had a higher descent rate than the ones against the strong wind:[152] "If we had a long straight ridge and a suitable wind blowing at right angles thereto, we would have attempted to have sailed horizontally along the top of the ridge, transversely to the ascending current." The concept of soaring, as opposed to "aspiration" was becoming better understood.

To share the excitement and beauty of flying, Chanute wrote another article "Recent Experiments in Gliding Flight" and Herring mailed his version to be published in Means' third *Annual* in 1897.[153] Chanute also submitted the second part of his article on "Sailing Flight," first reviewing theories as advanced by other researchers, followed by his theoretical interpretations. The Russian engineer Stephane Drzewiecki had mentioned at the 1889 Paris Aeronautical Conference that "lift" and "drift" needed to be recalculated for each curved wing shape. Seeing Lilienthal's table of air pressures in Moedebeck's 1895 *Taschenbuch*, Chanute succeeded to calculate lift and drift on wing surfaces of 1/12th curvature, so he included Lilienthal's table in his article. In closing, Chanute presented his contemporary understanding of soaring: "The simplest and most satisfactory explanation thus far is that which assumes ascending columns or trends of wind to exist at opportune times and places, but that it does not account for the cases in which all observers are agreed that the wind is horizontal." The information presented in this article, especially the translated table from Lilienthal, contributed significantly in the evolution of the aeroplane during the next decade.

Means' three volumes of *Aeronautical Annual* did much to encourage enthusiasts at this time of growing interest. "It was portentous service to aeronautics
which you rendered in 1895, 1896 & 1897 by republishing some of the classics,
describing promising experiments and formally giving data which have led to
final success,"[154] Chanute complimented Means a decade later.

How it Feels to Fly: Reporter Tries an Aerocurve

One of the nicest reports about flying a Chanute-type glider was written by
Harry Macbeth, who had tried the sport in 1896 and again in 1897. His report
appeared on September 8, 1897, in the *Chicago Times-Herald*:

> Any man endowed with an average amount of nerve, a cool head, a quick eye
> and a fair muscular development, can soar through the air nowadays, provided
> he is equipped with a machine like the one being used by A. M. Herring among
> the sand dunes near Dune Park, Ind. All that is necessary for him to do is to
> seize the machine with a firm grasp, say a prayer, take a running jump into
> space and trust to luck for finding a soft place when he alights. His chances of
> getting hurt are about one in a thousand in his favor, while having more sport
> to the second than he ever dreamed possible.
>
> Mr. Herring has been making flights with his machine for more than a
> week. Last year with a machine almost identical in construction he made daily
> experiments in flying for over a month. He made several hundred ventures
> into the air, and in none of them did he receive as much as a scratch. William
> Avery, who was one of the party to which Mr. Herring belonged, took the same
> risk fully as many times, and he, too, has yet to spend a cent for arnica or court
> plaster as a result of his seeming recklessness.
>
> With all these arguments before him in favor of the docility of the flying
> machine, a reporter for the TIMES-HERALD persuaded himself yesterday af-
> ternoon that he would like to hitch himself to the airy steed and try conclusions
> with a fish eagle that circled over his head a mile or so. It looked so very easy
> and it was a thousand per cent better fun to look at than shooting the chutes.
>
> One will never know what it is to sail through the air at a speed of thirty or
> forty miles an hour, sometimes at a height of ten feet and at the next moment
> three times as high, until he has tackled the aerocurve, or gliding machine.
> The first step is to get under the apparatus, and this is the most difficult part
> of the performance. The machine weighs only twenty-three pounds, but it is
> as big as the bay window of a cottage and has an alarming tendency to topple
> over on a man's head at a critical moment. With but two small upright sticks
> to grasp and a frail wooden bar under each arm on which to support the weight
> of the body one is not deeply impressed with the stability of the machine on
> coming into actual contact.
>
> Once underneath the machine one finds himself standing on a wide plank,
> which rests on the sloping side of a sand hill. The hill is about 100 feet high

and steep enough to test the lungs and legs of the strongest man. You face the wind as squarely as possible and shift the machine to and fro until you feel that it is balanced fairly on your arms. You are suddenly aware that the broad expanse of varnished silk above your head is pulling on your arms and trying to get away from you with each gust of the freshening wind. At the same time you remember the caution to keep the front edge of the machine depressed until the instant of your departure from earth.

It becomes necessary to start, of course, if one wants to fly. In the meantime a sickening fear comes over one that he may lose his balance and plow a long and deep furrow in the sand with his nose. The wind grows stronger, and blows with surprising steadiness. Somebody—one can't turn his head to see, who it is—mutters that the wind is just right, and that it is a good time to start. Grasping the uprights with a grim determination to never let loose, and drawing a deep breath, one takes four or five running steps down the plank and jumps off, expecting to drop like a stone to the sand. To his surprise and pleasure he experiences about the same sensations felt by a man when taking his first ascension in an elevator. There is the same queer feeling of being lifted from beneath and a corresponding exhilaration as the sense of motion is realized.

As the machine mounts in the air one sees the ground sinking beneath. He imagines he is a hundred feet in the air, and begins to wonder if he will ever come down and be able to see his folks again in this world. The thought no sooner comes, when the machine suddenly begins to descend with lightning speed. The wind rushes in the face of the operator like a hurricane and hums through the network of fine wire that forms part of the framework with a high, shrill note. There is a rustling sound, as of sand rushing over the white silk surfaces that sustain the machine in the air. All of these things are noted in a moment of dread, for the earth is rising all the time, as though to strike one. Just as one stretches his legs out expecting to plant his feet on something solid, the wind suddenly lifts the machine again toward the sky. As it mounts upward, one's confidence returns. It is not so dangerous after all, just as Mr. Chanute and Mr. Herring and Mr. Avery said, and the possibility of flying across the valley and returning to the starting point is mentally revolved. The machine settles down slowly and steadily, and to the disappointment of the operator his feet strike the sand. His experience in the air is over.

He turns around and looks up the side of the hill, feeling that he has traveled at least a thousand yards. When the tapeline is brought out, he is somewhat disgusted to find that he is only 110 feet away from his starting point. He wonders how this can be, when he was up in the air at least ten minutes. Then he receives another shock, when he is told that his flight lasted just five seconds. He still fails to understand, knowing positively that he was at least 100 feet up in the air, but some of the observers tell him that he was never more than thirty feet above the earth. This is the funny part of coasting on the air when one is beginning. It's different when you know how.

Mr. Herring expects to break camp today. He is satisfied that nothing more can be accomplished for the benefit of science by continuing the experiments with the machine which he and Octave Chanute have jointly invented.[155]

Elegant Engineering:
The Chanute-Type Biplane Glider

1896 was a turning point in aeronautics, as earlier flying machine inventors were overshadowed. In May Samuel Langley's aerodrome demonstrated the technical feasibility of heavier-than-air flight. Otto Lilienthal, who had shown that an aircraft with a man onboard could fly, died in August, ending a promising aeronautical career. Later that month, Chanute introduced the Pratt-trussed biplane configuration to aviation. In the 1930s, when soaring became a nationalistic and competitive sport, engineers and glider pilots including Robert Kronfeld,[156] but also today's historians, including Tom Crouch, agree that this strong, light, and straightforward rectangular structure, with its vertical supports in compression and its diagonals in tension, was the first modern aircraft structure. Its origin, however, soon became clouded in controversy, as Herring also laid claim to its invention.

Over the next few years, Chanute discussed the biplane's performance and the excitement of flight, experienced by engineers and reporters with no prior knowledge of flying, in many articles, mentioning Herring and Avery as part of the story.[157] This upset Herring, who considered himself the leading American flying machine experimenter and the sole designer of what became known as the Chanute-type biplane.[158] He firmly believed that the biplane's success was based on his elastic regulator, attached to its tail. Interestingly, Herring also stated that "the success of Langley's model in May 1896 depended upon the use of curved surfaces and on the elastic hinging of the tail, an invention of mine."[159] Opinions differ on the value of Herring's "regulator" among the participants in building either machine.

The editors of the *London Times* wanted an updated entry on aeronautics for the next supplement to the *Encyclopedia Britannica*. To bring new material, Chanute assembled a brief history of manned flight and then gave engineering details on designing a glider and how to calculate the performance of a machine.[160] Just hearing of this literary project upset Herring, as he assumed that Chanute would not give him the credit he thought he deserved. Right away, he wrote and complained to his financial supporter Arnot, who contacted the *London Times* editors requesting to review Chanute's essay. They instead forwarded Arnot's letter to Chanute, who replied that there was no reason to alter his article. A week later, Chanute asked Herring to clarify his concerns.

It took Herring two weeks to respond. "I have never seen why you should claim the whole credit for the invention of the two surface gliding machine, since its success depends wholly upon the efficiency of the regulating mechanism which was my work and furthermore I made the design for the original two surface machine alone at my home & put them on paper as a scale drawing in the course of one afternoon in your study. This first machine, it is true, was built with your

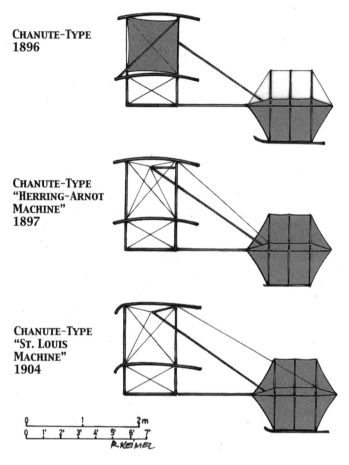

CHANUTE-TYPE
1896

CHANUTE-TYPE
"HERRING-ARNOT
MACHINE"
1897

CHANUTE-TYPE
"ST. LOUIS
MACHINE"
1904

Engineering development of the Chanute-type gliders, side view in comparison (1896, 1897, and 1904). Drawings by Reinhard Keimel.

money and you paid me for my time. And you did suggest superimposing & trussing the surfaces together. But I do not think this is sufficient for you to take entire credit for the invention of the two surface gliding machine."[161]

Clearly, Herring and Chanute held fundamentally opposite views on what had happened in 1896. After reading Herring's letter, Chanute replied with Old World courtesy: "It is natural for every man to overvalue his own achievements. I do it myself." He then recalled the facts as he saw them, point by point, and closed his letter stating, "Go on and demonstrate some important results and there will be no lack of appreciation in my writings."[162] In a separate letter, Chanute suggested that Herring find someone to pay for the renewal of the joined British patent or let it lapse. There was no reply to either letter, and the British patent expired soon thereafter.

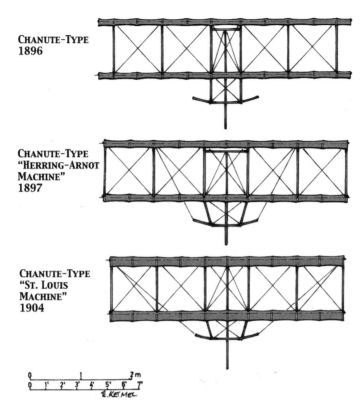

Engineering development of the Chanute-type gliders, front view in comparison (1896, 1897, and 1904). Drawings by Reinhard Keimel.

In the next few years, the public became more curious about aviation and the people who had reportedly flown. Carl Dienstbach, a German correspondent living in New York, discussed "Herring's Work" in the *American Aeronaut* in May 1906. Chanute thought that the facts were a bit distorted but did not comment. The next issue carried an article by Zahm, discussing who invented this machine: "It is questionable whether the type often described as the 'Chanute glider' can be properly regarded as the invention of one man. It seems rather the development of several minds, and in that evolution Wenham seems to stand first, followed by other Englishmen as Stringfellow, Phillips and Hargrave, not to mention the continental experimenters. However Mr. Chanute was the first to build such a glider, to operate it and to patent it."[163]

Two years later, Chanute agreed to document the evolution of his biplane. He credited Wenham as the originator of superposed surfaces in an aeroplane, patented in 1866. Two years after that, Stringfellow produced a model based on Wenham's principles. In 1878, Linfield tested an apparatus with twenty-five superposed surfaces, towed by a locomotive, but it proved unstable. Chanute witnessed Renard's gliding model at the 1889 Aeronautical Congress in Paris, traveling against the wind and changing direction by using a rudder. In 1893, Horatio Phillips patented variously shaped wing sections, deducing what he thought was the best shape to provide lift. Hargrave invented the cellular kite, described at the 1893 aeronautical conference, and Chanute had experimented with a similar gliding model, which did not perform to his satisfaction. Each of these were experiments in the evolution of the biplane glider.

After flying and then discarding the Lilienthal-type glider in 1896, Chanute and his men experimented with the *Katydid*, learning with each alteration of the wing position. Any engineer had to foresee the variety of ways in which technology could fail, so Chanute calculated the air resistance in front of Herring, sketched a triplane design on a sheet of cross-barred paper, and asked Herring to build the glider in Avery's shop. "Being a builder of bridges, I trussed these surfaces together to obtain strength and stiffness. When tested in gliding flight, the lower surface was found too near the ground. It was taken off and the remaining apparatus consisted of two surfaces connected together by a girder composed of vertical posts and diagonal ties, known as a 'Pratt truss.' Then Mr. Herring and Mr. Avery together devised and put on an elastic attachment to the tail. This machine proved a success. Over 700 glides were made at angles of descent of one in six or one in seven."[164] One year later, Chanute further clarified: "In point of fact, I made the design and Herring made the working drawing under my instructions, but what is the use of entering into a controversy about it now."[165]

There is no doubt that Chanute's engineering and mathematical knowledge in iron bridgework formed the basis for the biplane glider. The truss bracing and the tail, consisting of a fixed vertical tail and horizontal stabilizer, originated with Chanute. Herring introduced the "regulator" and Avery shaped the uprights aerodynamically, always stating that he built the machine at a carpenter's salary of $0.60 an hour.[166] One can hypothesize that the overall design process was a team effort in which one word triggered the thought process in the mind of the others.

When Means asked to use the biplane design, Chanute explained: "You are mistaken in assuming that I control the whole of the machine which you saw in gliding flight. I originally made the general design, it is true, but Mr. Herring supplied the automatic regulator and I should not feel justified in using it without his consent, particularly now that the patent office has again rejected his application. He is so sensitive and has so many illusions as to the value of the device, that the praise which you unduly bestow upon me, would cause him to boil over."[167]

After the 1896 flying season, the biplane was stored in the rafters above Avery's shop. When Matthias Arnot ordered a glider, Herring and Avery redesigned the wings and tail and Avery used parts of the original structure for the "Arnot machine." After two weeks of flying in September 1897, the glider went back into storage, because Arnot had become interested in powered aeroplanes. When the newly formed Aero Club of America planned an exhibit, Herring wanted to show "his" biplane, so the glider, eaten by moths and mice, was removed from Avery's shop in December 1905, shipped to New York, and refurbished by Herring. After the exhibition closed, the glider was shipped to the National Museum in Washington.[168]

The little 1896 glider was a step forward in the evolution of the flying machine. It had proven the ideas that preceded it and embodied the significant progress toward maintaining equilibrium and exerting limited control. Nevertheless, the application of propulsion and effective control remained for the next generation of experimenters.

Flying Flivver is Assured, Opening the Field of Interest

During the last decade of the nineteenth century, able men like Langley, Maxim, Pilcher, Hargrave, Lilienthal, and Chanute, made significant advances in the "inchoate art of aeronautics," but to fly higher, faster, and farther with a practical aeroplane remained futuristic. Believing that final success could only come when different people worked the same problem from various angles, Chanute established awards to encourage research, publishing, and the construction of kites or aeroplanes. Being a realist, he knew that he would not be the one to perfect the successful aeroplane, but he hoped to be remembered as the one who furnished a prototype.[169]

Sitting in his shirtsleeves at his big flat desk and smoking his favorite cigars, Chanute did his best work in the evening, when he could think of new ideas and write. His letterpress books contain copies of thousands of letters, not only to well-known aeronautical researchers, but also to many faltering flying machine builders whose experiments had made even the slightest news. Receiving letters from inventors who declared to have solved the problem, Chanute may have smiled while reading, but his reply was usually supportive; he critiqued and shared his knowledge for the purpose of encouraging those who wrote him. The last sentence in his letters frequently read something like: "I shall be glad to place at your disposal the information I have gained about the travel of the center of pressure, the effect of wind gusts, or the best shapes for sustaining surfaces. I shall be glad to answer further questions."

Traveling across the country on his various consulting assignments, he often visited bookshops and libraries to obtain aeronautical material, and his personal library soon became an information center that contained more data on mankind's efforts to fly than had ever before existed in one place. When the Chanute estate donated his engineering library to the John Crerar Library in April 1911, the subsection on aviation in the Dewey Decimal System had to be expanded to handle the 1,500 items from his collection.

In the mid-1890s Octave Chanute developed the most modern and successful glider up to that time. He had systematically approached the problems
necessary for flight and had moved solutions forward through experimentation,
observation, and communication. Chanute had also begun an international
conduit, vital for the exchange of ideas and encouragement as the invention of
the airplane evolved. Even though he would forego further serious experiments
himself, the effect of Chanute's exchange of knowledge became obvious during
the next decade.

CHAPTER 8

Encouraging Progress
in Flying Machines

BY THE EARLY DAYS of the twentieth century, the ancient taboo of flight had slowly emerged as a thoroughly modern inquiry of science. Although doubters still existed and the press still glorified experiments gone awry, most engineers fully anticipated the arrival of a powered flying machine.

To realize his personal goal of witnessing sustained mechanical flight, Octave Chanute freely shared what he had learned; for him, technical information was a public commodity and he impressed on his correspondents the need to share what they had discovered so that future investigators could avail themselves of the known and take problems to the next level. When newcomers contacted him for feedback, he studied their ideas and gave them input, but he also reminded them to work without expectation of reward other than being remembered: "For, in the usual course of such things, it would be the manufacturers who would reap the pecuniary benefits when commercial flying machines were finally evolved."[1]

By linking his many correspondents into an informal network, Chanute provided motivation and became the focal point of a far-reaching international community of flying machine experimenters. Unfortunately, however, after spending more than two years and $10,000 on practical flying and encouraging others to experiment with gliding machines, Chanute became acutely aware that his two tie preserving works were losing money (see chapter 6). To return his business to profitability, he elected to visit Europe late in 1899 to learn more about their timber treating procedures, but also to meet some of his many aeronautical correspondents.

The German military balloonist Hermann Moedebeck was high on his list of aeronautical people to meet. Two years earlier, Moedebeck had introduced a monthly, the *Illustrirte Mittheilungen*, to carry aeronautical news from around the world. Moedebeck had then published Chanute's article "American Gliding Experiments"[2] in 1898 with a German translation, and more recently the article "Conditions of Success in the Design of Flying Machines."[3] Now the two men talked about updating Moedebeck's earlier *Taschenbuch*, because so much had happened since its publication in 1895. Chanute offered to rewrite the chapter on artificial flight to include an appreciation of Otto Lilienthal and an account of gliding experiments made by others.[4]

After a stop in Paris, Chanute traveled to London and finally met the seventy-five-year-old Francis Wenham, with whom he had corresponded for almost a decade. In a follow-up letter written from Chicago, Chanute complimented the British civil engineer on "shedding lucid light on aerial navigation 35 years earlier" and stated sincerely, "I am eight years your junior and hope that I may be able to advance the question equally with you before I step out."[5]

While in London, Chanute met with Sir Hiram Maxim and also with Thomas Moy, who had just submitted his claims for an "Aerial Vessel" to the British Patent Office. A few months later, Moy sought funding for his machine, and Chanute replied: "I am well disposed to assist you, but my ready money is limited. . . . Singular as it may seem, I do not care to increase my present fortune, I have divided it up among my children and wife, so as to leave each a moderate competence, which I deem the happiest and safest. I am prepared to spend a couple of thousand pounds upon devices of my own without hope of return, but if I spend it on another man's devices I should want to get it back."[6] A few months later, Chanute showed his support and mailed Moy £20.

Collaboration for Progress

Since the mid-1880s, the United States Patent Office had granted more than one hundred patents on various approaches to navigate the air.[7] Chanute had studied each claim critically and determined that there would be enough money to solve the problem, if one could have all the money that had been spent on patents for aeronautical devices.

To produce a practical flying machine, he had determined the ten most critical problems needing solutions, with equilibrium in flight the most important one. Knowing that the angle of the air shifted on the wings in flight, the veteran engineer now wondered if rocking or oscillating wings could supply automatic equilibrium, but he needed to find a young engineer to experiment and verify the merits of such a device.

The 1893 Conference on Aerial Navigation had debuted several experimenters, one of whom was Edward Huffaker; he had submitted his essay on "Soaring Flight" and then began a correspondence with Chanute. A year later, the two men met in Chuckey City, Tennessee, to watch the soaring birds: "The bird was not simply gliding, utilizing gravity or acquired momentum, he was actually circling horizontally in defiance of physics and mathematics."[8] Chanute did not yet grasp that a parcel of warm air, or a "thermal," could provide an upward movement; he only thought of "aspiration," by which experimenters believed the bird encountered a current and was then drawn forward into that current to achieve soaring flight.

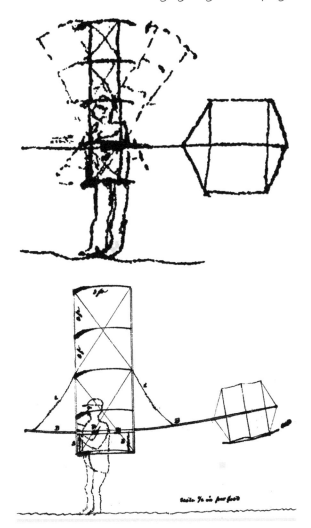

Chanute's ideas on how to achieve automatic stability: Top: multiplane with "rocking" wings, letter to F. Wenham, September 24, 1897. Bottom: multiplane with "oscillating" wings, letter to E. C. Huffaker, December 22, 1898.

When Huffaker expressed an interest in working for the Smithsonian Institution, Chanute arranged an interview for him to demonstrate his model with cambered wings, and Langley hired the Tennessean in December 1894. Working conditions were not always pleasant, so Huffaker left the Smithsonian in 1898. Not finding a job back in Tennessee, he accepted Chanute's offer of $50 a month to experiment with flight. In December 1898 Chanute mailed him sketches of his "rocking wing" glider and explained how he envisioned the functioning of the wings. In the next three years, Huffaker built and tested a variety of flying models.

Several enthusiasts formed the Boston Aeronautical Society in May 1895. One of its members, Charles Lamson from Portland, Maine, had built and flown a Lilienthal-type glider that crashed on its first flight. Looking for a safer way to fly, he contacted Chanute in March 1896. Because Hargrave's kites were reasonably steady, Lamson started to build manned kites, and Boston newspapers had a grand time reporting on these "mammoth kites" in flight.[9]

After leaving Chanute's employ in July 1897, Augustus Herring joined the Truscott Boat Yard in St. Joseph, Michigan. Building flying machines for himself in his spare time seemed more congenial than experimenting under the direction of others, so he worked on his powered aeroplane, writing periodically. In July 1898 Chanute reported to Means that Herring had assembled his aeroplane to be photographed, but found it unmanageable. "It struck me that he was weakening in his faith of success, and a little afraid to try it."[10] On October 10, 1898, Herring wired Chanute: "About ready today. If you come, don't bring any strangers." Anxious to see an aeroplane fly, Chanute took the night boat of the Graham & Morton Transportation Company, leaving Chicago at 11:30 P.M. and arriving in St. Joseph at 4 A.M. on October 11. A few hours later, he met Herring, who described his flight on the previous day, but on the present day he could not repeat the performance. Disillusioned, Chanute returned to Chicago and reported to Langley: "Herring invited me to St. Joseph Oct 10 to see him fly. He was unable to perform the feat in my presence, but said that on the preceding day he had flown ~50 ft by running against a strong wind and simultaneously setting his propellers in motion with a compressed air motor."[11] On October 22 Herring reported a second flight. In the words of Tom Crouch, maybe it is not important if he flew once or twice or not at all, because these hops in a powered Chanute-type biplane with weight shifting as control were not a significant contribution to the development of the airplane.[12]

When the eminent physicist Langley puzzled about the efficiency of wings late in 1897, he contacted the man whom he considered the most knowledgeable person on the subject. Chanute gladly obliged and sent a table of air pressures at various angles on flat plates and on curved surfaces, including his own interpretation and calculations on drift versus lift.[13] Studying the data, Langley then confided to Chanute: "You and I have been working at the same problem from opposite ends, and I should express the hope that we might meet in the middle, were I sure of going on; but the next step in my own experiments would be an expensive one, the construction of an aerodrome capable of carrying a man for a long flight. . . . If you hear of anyone who is disposed to give the means to such an unselfish end, I should be very glad to meet him."[14] Chanute replied with customary promptness that he would "truly like to meet in the middle before either of them flew off into the hereafter," but he did not know of anyone interested in funding a scientific experiment, "nor do I see what promise of financial profit, or fame, could be made to a rich man furnishing the funds, while I do see a decided danger of an experimenter being injured, if not killed. It is for this reason that I have confined myself to the question of promoting safety, and have experimented at my own expense."[15]

Approaching the twentieth century, the military found itself between the antiquity of its role as a frontier constabulary and the demands of a rapidly industrialized world. Believing that the aeroplane could revolutionize war, the assistant secretary of the navy, Theodore Roosevelt, recommended in March 1898 that the government should investigate Langley's aerodrome, but also consult with outside experts such as Robert Thurston and Octave Chanute,[16] Roosevelt having known both engineers for several years. A board of U.S. Army and Navy officers then recommended developing a full-size, man-carrying machine for war purposes. Half a year later, the Board of Ordnance allotted $25,000 to Langley to develop, construct, and test an airship, with additional funding supplied in 1899. Langley encountered many mishaps in the next five years, designing and constructing his "Great *Aerodrome*."

Another enthusiast began corresponding with Chanute in the late 1890s. Twenty-eight-year-old George Spratt from Pennsylvania, a former medical doctor who now worked his father's farm, wanted to determine the most efficient wing shape. Chanute did not encourage him to study this topic: "Such experiments have been found to be very delicate and perplexing, as they are affected by the sides of the vessel and the varying velocity and directions of the currents. Also, many experiments were tried by Horatio Phillips in England in 1884/5."[17] But Spratt was energized and built various wing shapes, then tested them in the natural wind or in a conduit placed into the line of the wind (to straighten the wind's whirls). He then made what he considered a perceptive discovery about the lifting performance of cambered wings. To prove his theories, Spratt considered building a full-size glider, and Chanute suggested, "If you can find a good hillside for preliminary experiments, I will send you a sketch of a gliding machine which I have in mind to build, on my third plan of securing automatic equilibrium, and you can build that. In that case I would be willing to pay for the labor, at an agreed price, and the machine would belong to me—until broken."[18]

While Chanute and others worked on developing a fixed-wing aeroplane, Count Ferdinand von Zeppelin launched his dirigible in July 1900. His *Luftschiff Zeppelin*, or LZ-1, cruised over Lake Constance in southern Germany for almost twenty minutes, reviving public interest in aerial navigation. Chicago reporters interviewed Chanute, who stated that the flight of the LZ-1 was a notable improvement over its predecessors, but "the disadvantages of the balloon ship were its frailty, its great size and its slow speed."[19]

Lilienthal's Approach Points the Way

To remain airborne for a period of time, one could install an engine to push or pull the flying machine through the air as Maxim and Langley envisioned,[20] or one could do what Lilienthal had advocated: use the uplifts created by nature to fly like the soaring birds. Chanute partially subscribed to the philosophy of the "School of Lilienthal,"[21] which also suggested that practical flying should follow logical, safe steps. In Chanute's opinion, a novice should construct a simple machine and select a sandy hill without bushes or trees, then try short glides and learn the tricks in a step-by-step process. "Make improvements to the machine and the flying skills of the operator as needed and start over again." Eventually, the glider could be flown higher and the pilot could attempt soaring flight, but "Even when man produces a perfect flying machine, he must acquire the skill to use it. The key to success is practice, practice, practice."[22]

Reading of Lilienthal's death in 1896, the bicycle maker Wilbur Wright from Dayton, Ohio, became more interested in manned flight. After studying the available literature, he joined a long line of want-to-be aeronauts who had come to Chanute for advice and wrote a five-page letter to the distinguished engineer. "For some years I have been afflicted with the belief that flight is possible to man." He then described the glider he intended to build: "In appearance it is very similar to the 'double deck' machine with which the experiments of yourself and Mr. Herring were conducted in 1896–7. The point at which it differs in principle is that the cross stays which prevent the upper plane from moving forward and backward are removed, and each end of the upper plane is independently moved forward or backward with respect to the lower plane by a suitable lever [or] other arrangement. Lateral equilibrium is gained by moving one end more than the other or by moving them in opposite directions."[23] Writing this historic letter, the budding aeronaut tried to make sure that there was no question about what he knew and what he planned to do.

May 1900 was a busy month for Chanute; the start-up of his two tie treating plants required much attention, but almost every day someone also inquired about aeronautics. Chanute responded a few days after receiving Wilbur Wright's letter that he was in sympathy with his proposal and agreed that no financial profit could probably be made in the investigation. Chanute had hoped for some time that someone might copy his patented structure and improve on it. He now suggested to his newest correspondent further literature to study and closed his letter with: "If you have occasion to be in this city, I shall be glad to have you call on me, and can perhaps better answer questions that have occurred to you. I intend to make further experiments myself, when my business and means will allow. If you do not expect to come to Chicago soon, I shall be pleased to correspond with you and have a detailed account of your proposal."[24]

Half a year later, Wright reported on their flying activity at Kitty Hawk, North Carolina, and Chanute read with surprise "how they had disregarded the fashion which prevails among birds, have placed the tail in front of their apparatus and called it a front rudder, besides placing the operator in horizontal position [in their 17-foot wingspan glider] instead of upright."[25] Chanute replied a week later, "It is a magnificent showing, provided that you do not plow the ground with your noses." Because their experiences would be of interest to other enthusiasts, he asked, "I should like your permission to allude to your experiments in a brief and guarded way as you may indicate." Wright responded right away that they did not want to keep their machine a secret but wanted to test its possibilities

Chanute-type glider at the dunes along Lake Michigan, Indiana, in September 1897. Chanute photo album, Library of Congress, Washington, D.C. WB039P-34-009.

Wilbur Wright in 1901 glider, Kitty Hawk, North Carolina. Library of Congress, Prints and Photographs Division, Washington, D.C. (LC-DIG-PPPRS-00570).

first. Continuing his letter, Wright also asked Chanute if he could stop in Dayton, because he wanted his opinion on the control principle for their glider. Using the analogy of twisting a cardboard box, the Wrights thought of warping the structure so that the trailing edge of one wingtip would flex upward while the other one flexed downward. This warping caused the lift on one side of the plane to be greater than on the other side, thus banking the aircraft toward the side with the upward twist.

As the Wrights delved deeper into the unknowns of manned flight, they quickly discovered that Chanute was a great source of information and that if he did not have an answer, he knew someone who might. In the first few years, Chanute provided a number of insights that proved critical to the Wrights' eventual success. Usually Wilbur wrote lengthy letters explaining his thoughts and Chanute often responded like a good mentor, trying to make his student think harder. In other cases, he may not have fully comprehended what Wilbur tried to explain, but being part of their developmental process gave the elder engineer the opportunity to refine his own aeronautical knowledge, and he appreciated that.

1901: An Eventful Year for Research

With the holiday season approaching, Chanute's note to his friend James Means in late December 1900 was typical: "I meant to write you yesterday to wish you a Merry Christmas, as well as a happy New Year, but a business bore took up all my time, so that only my good wishes went to you by wireless telepathy. I hope they were effective. My wife has been ailing and goes to California in a few days. I will follow her about the middle of January. The wood preserving affairs have kept me closely at work all year. But once in San Diego, I will take up aviation again and make some experiments, if I can."[26]

Prior to leaving for his winter sojourn in San Diego, Chanute mailed Huffaker specifics for the glider he wanted to have built. Chanute's design intended the structure to be lightweight but strong and foldable, with the wing surfaces rocking back and forth and the wing curvature changing automatically with the wind. Anxious for results, Chanute followed-up a few days later and wrote, "Let me know how soon you expect to have the machine ready for trial and whether you think it best to try at home or to come to Chicago to do it. In the latter you will have the benefit of Avery's instructions about its use which you will find awkward at first."[27]

In early May, Wilbur Wright described their future plans to Chanute and made an interesting offer: "It is scarcely necessary to say that it would give us the greatest pleasure to have you visit us while in camp if you should find it possible to do so."[28] Curious about his correspondent, Chanute arranged a two-day stopover in Dayton on his next trip east, later in June 1901.

Little is known of this first personal meeting, but Wilbur described their visitor in a letter to his brother Reuchlin as a recognized "leading authority of the world on aeronautics. . . . He was very much astonished at our views, methods and results, and after studying the matter overnight said that he had reached the conclusion that we would probably reach results before he did. He will visit us at Kitty Hawk and will send one of his machines for trial."[29] Chanute indeed had a different view on how the future airplane should be operated. Warping to create lift, as proposed by the Wrights, and braking to create drag, as proposed by Chanute and Mouillard, were opposite approaches, but at that time no one had proven which method would eventually succeed.

Continuing his travel, Chanute stopped in Chuckey City, Tennessee, to see the glider that Huffaker was building for him. Incorporating knowledge gained under Langley, Huffaker used paper tubes, strengthened with shellac, for the framework. "I fear they will not stand long enough to test the efficiency of the ideas in its design," the disappointed Chanute wrote Wilbur later the same evening.[30] Mailing his next monthly paycheck to Huffaker, Chanute reiterated that the glider needed to be completed by July 1. "I propose to have several experts go out to the sand dunes with me, so as to have a sort of tournament."[31] Maybe some useful information could be extracted from this "rocking wing" glider design, but it did not look promising.

Chanute's other protégé, George Spratt, continued studying the movement of the center of pressure. Reading his latest report, Chanute responded: "The results show a reversal of the movement of the C. P. for the sections less than a semi-circle, but agree with my general understanding of results obtained by Mr. Huffaker for Prof. Langley. Of course, the 'Plin's curve,' or birds curve, is the most stable and was the one used by Messrs. Wright in their experiments." Because Spratt had medical training, and assuming that the Wrights needed someone in case of emergency, he continued, "These gentlemen leave on the 8th for Kitty Hawk, off the North Carolina coast, and Mr. Huffaker is to join them with a gliding machine, which he has been building for me. This will prove a failure, but I think all the experiments will be instructive. The spot selected is a wilderness and will require the erection of a shed, so that it will be about 1 August before serious gliding will begin. If your farm work will admit of your leaving, and you want to visit us, I will defray your expenses and perhaps you can assist in putting the machines together and in gliding."[32] Spratt happily accepted the invitation.

The Wrights arrived at Kitty Hawk on July 10. Their first visitor, Huffaker, whom Chanute had described as a "trained experimenter who lacks mechanical instinct," arrived a week later and personal difficulties developed quickly. Wilbur wrote home that their guest was intelligent but lazy, and that the glider he had built for Chanute was a total mechanical failure.[33] The culture clash between the men deepened with every day. The forty-five-year-old Huffaker had grown up in the Appalachian Mountains, while thirty-year-old Orville and thirty-four-year-old Wilbur were raised in a midwestern city. Huffaker had an academic education, while the Wrights had none. One man believed it unimportant to dress in the field as one would in the city, while the others believed the opposite; and each man felt convinced he knew the correct answers. Huffaker, with his academic education, also had a hard time taking the Wrights as seriously as they would have liked.

Spratt, whom Chanute described as an amateur, arrived the following week and quickly became friends with everyone; he soon helped launch the Wright's

22-foot span glider with its 140-pound pilot. Chanute described the launching in a later article: "The machine was taken upon a sand hill about 100 feet high, sloping at an angle of about 10°; two assistants, each grasping one corner, ran forward against the wind until a sustaining pressure was obtained. When they were told by the operator to 'let go,' the latter then guided himself, landing at the bottom of the hill by striking the sand at a flat angle and sliding thereon on the shoes attached to the machine."[34]

On August 2, a heavy rain soaked the just-assembled, Huffaker-built glider and it collapsed before it was flown. Wilbur was quick in photographing the remnants of what he called the "Thousand Dollar Beauty." Huffaker replaced the uprights to make the glider at least presentable, but Chanute, arriving two days later, was disappointed, because it could surely not function as he had envisioned.

Having brought along his various instruments, Chanute carefully logged Wilbur's flight times (down to the split second) and distances flown, the wind conditions, and the slope of the dune, just as he had done for his own experiments five years earlier. Wilbur's longest flight covered almost four hundred feet, farther than Avery and Herring had flown in 1896, but the 1901 Wright glider displayed a temperamental, almost unpredictable, behavior. Watching the pitching movements, Spratt believed that this resulted from the sudden reversal of the center of pressure as he had observed on his models, and he explained how he had measured the balanced lift against drift (or drag).

The Kodak camera was another tool Chanute had brought along to document the various stages of flight. Some snapshots were double-exposed and others out of focus,[35] but the group photos were perfect, capturing the human face of the airplane's development. They showed the sixty-nine-year-old Chanute with his distinguished small, white beard, wearing a striped starched dress shirt with winged collar, black bow tie, and felt hat, sitting with Huffaker on a cot. They show Spratt with his tight-fitting cap, the clean-shaven Wilbur, and finally Orville, with his well-groomed moustache. Both Wrights wore white shirts with stiff collars and dark ties; in their mode of dress, the brothers were more closely aligned with Chanute than the others.

After spending one week in camp, Chanute left on August 11. This marked the end of Huffaker's employment, so he handed his final report, a 114-page document,[36] of the experiments during the past three years to Chanute, along with his diary of the activity at the camp.[37] With everyone gone, Wilbur flew their glider again, confirming that it did not produce the lift they had calculated and that it went into an undesirable side slip and yaw toward the opposite direction when he applied wing warping.[38] The disappointed brothers were ready to give up.

Seeing the temperamental glider in flight energized Spratt. He built a new wind tunnel to test wing profiles and was proud to send a photo with his latest research data to Chanute and the Wrights. "I suppose you will test a current from your fan and this may settle the question whether this gives the same results as

Wright Camp in 1901. From "The Recovered Legacy." Chanute Papers, Manuscript Division, Library of Congress, Washington, D.C. No. 6885.

the natural wind. If you succeed and get good results you will deserve a medal," was Chanute's encouraging reply to Spratt.[39]

The experiences in camp needed to be shared, so Chanute wrote to Moedebeck:

Just ret'd from North Carolina where Messrs W. & O. Wright are making new experiments in gliding flight. These gentlemen have been bold enough to build a two surfaced machine, 22 feet from tip to tip, 7 feet wide, dimensions, which I have not dared to employ. . . . They place the man head first, flat on his belly in a two-surfaced machine of 290 square feet of supporting surfaces, superposed 4 feet 8 inches apart, and thus far without accident. Their angles of descent have been flatter than mine, I think they will do better soon. . . . I write principally to tell you that we have not found the lift to be anywhere near as great as that indicated by Lilienthal in his table on page 106 of your book, but more nearly to conform to the lift of planes as indicated by the Duchemin formula. I would ask you therefore whether Lilienthal's coefficients in the table have been confirmed by other experimenters and are now generally accepted?[40]

One of the responsibilities of the president of the Western Society of Engineers (WSE) was to solicit speakers for their twice-monthly meetings. Chanute, the president in 1901, had sensed that the Wrights were at a crisis point, so he invited Wilbur to deliver a paper at the WSE's second September meeting. At first Wilbur was not in favor of "performing," but his sister nagged him into going and he arrived in Chicago late morning on September 18. He spent the afternoon at Chanute's home, later telling his sister that his mentor's den was ten times dirtier and more "cluttered up" than their father's ever was. There were so many models of flying machines and stuffed birds hanging that he could not even see the ceiling.[41] Katharine, however, was not sure if she should believe this story.

Dinner was served at the Chanute home at 6:30, after which the two men took a carriage to the Monadnock Building to arrive at the WSE headquarters at 8:00 sharp; about fifty-five members and guests attended the "Special Meeting" in the society library. In his introduction, President Chanute stated that man would eventually fly, but needed to perfect safety and stability prior to adding a motor to any flying machine. Beginning his presentation, Wilbur decided to follow-up on Chanute's statement, recognizing the safety and control aspect, and then talking about flying in general, using some of Chanute's slides. He then described his glider and how he achieved lateral control. Comparing their experiences at Kitty Hawk with earlier investigators, he indicated that previously published data relating to air pressures on curved surfaces seemed erroneous. Wilbur then tried to explain soaring flight as demonstrated by the birds. "What sustains them is not definitely known, though it is almost certain that it is a rising current of air. But whether it is a rising current or something else, it is as well able to support a flying machine as a bird." To utilize this "updraft" would help them prolong flight to gain experience and better understand the principles of flight. Wilbur's lecture was published in the December WSE journal[42] and Chanute ordered two hundred reprints for distribution to his correspondents.

Preparing for the talk and listening to the comments of attending engineers served as turning points in the Wrights' development; with Chanute's consistent moral support, mentoring, and prodding, they decided to continue. The brothers had used the Lilienthal table from Chanute's article in the 1897 *Annual* to calculate the supporting power and the resistance of the surfaces to determine the aerodynamic performance of their glider, but something went awry. They now needed help to understand the unexpected complexity of controlled flight. The only person they trusted at that time was their friend Chanute, so Wilbur compiled their current knowledge: "We have found that the apparent discrepancies between our calculations based on his table and our actual experiences may be brought under the following heads: (1) Errors of Lilienthal's formula & tables; (2) Errors in the use of them by others; (3) Errors in anemometers and other estimates of velocities; (4) Errors of our own in overlooking or improperly applying certain things."[43]

To understand why the performance of the Wright 1901 glider did not match their calculations, Chanute made his own inquiry and rediscovered "that in applying coefficients to an aeroplane, not only should the arching be taken into account, but also the outline or form in plan, its proportions of length to breadth, its aspect or direction of advance, and the movements of the center of pressure, which will be found to vary greatly on different arched surfaces," Chanute wrote to Moedebeck.[44] But he also questioned the accuracy of his anemometer, so he asked Charles Marvin from the Weather Bureau to calibrate it on the whirling arm. Because Marvin had won the "Chanute Prize" for his research with kites, Chanute wondered if he might have an explanation of the movement of the center of pressure in gliding flight.[45] Marvin promised to look into this intricate problem.

On January 7, 1902, Chanute presented his address as the retiring president at the 32nd WSE Annual Meeting, followed by several entertaining speeches by members. William Karner of the Illinois Central recited a poem without title, as he thought back to his flying in the dunes in 1897:

> . . . Well, I frankly confess there was a time
> In days of yore, when time was young,
> And birds conversed as well as sung,
> On a Chanute glide I soared aloft
> Until some people said that my brain was soft . . .

Research and Flying in 1902

There had to be a good explanation why the lift and drift data, as calculated by the Wrights, did not match the observed data when flying their glider. To succeed in solving the problem, the brothers had to crack the secret codes of flight. Maybe remembering Spratt's explanation, they decided to build their own device to measure "rectangular pressures" on various wing shapes that Wilbur described in a stream of lengthy letters to Chanute. "It is perfectly marvelous how quickly you get results with your testing machine. You are evidently better equipped to test the endless variety of curved surfaces than any body has ever been," and Chanute continued, writing, "I hope that you will prepare a paper at your leisure, so that your discoveries shall not be lost to others. I would take pleasure in presenting such a paper either to the Western Society of Engineers or the American Association for the Advancement Science of which I happen to be a member."[46]

Just prior to leaving for his winter sojourn, Chanute received a wire from St. Louis World's Fair organizers that they considered an aeronautic contest and wanted him and Samuel Langley to participate. So Chanute stopped in St. Louis on his way to California and agreed to help with arrangements for a tournament and an aeronautical congress in 1904.

Soon after arriving in Pasadena in mid-January 1902, Chanute received a letter from Spratt telling of his impending marriage. Chanute replied, "I congratulate you upon having found a nice girl to go into partnership for life. This will make a great change in your purposes. You will have to devote more thought and energy to material cares and to the accumulating of a competence, and any scientific experiments must be relegated to your leisure. If you need a little help, I shall prefer to furnish it, imposing no obligation whatever upon you, providing you eventually publish whatever you find, so as to help others."[47]

A few days later, Charles Lamson mailed a letter with photos of his just-patented kite. Because automatic stability remained on Chanute's mind, he wondered if Lamson could incorporate his third idea of "oscillating wings" into his kite design, so he replied: "The triplane surfaced one is especially interesting to me, as it is closely alike in outlines to a gliding machine with five surfaces which I have designed. I wish I had your skillful fingers to build a full-sized machine. Can you suggest some way to bring this about?"[48] Lamson agreed to build a foldable triplane with oscillating wings for $234. Given Chanute's engineering knowledge, it is surprising that he continued thinking of multiplanes instead of improving his far more stable biplane glider.

Looking forward to the next camp with the Wrights and thinking of improvements, Chanute asked Wilbur if they could rebuild, at his expense, his "multiple wing" and his "two surfaced," in which the operator hung in an upright position.[49] In Chanute's opinion, it would be easy to compare the merits of each design in flight, and the knowledge gained would help all in designing future machines. Wilbur, however, did not favor such work or a competition; he was convinced that they were far ahead of everyone else.

Even though it may have been expected, Chanute's wife Annie passed away early in April 1902 (see "1902: A Year in Change" in chapter 6). Trying to cope with his new life, the lonesome seventy-year-old widower briefly considered stopping all aeronautical involvements, but this did not last long. Only one month after Annie's burial, in the midst of the controversy over who had actually "invented" the biplane glider, the thirty-five-year-old Herring had the nerve to again contact the ever-tolerant Chanute for a job. Having lost his income after the Truscott Boat Yard burned and Matthias Arnot had died, Herring proposed to build a glider; Chanute reluctantly agreed to pay $150 to build new wings for the *Katydid*, using the frame and other parts still in Avery's shop.[50] The following month, Chanute stopped at St. Joseph and thought Herring's workmanship was excellent.

Sharing news about the upcoming activities in camp with the Wrights, Chanute wrote to Spratt: "My machine is completed in CA. Have also a new multiple winged machine to be tested. If your engagements permit of your being present, you may see how you can improve on the design."[51] Wilbur also invited Spratt to join them in camp.

The Wrights arrived in Kitty Hawk on August 28. This year's glider, with its thirty-two-foot wingspan, was significantly bigger than their 1900 or 1901 machines or Chanute's 1896 biplane. Following Lilienthal's footsteps, the brothers made practice accompany theory, isolating problems as they appeared. To control the big glider in flight, they had developed a hip-cradle to activate the wing warping, but the fixed tail still allowed the craft to sideslip when warping the wings. Orville then suggested changing the vertical tail to a moveable rudder,[52] which was the needed breakthrough for a working three-axis control system. Now both brothers shared the flying.

The Wrights' older brother Lorin arrived on September 30, Spratt came on October 1, and Chanute arrived with Herring as his assistant on Sunday, October 5. Early next morning, Herring assembled the *Katydid*, or the "swinging wings," and made a short glide in the afternoon, but the wings buckled and twisted in the wind. A second attempt ended after a twenty-foot glide, breaking the crossbar on landing. After a quick repair, Herring flew the glider as a manned kite,[53] proving that it did not produce sufficient lift. Chanute had expected better results.

The following day, the boxes with the Lamson-built "oscillating wing" glider were uncrated and Herring, Spratt, and Orville assembled the twenty-foot-wingspan triplane under Chanute's supervision. After flying it as a kite, Herring made a brief gliding flight, but could not adjust the springs easily to activate the oscillating mechanism to enhance longitudinal stability. In short, this glider also did not perform as hoped. When Lamson inquired, Chanute responded:

Augustus M. Herring flies Lamson-built "oscillating wings" at Little Hill (October 1902). From Hermann W. L. Moedebeck, *Taschenbuch für Flugtechniker und Luftschiffer* (1904).

"Workmanship is superb, but the joints frail. . . . I am changing my mind about the expediency of adding a motor, as the gliding machine (of the Wrights) has been so much improved that it seems safe to advance forward."[54] Hoping that his gliders might provide some insight, Chanute gave them to the Wrights as a present; a storm destroyed them and the shed a few years later.

With both his gliders unusable, Chanute watched Orville and Wilbur fly over longer distances. Despite his Kodak's sluggish shutter speed, Chanute again documented the gliding flights.[55] The obligatory group picture shows the participants sitting in the sand in front of the glider. Coupled with the passing of his wife, Annie, the unsuccessful showing of his gliders at Kitty Hawk had made the year 1902 a less than happy time for the 71-year-old Chanute.

Chanute and Herring left camp on October 14; Herring's employment with Chanute had ended and both men headed for Washington independently. Even though Chanute did not have an appointment, he stopped at the Smithsonian castle and saw the secretary for an instant. Langley's assistant, Charles Manly, then received the main points of the Wrights' experiments. Next, Chanute had a good talk with Albert Zahm, now professor at Catholic University, performing aerodynamic research in his new wind tunnel, and with Charles Marvin at the Weather Bureau. There was so much to learn and share!

Looking for a job, Herring also stopped at the Smithsonian, but the secretary was not available. A few days later, Langley wrote Chanute, who replied confidentially:

> I have lately gotten out of conceit with Mr. Herring, and I fear that he is a bungler. He came to me, said that he was out of employment and urged that I let him rebuild gliding machines 'to beat Mr. Wright.' I consented to building new wings for the multiple wing machine, but could give it no attention, as the work was done at St. Joseph, Mich. Herring adopted new forms of wings, but when the machine was tried in North Carolina, it proved a failure and he said that he did not know what was the matter. Doubtless he got some valuable ideas by seeing Mr. Wright's machine, and he announced that he wanted to return via Washington. I suspected, although he did not say, that this was to revive his former unsuccessful application for a patent."[56]

Back in Chicago, Chanute reported to his many correspondents about his latest trip to the Outer Banks. "Have just ret'd from Kitty Hawk where Mr. Wright is experimenting. He has built a new gliding machine. The machine is perfectly steady and under full control, gliding at angles of 6 to 7 and sustaining 125 pounds per horsepower."[57] In another letter, he wrote: "Wright has made a great advance over last year. As soon as practice has been obtained to learn all the tricks of the wind, I believe that soaring flight will be performed under favorable conditions, and that the time has nearly come to introduce a motor. I am glad that this type ('type Chanute' as Captain Ferber is good enough to call it) is likely to prove to be one of the types of machine with which safety will be achieved."[58] Writing to B. F. S. Baden-Powell, president of the Aeronautical Society of Great Britain and editor of their quarterly journal, Chanute gave the dimensions of the Wrights' latest glider, and explained: "It was equipped with a horizontal rudder at the front, and a vertical rudder at the rear, and it proved quite manageable, so that skill was quickly acquired to steer it up or down or sideways and to meet the turmoils of the wind."[59] The following month, Baden-Powell briefly mentioned the progress made by the Wrights in his presidential address, but not the use of the horizontal and vertical rudders.

This was Chanute's last season of experimenting with glider designs; neither of his approaches to automatic stability had yielded the hoped-for result. In the years to come, he provided encouragement to other researchers and always hoped to see an aeroplane in flight. Lecturing at Washington University in St. Louis in October 1902, Chanute told his listeners optimistically that they would see machines travel through the air in the not too distant future and prophesied that the flying machine would soon be in use for sport and recreation.[60]

The Aeronautical Mission in Europe

In an effort to regain his mental poise after his wife's death, Chanute delved deeper into his tie preserving business, but he also sought more involvement with aeronautics, which now was no longer a "side issue." When St. Louis World's Fair organizers looked for a goodwill ambassador to promote their aeronautical contest in Europe, Chanute happily accepted the appointment, because this would allow him to meet his many correspondents in an official capacity.

Chanute and his three daughters spent New Year's Day 1903 in Boston and then traveled to Egypt. Vacationing for almost two weeks in Cairo, he also stopped at the French embassy and inquired unsuccessfully on the whereabouts of Mouillard's belongings and his heirs.

The Chanutes then ferried to Italy, where he met Cosimo Canovetti, a civil engineer from Milan, who had won a $200 award from the Smithsonian for his experiments with the resistance of variously shaped solids in air and water. The family then stopped in Nice, France, to meet the French artillery officer Captain Louis Ferdinand Ferber, who had also experimented with gliders. During their discussions, the two men realized that they both believed in Lilienthal's philosophy. Seeing Ferber's glider, Chanute detected some resemblance to the Wrights' glider, so he promptly wrote to Wilbur about his visit. "Ferber has an abominable practice ground, it being a series of rocky slopes with precipices above and below. He has accordingly built what he calls an 'aerodrome' with a mast, 59 ft high and a rotating arm 98 ft across, running on ball bearings. He has a gasoline engine of 6 hp, which weighs complete with the shaft less than 200 lbs, and he estimates that the screws will weigh about 22 lbs or more."[61]

On Friday, March 13, the sun shone brightly. Chanute's three daughters traveled by train to Paris, while he went to Austria to begin his aeronautical mission for the St. Louis World's Fair. Arriving in Vienna in the midst of a furious snowstorm, Chanute spent the afternoon with Wilhelm Kress, with whom he had corresponded since 1894. His large flying boat *Drachenflieger I* was constructed from steel tubing, and a heavy Daimler engine provided power. Kress had tried to launch it from the shores of the Wienerwaldsee, but the pontoons filled with water and it sank; he then added a fourth wing to provide more lift and looked for a lighter motor. Seeing the structure, Chanute wrote to Wilbur that "it seems to possess some excellent points in construction, and that it may actually fly if a motor lighter than the present one can be obtained."[62] But two days later, Chanute reportedly told Viktor Silberer, president of the Aero Club in Vienna, that he did not think the approach that Kress took would succeed.[63] That evening, Kress hosted a dinner at the Hotel Bristol, with eleven influential club members attending.

On Sunday morning, Chanute talked at an informal meeting about the St. Louis events, followed by a general discussion of aeronautics in Austria and the United States. Later in the afternoon, Chanute met with the eighty-six-year-old Friedrich Ritter von Lößl, a senior board member of the Aero Club, who had also studied the air resistance on locomotives and bridge members while developing the railroad system in the Austro-Hungarian Empire. Chanute then took a train to Breslau in southeastern Germany, where he met Moedebeck, now stationed in Graudenz. The updated *Taschenbuch* with Chanute's contribution on artificial flight[64] was ready to go to the printer and the two men discussed a possible translation of the book into the English language.

Continuing his travel to Berlin, Chanute first took care of some tie preserving business and then met with the German commissioner for the St. Louis Fair and members of the *Berliner Luftschiffer-Verein*. They invited him to the officer corps of the Airship Battalion at the military airfield at Reinickendorf-West, where a special breakfast ensued[65] and Chanute shared his excitement of flying heavier-than-air equipment. After a brief stop in Strasbourg, meeting members of the *Oberrheinischer Verein für Luftschiffahrt*, Chanute arrived in Paris on March 22, catching up with his daughters and enjoying life in this excitingly social city.

Little aeronautical progress had been made in France after the Montgolfier brothers had ascended in their balloon a century earlier, but in October 1901 a dapper young Brazilian, Alberto Santos-Dumont, had dazzled Parisians by flying his dirigible around the Eiffel tower, winning the Deutsch prize and 50,000 francs. This rekindled interest in flying, and wealthy members of the Aéro Club de France became interested in spreading their wings beyond ballooning.

Late in 1902 Chanute had written to his many European correspondents about the various American aeronautical activities, and these reports now appeared in print as he appeared in person.[66] Naturally, Parisians wanted to meet the French-born civil engineer and hear him talk about flying in the New World. In a well-attended after-dinner lecture at the Aéro Club on April 2, Chanute described with some humor his own experiments, and followed with more serious reporting of Orville and Wilbur Wright's gliding in their machine. Combining his enthusiasm with spectacular lanternslides, Chanute made a point that flying was exciting, leading several Paris papers to mention the talk of the *célèbre aviateur américain* the next day.

Two days later, Chanute wrote to Wilbur Wright: "It seems very queer that after having ignored all the gliding experiments for several years, the French should now be over-enthusiastic about them. The German and the English had taken more notice, and it does not come as a surprise to them that men actually take toboggan rides on the air."[67]

The first report of Chanute's *dîner-conférence* talk was published in *La Locomotion*. Ernest Archdeacon described the leading figure in American aeronautics as being different from other inventors, because Chanute believed in sharing information to help others. Perhaps taking license with what he heard in private conversations, Archdeacon described how the Wrights operated their glider: "To regulate the equilibrium in the transverse direction, he moves two cords, which operate by warping the right side or the left side of the wing, and, simultaneously, by the displacement of the vertical tail in the back."[68] Archdeacon also reported that Chanute would send drawings of all gliders so that similar ones could be built in France; he reminded club members that a Frenchman should perfect aviation. The secretary of the Aéro Club, Georges Besançon, submitted a similar report to *L'Aérophile*, published later in April.[69]

Continuing his mission, Chanute met the French commissioner Michael Lagrave and his committee and discussed the rules of the St. Louis aeronautical contest and how to grow aviation. One committee member, Jacques Balsan, was now interested in flying gliders after having flown his balloon *Saint Louis* in late 1900 from Paris to Danzig, a 950-mile distance.

On April 10 the Chanutes left for London. Having read John Bacon's recently published book *Dominion of the Air*, he arranged to meet the author and spent an afternoon at his home in Berkshire. Bacon's daughter Gertrude later wrote about the

> visit of the courteous, keen-eyed, white-haired American gentleman with a pleasant smile and eyes that looked into the future, sitting at our table and telling us, in modest, matter-of-fact tones, enthusiastically of certain wondrous experiments with kites and gliding machines that he was making in his own country, and of the rather striking work that was being done by two young American brothers of Dayton, Wilbur and Orville Wright. They had chosen as their experimental ground a remote stretch of the North Carolina coast where a steady wind blows constantly from the Atlantic, and the sand dunes gave admirable jumping-off places. He told us how he himself had gone into Robinson Crusoe-like camp with them and had watched them fly. We liked the old gentleman and we were mildly interested in what he told us.[70]

During the afternoon, Gertrude showed her collection of "Blindfold Pigs" and cajoled Chanute into adding his work. Seeing Maxim's contribution, Chanute agreed and was blindfolded, given a pencil and paper, and asked to draw his rendition of a pig, an exercise intended to give psychological insight into one's character. Removing his blindfold, Chanute probably chuckled, but added his autograph.[71]

After this pleasant interlude, Chanute presented a final slide-talk at the Aeronautical Institute in London and the family returned to Chicago late in April. Traveling through central Europe for six weeks provided many opportunities to meet prominent aeronautical personalities, which convinced Chanute that America was far ahead of Europe in regard to aviation. In the next few months, publications on either side of the Atlantic included colorful reports of Chanute's visit. His talks on aeronautical activities in America stirred controversy but also precipitated a revival in aeronautics in France. Over the next few years, the Voisin brothers, Blériot, Farman, Delagrange, and Archdeacon built gliders from the drawings supplied by Chanute. In later years, the Wright machine was dubbed "*type du biplan américain Wright*," while Chanute's biplane with the tail in the back became known as "*type du Chanute*."

As promised, Chanute mailed the text of his Paris talk, with photos and simplified scale drawings of his three gliders and the Wright's 1902 machine, to the editor of *L'Aérophile*. Describing his own gliding experiments, he reminded his readers that all went without accident. He added that his open invitation to improve on his work had received no answer until 1900, when the Wrights took up the problem, and the remainder of the article discussed their work. Even though Chanute had reportedly described the wing-warping concept at the Aéro Club meeting in April 1903, his French-language text only mentioned the "warping of the wings to steer to the right or left," without noting a corresponding movement of a vertical rudder. In closing his *L'Aérophile* article, Chanute reiterated: "It is necessary to use a great deal of prudence while conducting gliding flights. Do not try to beat existing records, it only prompts foolhardy adventures and causes accidents. Competitive runs where amateurs practice together are fruitful because the fliers learn from one another. . . . One must never forget that balance, control and safe landing are the main things, no matter the length of the distance flown. . . . It can be hoped that when many fliers start to work seriously, progress will happen so fast that the time will be near when aviation becomes practical."[72]

When the Airplane was Born: The Wrights Did It!

Shortly after returning from Europe, Chanute visited the Wrights in Dayton, bringing along a Richard anemometer as a gift. He shared stories about his trip and listened to the improvements the brothers had made to their "whopper flying machine." Chanute then arranged for Wilbur to address the Western Society a second time in late June. As part of this WSE talk, Wilbur admitted that progress was slow and that they had acquired only four hours' flight time in about one thousand glides. The brothers needed much more experience in steady soaring flight to become proficient.

Preparing an article for the *Revue Générale des Sciences,* Chanute wrote to Wilbur and asked if he should mention the warping of the wings.[73] Wilbur responded quickly: "It is not our wish that any description of this feature of our machine be given at present." He also stated that the control was too complex and that experimenters should use machines of less than twenty feet span before trying bigger machines. As Lilienthal had advocated, "one thing at a time is a safe rule."[74]

Chanute mailed the English text of his *Revue* article to Wilbur for final comment, who pointed out that it was news to him that "the tail is operated by twines leading to the hand of the aviator.[75] He wanted this sentence removed. Not wanting to divulge confidential information, Chanute asked how the tail was operated, and Wilbur explained, "The vertical tail is operated by wires leading to the wires which connect with the wing tips. Thus the movement of the wing tips operates the rudder. This statement is not for publication, but merely to correct the misapprehension in your own mind."[76] Reading Wilbur's letter, Chanute sensed apprehension, and responded, "I was puzzled by the way you put things in your former letter. You were sarcastic and I did not catch the idea that you feared that the description might forestall a patent. Now that I know it, I take pleasure in suppressing the passage all together. I believe however that it would have proved quite harmless as the construction is ancient and well known."[77] The article was published in November 1903, without mention of the rudder's operation.[78]

Having worked in business and technologically advanced settings throughout his life, Chanute knew how to balance the excitement of new technology with one's need to maintain confidentiality. In his other life, wood preservation, he always wanted to learn and talk about new processes, but that did not stop him from withholding pertinent information until he had it patented. Chanute understood the need for protecting potentially valuable "trade secrets," but he sincerely doubted that anyone would make a fortune from the airplane.

While the Wrights worked on their *Flyer,* their nearest competitor, Langley, had similar goals. Reading in the press of the various attempts to launch Langley's "misbehaving airship," Chanute hoped to receive an invitation to attend a launch, but nothing came. In early December 1903, Langley's *Aerodrome* and its pilot tumbled into the Potomac a second time, and the press ridiculed the secretary for wasting government money. Chanute felt sorry for Langley that pesky reporters labeled him a bungler, but on the other hand, he thought that the physicist had undertaken too much at once to produce a full-fledged, man-carrying flying machine.

In late September 1903 the Wrights left for Kitty Hawk, sure of interesting results. Again they invited Chanute to visit their camp, but he had just lost the tie preserving contract with the Rock Island and had to locate another customer before he could have fun. In early November Chanute traveled to Washington on tie treating business, visited Langley briefly, and then continued to Kitty Hawk, arriving on November 6. The weather was bad for most of his weeklong stay, so the Wrights and Chanute, possibly the most knowledgeable aeronautical enthusiasts of the day, huddled around the camp stove, talking about flying and drinking coffee prepared in Orville's newly fashioned French drip coffee pot.

The weather improved after Chanute left, but the *Flyer* developed more technical problems. Waiting for parts to come from Dayton, Orville wrote home: "If we should succeed in making a flight, and telegraph, we will expect Lorin as our press agent (!) to notify the papers and the associated press."[79] On November 28 the propeller shaft cracked a second time, and this time Orville took it to Dayton. Traveling back to Kitty Hawk, he heard of Langley's latest attempt to launch his aerodrome. The brothers now wondered about their luck, but they were determined to stay until they had flown. Everything was ready on December 14, but the *Flyer* stalled after 3½ seconds in the air. Three days later, they tried again, and the first sustained, controlled, powered flight became reality; they wired home to set the prearranged news release in motion.

Reading the telegram, Katharine decided to share the news with Chanute: "Boys report four successful flights today from level against twenty-one mile wind. Average speed through air thirty-one miles. Longest flight 57 seconds." The telegram arrived in Chicago shortly after 8 P.M. and Chanute replied right away: "I am deeply grateful for your telegram of this date advising me of the first successful flights of your brothers. It fills me with pleasure."[80] He also wired the Wrights at Kitty Hawk: "Immensely pleased at your success. When ready to make it public, please advise me." In the following days, Chanute shared the text from Katharine's telegram with some close friends. In a letter to Spratt, he commented: "It would have been safer to test the machine thoroughly as a glider before launching it with the motor on board, but they took the risk and won."

Chanute's next talk at the AAAS meeting in St. Louis was scheduled for the end of December. He had received confidential information from Charles Manly on Langley's latest attempt to launch his aerodrome but had no reportable news from the Wrights. Several Chicago papers had discussed the Wrights' flights and Chanute wondered if they had really flown more than three miles at an altitude of sixty feet, so he wrote Wilbur on December 27: "I have had no letter from you since I left your camp, but your sister kindly wired me the results of your test of Dec 17. Did you write? It is fitting that you should be the first to give the Association the first scientific account of your performance. Will you do so? Please wire me at St. Louis." Wilbur wired back: "We are giving no pictures nor description of machine or methods at present."[81]

Louisiana Purchase Exposition in St. Louis, 1904

To promote the aeronautical happenings at the upcoming World's Fair, Chanute told members of the AAAS's Section D in St. Louis on December 30, 1903
that, "The empire of the air is still to be conquered, but we have certainly got
a further glance into the Promised Land than we have ever had before."[82] He
described the progress made with balloons and aeroplanes but also their limits,
and talked about the failures of Langley's *Aerodrome*, which only proved that a
better flying machine had to be invented. He mentioned that the Wrights had
acquired the necessary proficiency and that they had flown their airplane two
weeks earlier. In closing, he discussed the prospect of anyone winning the prize
money offered by the World's Fair Commission.

Listening to the talk, the editor of the widely read *Popular Science Monthly*
asked to publish it in his magazine. It appeared in March, and Langley included
it in the *Report* of the Smithsonian for 1903.[83]

After wiring Chanute that the brothers would provide no information, Wilbur
wrote him a five-page letter, causing the recipient to puzzle why the brothers were
so hesitant in sharing such exciting news. The Wrights then submitted a detailed
statement to the press, which none of the major Chicago papers printed, and
mailed a clipping to Chanute. The last sentence puzzled him: "All the experiments
have been conducted at our own expense, without assistance from any individual
or institution." He thought he had given them assistance, just as he had helped
other budding experimenters, but the brothers apparently thought differently.

St. Louis World's Fair organizers believed that liberal prizes would stimulate
progress in aeronautics, and many people showed interest in the prize money,
but one after the other dropped out. Santos Dumont came and went away with
his gasbag damaged—no one knew how it happened. The Wrights also found
the grand prize tempting. Maybe Chanute was right that by flying in public
and revealing the extent of their success, they would ultimately profit, but they
continued working on their *Flyer* with as little publicity as possible.

Even though Chanute was not interested in experimenting with automatic
stability in a glider anymore, he liked Avery's suggestion to enter a reengineered
biplane glider in the aeronautical contest. The wings were now shaped like
"Plin's curve," resembling the profile of a bird wing, with the highest part of the
curvature being about 1/12th aft of the leading edge. It had a fifteen-foot, three-
inch wingspan with a six-foot chord, and the aerodynamically shaped uprights
consisted of aluminum tubing. Avery felt sure that he could win the prizes for
gliders and make twenty flights of four hundred feet in length, or forty flights
showing automatic stability; he told Chanute that this would be "like picking
up money from the street."

Because there were no hills on the fairgrounds, some device was needed to
launch the glider. The Wrights had built a wooden track to assist takeoff at a

total cost of $4, which amused Chanute.[84] Percy Pilcher had attached a towline
to his glider and had horses or people pull him into the air; but this did not
prove to be safe, as his glider crashed and killed the pilot after a rope break.
Early in 1902, Albert Merrill from Boston had suggested using a derrick with a
falling weight to launch his glider, an idea later picked up by the Wrights, but
Chanute was concerned about the connection, and replied: "Not knowing the
construction, I can not judge the safety of your experiments. Remember, while
your backer will only risk money, you will be risking your limbs in the attempt to
advance knowledge."[85] A decade earlier, Chanute had discussed a high-wheeled
carriage with a drum, used by the Italian military to control and retrieve their
captive balloons. Combining such a windlass with an electric motor and using
a light railway track with a flatcar for Avery to stand on while holding the glider
seemed the most sensible choice. Early in September, Chanute submitted his
claims for a glider launcher to the U.S. Patent Office, Avery began the installation of the rope-reeling drum onto the electric motor, and Chanute withdrew
from the World's Fair jury to avoid a conflict of interest.

On Friday, September 23, Avery experienced his first glider launch. Because
the winch was not ready, he used manpower. One end of the rope was connected
with a tow hook to the glider, held upright by Avery, while three men took the
other end of the rope and started to run. With the men pulling, Avery ran down
the field, holding the glider steady, until he felt himself lifted off the ground.
Reaching a height of about thirty feet (the length of the rope), he released and
glided down.[86] Having spent about $600 for equipment so far, Chanute mailed

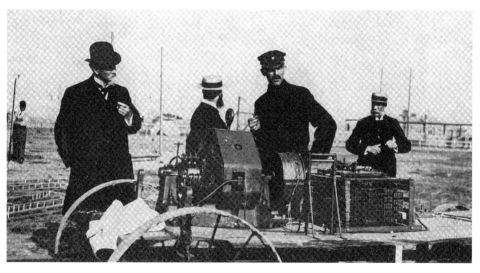

Electric winch at the St. Louis World's Fair (October 1904), showing Chanute and
Avery in the foreground. From *Cherry Circle*, January 1908.

Avery his next instructions: "Make your record by 30 Sep. Test efficiency of making line to the car instead of machine. Also put rubber soles on your shoes to ease alighting," and "Be very careful in experimenting!"[87]

But to fly the 400-foot distance and win the prize for gliding, Avery needed more speed and a higher launch. On October 6 the electric winch, driven by a ten-horsepower motor, pulled Avery and his glider into the air for the first time. From then on, he launched almost daily, usually reaching an altitude of seventy feet and flying over a distance of 300 to 350 feet, at times turning in flight before gliding back to earth. Avery's flights were impressive, the crowds cheered, and Jacques Balsan decided to buy the glider.

Originally, the jury had stated that they would award the prize for gliders for a flight of 400 feet; they now clarified that the distance was not from the point when air supported the glider but after the line was released. This was impossible to achieve on the stadium grounds, and Chanute withdrew the machine from the contest and wrote Avery: "I regret very much that you should have no chance to win a prize, but it is better for me to lose my money than for you to lose your life."[88] Talking things over, the two men then agreed that they would jointly decide when to stop the flying.

William Avery performing winch launch in St. Louis (October 1904). Chanute photo album, Manuscript Division, Library of Congress, Washington, D.C.

Meanwhile, exhibition management realized that the entrance-paying public wanted to see the only heavier-than-air machine fly, so they suggested moving the winch to the Plaza of St. Louis, which provided more space and less-turbulent air. Using a one hundred–foot line allowed higher and longer flights, but Chanute was concerned and sent further instructions to Avery: "You need to examine your apparatus carefully after each glide, to make sure nothing is broken or loose. Take no chances in tumultuous winds. When you have made the prescribed number of glides to win the prize, stop, and make your own arrangements with the administration. It probably can afford to pay you a good price to secure large attendance."[89] Even though Avery used a longer towline to reach a higher altitude, the glider did not have the performance to extend its glide to the required four hundred–foot distance to win the prize.

After being launched and landing on his 1¾-inch thick rubber-soled shoes for almost a month, the inevitable happened. On October 25, the hemp towrope broke, with Avery and his glider still climbing at about forty feet. He landed hard, broke his right foot, and dislocated several tendons. The glider received only minor damage and Avery repaired it as soon as he could move without crutches. In February 1905 Chanute sold the glider for three hundred francs to Balsan and shipped it to France; after flying it for several years, Balsan donated the glider to the Musee de l'Air in Paris.

Chanute's "Means for Aerial Flight," or glider launcher, received a patent on October 30, 1906;[90] little imagination is needed to recognize the ancestry of modern glider winches and towing hooks in this 1904 design. No doubt, Chanute knew that experimenters would achieve reliable mechanical flight in time, but the St. Louis prizes, big as they were, failed to bring forth much evidence of progress.

To Get into the Air: Powered or Unpowered

While attending the 1905 International Railway Congress in Washington, Chanute was one of the guests at Alexander Graham Bell's "Wednesday Evening" gathering. On May 10 about twenty-five to thirty scientific friends, including Langley and his assistant Manly and Emile Berliner of telephone fame, congregated at Bell's mansion on Connecticut Avenue. The discussion soon turned to aviation and Chanute spoke of several aeronauts, including Kress, Montgomery, and the Wrights, but Bell had only one question: "Had the Wrights really flown?" Chanute's simple reply left a lasting impression: "I have seen them do it." He had indeed seen Orville fly on October 15, 1904, at Huffman Prairie, Ohio, even though this had been only a brief flight.

Believing that they had almost perfected their invention, the Wrights decided in January 1905 to offer their *Flyer* as a military weapon to the U.S. government via their local congressman. The Board of Ordnance responded that they would be interested once the machine reached practical operation. Taking umbrage, the Wrights decided to contact foreign governments. Because they did not want to reveal details, they needed someone of worldwide eminence to testify to their efforts. Chanute was the only person with these qualifications, so Wilbur wired him on October 30, 1905, to come and "witness their machine breaking another record." The weather prevented a takeoff and Chanute returned to Chicago not having seen them fly that year. The Wrights then dismantled their airplane and Wilbur wrote Chanute that their friends could now talk about their flying.

The upcoming AAAS Meeting in December seemed like a good venue for publicity, so Chanute asked cautiously, "Would you like to prepare a short paper announcing what you have accomplished during the last two years or would you like me to do it? If so please indicate how much you would wish me to tell."[91] Wilbur replied that Chanute could discuss their progress but not the method of operation, so neither made anything public.

Chanute's hope for a national aeronautical society came true more than a decade after first suggesting it. The Aero Club of America formed in late summer of 1905 as an offshoot of the Automobile Club of America. Charles Manly, the mechanical engineer who had gained publicity in 1903 as the pilot of Langley's ill-fated *Aerodrome*, spoke at the first meeting. He provided background information on aeronautics, which he felt was inspired by the "great patron saint of aviation, Octave Chanute," and mentioned that the Wright brothers had successfully flown a new version of their *Flyer*. Manly's speech ignited enthusiasm, and the group decided to include "aerial technology" in the next auto show in January 1906. Trying to gather material on short notice, Augustus Post, the secretary of the Aero Club, contacted Chanute requesting artifacts to display at the upcoming show.

Anticipating a move from their rented house on Huron Street to a better neighborhood, Chanute decided that he could help himself and the Aero Club by cleaning up. He mailed Post a list of flying models that could be shipped for $30, but stated they were all in sad shape. Post replied that the club desired his material and Herring would restore them, so Chanute mailed $10 to join as a charter member. Avery pulled the glider and kites from the rafters above his shop and Chanute's daughters removed the kites and stuffed birds from the ceiling in his den. He also gathered twenty-eight large-size photos of the 1896 and 1897 gliding experiments and offered his duplicate books to start a library. Everything was packed into three large crates and shipped to New York.

The weeklong Aero Club Show opened with great fanfare on January 13, 1906. One wall of the exhibition room showed flying machine photos, divided under the headings: Lilienthal—Herring—Wright Bros.—Langley—Maxim—Pilcher; Chanute's name was mentioned in some photo captions under Herring's contributions. When Wilbur saw this, he wrote to Chanute: "I was under the impression that I had learned somewhere that you had conducted some experiments about 1896 or 1897. Possibly my memory is at fault."[92] Chanute was very much amused about Herring and how he had arranged the exhibit, but said nothing.

The exhibition was a good display of the current status in aeronautics. An original Lilienthal glider, looking a bit tattered, hung beside Chanute's biplane, described as the "Herring-Arnot Glider," and the latest creation of Santos-Dumont, all contrasting sharply in shape and construction with Bell's tetrahedral kite. Langley's *Aerodrome* and the propeller shaft of the Wright *Flyer* were also on display, among many other artifacts.

Chanute did not go to New York; he and two of his daughters spent the winter of 1906 in the New Orleans area at the rice plantation of his niece, Amelia Gueydan. Many years later, Amelia's daughter Pepilla recalled, probably with a smile, that her uncle Octave was fond of watching the buzzards, and the best place to do so was the city garbage dump.[93]

During the Chanutes' lengthy absence from Chicago, the servants usually forwarded the mail. Receiving a letter from an author who wanted to discuss Chanute's life in conjunction with other successful French descendents, he responded tersely: "I earnestly beg that you will have all mention of myself omitted from the proposed article on 'The French in America.' Feel honored that my name was suggested, but deprecate publicity, especially in a connection which seems to differentiate one nationality from another, while we are all American and have been absorbed in the mass."[94] He considered himself an American and was proud to be one.

Even though the Wrights now talked openly about their *Flyer*, the public still knew little about its details. To obtain this inaccessible information, the Aero Club in Vienna, Austria, tried a unique approach. They wrote Chanute in February 1906 and suggested giving him an honorary membership in exchange for drawings of the *Flyer*, "to create a diploma for the Wrights." Recognizing this as a roundabout honor, Chanute replied: "This is the first opportunity which I had to say that I have not been what you term 'the teacher' of the Wright Brothers, in the usual meaning of the word. I gave them, as I did to several others, such information and experiences as I had gathered by some investigations of my own; I did not feel to be a sufficient mechanic to finally develop a successful dynamic flying machine. If the Wright Brothers have progressed so very much further than others to whom similar information was imparted, it has been entirely the result of their own genius, perseverance and great mechanical ability. As to the construction of their present machine, I regret that I can tell nothing nor can I send a schematic sketch."[95]

Wanting America to be a strong contender in aeronautics, the Aero Club approached the Wrights to provide further evidence of their aeronautical achievements. Looking for publicity, Wilbur described their experiments and listed the names of several individuals who had seen them fly, but Chanute was not one of them. Reporters were now frantic for information, but the brothers continued to refuse giving technical details, which Chanute found amusing.

Again, the brothers needed Chanute with his worldwide reputation, because the representative of the French government was coming to Dayton to finalize negotiations for the purchase of a machine. The deal with the British government had failed earlier, and the Wrights really wanted to succeed with the French, so they wired Chanute. He hurriedly rearranged his business schedule and traveled to Dayton for one day, April 1, 1906, but the French official did not appear.

In the following months, Chanute warned the Wrights gently that they might be setting their sights a little too high, especially because French experimenters were making rapid progress. Wilbur, however, felt that $200,000 was not too high an asking price. "We are convinced that no one will be able to develop a practical flyer within five years. . . . Even you, Mr. Chanute, have little idea how difficult the flying problem really is. . . . We would be very glad to know exactly upon what grounds you base your opinion of the selling value of our machine."[96] Chanute's laconic answer came a few days later: "The value of an invention is what it costs to reproduce it, and I am by no means sure that persistent experimenting by others may not produce a practical flyer within five years. The important factor is that light motors have been developed. . . . I cheerfully acknowledge that I have little idea how difficult the flying problem really is and that its solution is beyond my powers, but are you not too cock-sure that yours is the only secret worth knowing and that others may not hit upon a solution in less than 'many times five years?' It took you much less than that and there are a few other able inventors in the world."[97]

Because powered flying was complex, would-be pilots frequently began with the simpler and less costly sport of gliding as the first step in learning the basic laws of flight. "Only pluck and address were needed for its manipulation, and a spice of danger is what the true sportsman loves, and nothing great is ever accomplished without perseverance, resource, and, often enough, risk."[98] Anyone could buy a glider for about $100 or build one for less money, because drawings for the Chanute-type biplane glider, with control by weight shifting and no concerns about litigation, were widely distributed in magazines, newspapers and pamphlets.[99]

Flying with lighter-than-air craft was also exciting and the press described the daring flights of balloonists in colorful language. Chanute had chatted with the enthusiastic thirty-year-old Roy Knabenshue, who flew Tom Baldwin's airship at the St. Louis World's Fair, to the delight of the crowds. Two years later Knabenshue proposed a publicity stunt that would take aviation's senior promoter along on a flight across Lake Erie in early January 1907. This sounded exciting and Chanute appreciated the offer, but at seventy-four years old, he did not think that he was up for such a ride in the middle of winter, and he sent his regrets.

The long series of experiments by various investigators had demonstrated the soundness of the principle of the aeroplane. More and more people became interested in aeronautics, including fifteen-year-old Larry Lesh. Searching for information on flying models in Chicago libraries, he read articles by Chanute and contacted him in August 1906, visiting him at his home a few days later and meeting Avery in his shop. Lesh later wrote: "I was relieved to find that Chanute was a very simple, cordial and delightful person. His head was high and broad, the repository of an amazing store of knowledge. . . . I left the Chanute mansion with an armful of books, papers and photographs. The next day, I received a letter with a check for $50 and instructions how to build my full size glider."[100] Chanute believed in training intelligent pupils to carry on his ideas, so Lesh continued visiting Chanute and Avery until his parents moved to Kansas City.

When the recently formed Aero Club of Chicago wanted to stage an air meet at White City in July 1907, Chanute, one of the charter members, suggested hiring Lesh to bring and fly his glider for $100. "Mr. Chanute visited the show and saw my glider for the first time. He tactfully suggested that I keep it in the tent, pointing out that it appeared too weak in the joints."[101] Owing to weather, Lesh made only one flight during this week but received his pay for flying the only fixed-wing craft.

The growing Kansas City community remained a favorite of Chanute, and he continued dealing with real estate there. In 1900 he bought half a city block on the corner of 8th and Bank Streets, and then partnered with the Maxwell-McClure Notion company to erect an office and warehouse building.[102] The resulting seven-story "Chanute Building," at that time one of three tall buildings in Kansas City, was erected in 1906 at a cost of almost $140,000. Curiously, and perhaps just for lyrical convenience in 1942, the musical producers Rodgers & Hammerstein selected the ultra-modern Kansas City of 1906 as a scene in their first musical, *Oklahoma!* Their fictional cowboy Will Parker boasted of life in the city, singing, "Ev'rythin's up to date in Kansas City. They've gone about as fur as they c'n go! They went and built a skyscraper seven stories high, about as high as a buildin' orta grow!" Could Parker's fictional seven-story skyscraper have been the just-erected Chanute Building, not far from the Union Stock Yards?

In October 1906 Chanute combined business with pleasure; he finalized the rental contract on his just-erected building in Kansas City and presented a talk for the Technological Society. The audience filled the Central High School auditorium to its limits, and Chanute's first sentence caught everyone's attention: "There is today in the world a practical flying machine. I have seen it fly with my own eyes and know. The secret of aerial navigation has been solved." The talk's highlights were slides of his and the Wrights' glider flying. "Now that we have successful airships, we are able to see their limitations. I do not believe they will be of commercial importance. They will be used in the art of war, in the explorations of swamps and countries of impassable forests and for the ascent of mountains that have not been climbed."[103]

Reading of this talk, the mayor of Chanute, Kansas, invited Chanute to lecture in his town in the middle of December. The social schedule of the family was full, but Nina and Alice wanted to see the town named after their father (see chapter 3), so he agreed. Members of the planning committee proudly showed their town of almost ten thousand citizens. "Dee-lighted" was the Rooseveltian phrase Chanute reportedly used when asked what he thought of the city named after him. "I was not expecting to see such a large and prosperous city. It is also a very pretty town, and the streets and buildings are all well kept. It is thirty-three years since I was here, then the city was a mushroom town of 800 inhabitants,"[104] and a railroad intersection. His lecture in the Opera House was well attended, and its citizens later stated proudly that their town was the first "high flyer" designed by Octave Chanute.

After a long illness, Samuel Langley died a disappointed man on February 27, 1906. A memorial meeting took place on Monday evening, December 3, in the lecture room of the National Museum in Washington. Chanute discussed Langley's contribution to aerial navigation and his faith in the possibility of success that was an inspiration to so many investigators. Chanute told his audience, "We are now in the possession of a solution of a problem, which had baffled the ingenuity of man since the dawn of history,"[105] but he also stressed that other people needed to continue Langley's work. Beyond that, Chanute also clarified some of statements reported in the press about Langley's experiments with his aerodromes. The facts, as Chanute chronicled them, now became a matter of history.

For the past several years, Chanute had worked with W. Mansergh Varley from England on translating Moedebeck's *Taschenbuch* of 1904. The attractive *Pocket-Book of Aeronautics* arrived late in April 1907 and Chanute liked seeing his contribution on "Artificial Flight."[106] Thanking Moedebeck, he also expressed his concern that the Wrights were in danger of being overtaken by their competitors. "It is possible that both you and myself are wrong in condemning their policy of not showing the machine until an agreement has been reached for its sale. The luck which experimenters will have within the next few months will determine the matter."[107]

Promoters thought that aviation should not remain a curiosity, and that the sport had to grow safely, so the Aero Club set rules for licensing pilots and established a correspondence school.[108] The club also published a book titled *Navigating the Air* in which President Cortlandt Field Bishop wrote the preface, giving information on the organization, and Chanute discussed "The Wright Brothers' Motor Flyer."[109] Other invited authors were Graham Bell, Herring, Lowe, Ludlow, Rotch, Stevens, the Wright brothers, and Zahm, each discussing a different aspect of flying.

The *New York Herald* promoted aviation by publishing interviews with prominent aeronauts that were then distributed widely by the Associated Press. Graham Bell, receiving an honorary degree of science at Oxford, stated in London: "The age of the flying machine is not one that belongs to the distant future. There is left only the question of improving the machine that has been invented. To Mr. Chanute should be given a great amount of credit for what has been accomplished in this direction in America. He started the experiments with the flying machine and reproduced the gliding machine of Lilienthal and induced several Americans, among them the Wright Brothers, to experiment."[110]

Early in 1906 the editor of the *New York Herald* had asked Chanute for facts on the Wrights flying experiments, and published his comments in mid May. "Having made some experiments of my own in gliding flights in 1896, I published a full description with photos and detailed plans in the Aeronautical Annual of 1897, and I invited other searchers to improve upon my practice. I was written to on May 13, 1900, by Mr. Wilbur Wright who stated that he was minded to experiment, although he believed no financial profit would follow, and who asked for information and suggestions. These were gladly furnished."[111] Chanute added that the Wrights had kept him informed of their progress and invited him to Kitty Hawk, and that he believed all that the Wrights had said about their accomplishments.

Early in June 1907, the mechanical engineer William Stout from Minneapolis contacted Chanute about flying and flying toys to use in teaching. Interested in helping but also knowing that he needed to clean up, Chanute replied that he would be glad to dispose of the toys he still owned, "now that the problem of artificial flight is solved."[112] Reading this, Stout decided to go to Chicago and pick up the toys. Sitting and talking in his library, Chanute mentioned that the Civil Engineers' Society in St. Paul, Minnesota wanted a talk on artificial flight, but his business kept him in Chicago. Naturally, Stout volunteered to give the talk and Chanute supplied slides, which helped both of them. "All in all I had myself a time with a very much interested audience, including some of my old Phi Beta classmates. That little ego of mine swelled up a bit and that occasion whetted my enthusiasm. With this start I went back to work with more intensity, scientific reading and a lot of engineering ideas."[113] Stout later formed the Stout Metal Airplane Company and remembered Chanute throughout his aeronautical career.

To satisfy public demand, several new aeronautical magazines appeared during 1907. The *American Magazine of Aeronautics*, edited by Ernest La Rue Jones from New York, emerged in July, and Chanute's "Conditions of Success with Flying Machines" was a feature article in the first issue.[114] Another magazine, the *American Aeronaut and Aerostatist*, was published in St. Louis and edited by A. Kaufmann; Chanute's article on "Pending European Experiments in Flying" appeared in October,[115] because the editor and the author wanted Americans to know that Europeans were catching up. Chanute described the "motor-equipped flying machines" being built in Europe and stated that the pilots learned from each mishap and profited by every failure.

The planning committee for the next Aeronautical Congress as part of the World's Fair in Jamestown, Virginia, met in the middle of November 1906. Chanute served on the Committee on Congress and Program and had already submitted a short article to *Engineering World*.[116] Concerned about attendance, the Aero Club decided to hold this 3rd American Aeronautical Congress on October 28–29, 1907, in New York. On opening day, Willis Moore, chief of the U.S. Weather Bureau and president of the Congress, read his address, ghostwritten by Chanute.[117] Then Chanute read his paper "Soaring Flight"; Spratt discussed his pet topic "Curvature, a Relative Term," and Lesh shared his experiences building and flying a Chanute-type glider. The next day, John J. Montgomery and Israel Lancaster were two of the many presenters.[118] Half a year later, the Aero Club of America elected Chanute an honorary member after he read his paper on "Future Uses of Aerial Navigation."[119]

In June 1908 Chanute became one of four honorary members of another just-formed association, the Aeronautic Society of New York. Still encouraging youth in the pursuit of aeronautics, he funded a silver "Chanute Cup" that was to be awarded at gliding competitions, sponsored by the society at their airfield at Morris Park, New York. At the March 1910 competition Larry Lesh took second place and another newcomer, the seventeen-year-old Ralph S. Barnaby, took third place flying his Lilienthal-type glider over a distance of 114 feet, marking the beginning of Barnaby's lifelong career in naval and civil aviation.

In spring of 1909 the Washington Aero Club formed, and Chanute was one of nine elected honorary members, which gratified him. Early in 1910, prominent Chicago businessmen organized the Aero Club of Illinois and members elected the seventy-eight-year-old Chanute its first president. Chanute appreciated all the honors.

Even though many exciting aeronautical events were happening around him, Chanute's advancing age and failing health slowed him down. He did not enjoy traveling anymore and was tired of writing. During the initial period of trying to solve the problem of manned flight, he filled a one thousand–page letterpress book with copies of letters in less than a year, but now he needed more than twice that time to fill a book with copied letters. Even though Chanute did not write articles as often as he did in the past, his name still frequently appeared in the press, as reporters sought his comments.

Possibly influenced by Chanute's constant promoting of aerial navigation,[120] Chicagoans wanted all atmospheric routes to center in their city and wanted every citizen to learn how to fly. The Chicago Athletic Association, with Avery as an active member, chartered an aero section in February 1909, and their first project was a large "wind craft" with an engine. Chanute had just read of Gustave Eiffel's latest experiments on staggering superposed planes to reduce the interference of the lower plane with the upper one, so Avery incorporated Eiffel's findings into the design of the *Cherry Circle* airplane. "The most radical feature will be a peculiar curvature of the surfaces. The planes will resemble the wings of an albatross and will look from the front somewhat like a cupid's bow, or an ideal sail on a boat."[121] This airplane was reportedly built and flown in Beloit, Wisconsin, later in the summer.

At about that same time, Larry Lesh designed his *Aeroplane No. 1* and sent drawings to his mentor for comments. Chanute suggested positioning the upper wing similarly to the *Cherry Circle*, sketched his ideas, and commented: "I have found that the downward position like that of the gull, is more effective than the upward position, like that of the buzzard. It occurred to me that it is possible to provide a substitute for the flexing of a birds wing by placing 'flap valves' at the outer ends, which, when opened by a cord, shall diminish the supporting power of that wing, and tilt the bird sideways. This idea reminded me of the fact that I am half owner of a clever soaring machine patent in which the surfaces at the outer ends of the wings are aided for steering purposes. I send you Mouillard's patent. I hope that you will study the patent and profit by it."[122] Even though Chanute was so eminently at the center of the world's aeronautical information web, he still may not have fully grasped the significance of the control system to turn and bank the airplane.

In early November 1908, Lesh took his latest glider to the Morris Park Race Track in the Bronx, and spectators smiled seeing the teenager launch his *Aeroplane No. 1*, using a horse to pull the glider into the air. But to win the Brooklyn Eagle gold medal, he asked for an automobile to pull him into the air more speedily. Unfortunately, the glider encountered a wind gust and crashed. Lesh broke his ankle and regretted that he provided grief for Chanute.

Larry Lesh and his *Aeroplane No.1*, designed with Chanute's input, auto towed over the Morris Race Tracks in November 1908. Chanute Papers, Manuscript Division, Library of Congress, Washington, D.C.

Chanute models, built and donated in 1909, in the Early Flight Gallery, Smithsonian Institution, Washington, D.C. (April 2008). Courtesy of the Chanute Family.

To create a new permanent exhibit, the National Museum asked several pioneers, including Chanute, for ¼-scale models later in October 1908. John Zimmerer, a local carpenter,[123] agreed to build two sets of the 1897 biplane glider, the 1896 *Katydid*, and the 1902 oscillating wing glider. One set was shipped to Washington[124] and Chanute received the second set. These models are now at the Museum of Science and Industry in Chicago.

With the knowledge gained building the models, Zimmerer and another Chicagoan, H. Wixon, built a full-size glider with much input from the aging aeronaut. Always concerned about accidents, Chanute warned them to be careful until "the tricks of the wind are learned; do not fasten yourself into the machine in any way, you may want to get out quickly. Do not start from a wagon, but select a hill. You will probably make some mistakes at first and break the machine, but be careful not to break your own bones."[125] On a hot, late summer day in 1909, Chanute took his Kodak and his anemometer and joined Wixon and Zimmerer in the Indiana dunes. The seventy-seven-year old Chanute came home thoroughly tired, having walked for miles among the same sand dunes where he had experimented thirteen years earlier.[126]

Researching soaring flight had been an ongoing project since the early 1890s. After many rewrites, *Aeronautics* finally published his article in April 1909:

> Now that dynamic flying machines have been evolved, it seems to be worth while to make the computations and the succeeding explanations known, so that some bold man will attempt the feat of soaring like a bird. The theory underlying the performance in rising wind is not new, but it has attracted little attention, because the exact data and the maneuvers required were not known and the feat had not yet been performed by a man. . . . The bird's flying apparatus and skill are as yet infinitely superior to those of man, but there are indications that within a few years the latter may evolve more accurately proportioned apparatus and obtain absolute control over it. It is hoped . . . that soaring flight is not inaccessible to man, as it promises great economies of motive power in favorable localities of rising winds.[127]

This latest article not only put to rest the old theory of "aspiration," it advanced the new theory that birds could soar in apparently calm conditions as long as the air was rising locally faster than the bird was descending.

The Arrival of the Practical Airplane

The era of human beings flying in airplanes seemed to have arrived and competition to advance was faster than expected. William Glassford, who had kept up his correspondence with Chanute, wrote in the summer of 1907 that the military leadership had taken more interest in aerial navigation, to which Chanute simply replied: "I think that the dirigible and the aeroplane should be taken up by our

Chanute biplane model, built in 1909. Wingspan of 1897 biplane is 47.5 inches. Museum of Science and Industry, Chicago (June 1996).

government." Half a year later, the chief Signal Corps officer of the U.S. Army requested proposals for a flying machine and a dirigible balloon. They received eight bids for dirigibles and twenty-four bids for airplanes. Tom Baldwin, with Glenn Curtiss as partner, submitted the winning dirigible bid, and the military accepted the two airplane bids from Herring and the Wrights.

The specifications required the airplane to be easily assembled and disassembled, carry two persons and sufficient fuel for a flight of 125 miles, and fly at a speed of at least 40 miles per hour. Chanute thought that these strict requirements marked a new era in the development of this inchoate craft. "They may bring such new elements into war as to render it far less frequent, and they may develop such new uses of their own as to make the world better and happier."[128] He also commented on the two competitors. Herring had been working toward a solution of the problem of aviation for the past fifteen years, and "the present undertaking is to bring either the fruition of all his endeavors or the defeat of all his hopes."[129] The Wright Brothers were "viewed with incredulity because of the mystery with which they have been surrounded in the hope of a rich money reward, yet it is now generally conceded that they have accomplished all that they have claimed."[130] In another article, he explained:

> They inaugurated negotiations for the sale of their invention to various governments for war purposes, asking, it must be confessed, very high prices. Being somewhat opinionated as well as straightforward, they made two mistakes: the

first that the principal market for flying machines would be for war purposes (where cost is no object), instead of for sporting purposes, as more correctly judged by the French, and the second that contracts could be obtained for a secret machine. Two years were spent in fruitless negotiations. Wright Brothers seem now to have changed their point of view, but meanwhile large numbers of French aviators have begun experimenting, operating in public, and teaching each other.[131]

At about the same time, Bell, Chanute, Means, and Lawrence Rotch of the Blue Hill Observatory publicized a plan for a $25,000 prize fund by means of hundred-dollar subscriptions for the "International Sport with Flying Machines." The announcement stated: "During the past sixty years, yachts have been designed and sailed in a spirit of generous rivalry, a direct result of the competition between the designers of the early racers. International rivalry arouses enthusiasm and thousands take pleasure in witnessing international contest. The devotees of pure science may well take the high ground that 'science knows no country,' but when it comes to applied science, international rivalry gives a needed stimulus and incites men to take the rich material which the pure scientists have furnished and to apply it under a waving flag."[132]

Chanute had predicted just prior to the 1907 Aeronautical Congress that a new generation of researchers would take aerial navigation to the next level. With the financial backing of Mabel Bell, wife of Alexander Graham Bell, the Aerial Experiment Association (AEA) formed on October 1, 1907, to design and build a practical airplane. "Any schoolboy could build a flying machine from Chanute's handbook,"[133] Bell had reportedly said. Avery offered his services to help build and teach how to fly their Chanute-type biplane. The *Hammondsport Glider* was built in January 1908, and each "Bell Boy" learned how to control the glider in flight and land. AEA members then designed, built, and flew their "aerodromes" in Hammondsport, New York. One member, the almost thirty-year-old champion motorcyclist Glenn Curtiss, was intrigued with speed and admired the stunning *Scientific American* silver trophy; to win it, he had to make an unassisted takeoff and a straightaway flight of one kilometer (or 3,281 feet) in public. After test-flying his *June Bug* briefly at Hammondsport, Curtiss took the next step.[134] On Independence Day 1908, after a couple of false starts, he made a prearranged public flight over one mile at a speed of thirty-nine miles per hour, winning the first leg of the coveted trophy, with its cash prize.

The Signal Corps had accepted the Wrights' bid and a French company had signed a contract to build their airplane, so the brothers returned to their old experimental grounds near Kitty Hawk to update their flying skills. Wilbur sailed for France in late May 1908 to demonstrate the *Flyer*'s capabilities, and two months later Orville took a similar *Flyer* to Fort Myer, Virginia, to fulfill the Signal Corps contract. Chanute arrived at the military field on September 12 and Spratt came the next day to watch the final trial flights.

One of the official observers, army Lieutenant Thomas E. Selfridge, a member of the AEA and the first army officer to pilot an aeroplane (the AEAs Drome No. 1 or *Red Wing*), asked on September 17 to be Orville's passenger "for the purpose of officially receiving instructions." After a few minutes in the air, the Wright *Flyer* nosed down and crashed, seriously injuring Orville and killing Selfridge. Standing about five hundred feet distant, Chanute watched the accident happen and saw one propeller blade snap off. Knowledgeable about wood, he later testified that the wood appeared to be brittle or overseasoned. Also knowing that Orville had installed a longer propeller to achieve higher speed, Chanute thought that the blade might have struck the upper guy wire of the rear rudder. Taking part in the investigation by courtesy,[135] Chanute explained that the cause of the accident was not faulty airplane design. Once the cause had been confirmed, he suggested that Katharine Wright request a contract extension. Orville flew an improved *Flyer* as part of the acceptance test the following year, and the military bought one airplane for $25,000, plus a $5,000 bonus.

On September 25, Selfridge was laid to rest in the National Cemetery at Arlington, Virginia, with an impressive military funeral. The honorary pallbearers represented the army and the navy, the AEA, and the Aero Club, with Chanute heading the latter group. In the following weeks, Chanute, as friend and ever-ready adviser, gave personal support and mental comfort to Orville in the hospital and to Katharine, but also to Wilbur, who was still flying in France. One month after Orville's accident, Herring made "technical delivery" of his airplane, but could not deliver an airplane; he was not an aeronautical contender anymore.

To compare the relative merits of the monoplane and the multiplane, Chanute assembled a listing of "First Steps in Aviation and Memorable Flights"[136] of various aircraft designs, listing Jean-Marie Le Bris as the first inventor to "warp the wings" and fly more than six hundred feet. The list continued with four entries for the Wrights, followed by the efforts of European pilots, but for some reason he did not include the one-mile public flight by Curtis half a year earlier. Chanute also stated that experimenters had achieved much success with biplanes, frequently called the "Chanute-type" aeroplane.

Aviation was a popular spectacle and the first "practical use" of the airplane was to provide thrilling exhibitions; aviators from around the world competed for fame and cash prizes, making fortunes while risking their lives. Every month someone flew higher, flew farther, flew faster, stayed up longer. "It all goes to show," Cortlandt Bishop reportedly said after Curtiss won the second leg of the *Scientific American* trophy, "that we are still in the experimental stage. What has been demonstrated is that a man can fly in a heavier than air machine. The perfect type of machine has not yet been devised."[137] Then the news flashed over the wires that Louis Blériot had crossed the English Channel in a monoplane on July 25, 1909, flying at nearly a mile a minute. Receiving many duplicate clippings, Chanute forwarded some with his dry comment to Glassford: "Now that Blériot has 'invaded England by his lonesome' the disquisitions will get more and more numerous." The big air show in Rheims, France, created more headlines, because it was a spectacle never before witnessed. The three-month-long ILA, the Internationale Luftschiffahrt-Ausstellung (International Aeronautic Exhibition) in Frankfurt, Germany, in the summer of 1909, included not only airplanes and airships but also featured a gliding competition.[138] Several Chanute-type gliders, built by August Euler and Oscar Ursinus, competed, and the longest flight reportedly covered more than 160 feet in distance.

Not all went well in the aviation business, with accidents happening much too often. In September 1909 Captain Ferdinand Ferber crashed fatally flying a Voisin airplane in France, and Chanute was saddened to report on the death of his friend in *Aeronautics*.[139]

Celebrating the centennial of St. Louis early in October 1909, promoters wondered if its citizens a hundred years later might consider "today's airships as amateurish as the poled flatboat of 1809 seemed to the present generation." Chanute discussed "Aviation in 2009" with a reporter from the *St. Louis Post-Dispatch*. He believed that the dirigible had probably reached its limits and that airplanes could not be built to carry more than five or six people, but he also thought that the next development should be motorless flight using updrafts, "as the soaring birds, without the action of motors, save that of the wind. That will only be practical in certain regions and under certain circumstances where there is an ascending trend in the wind, for it is those ascending winds produced by irregularities in the earth's surface or by the rising of heated air, which is utilized by the birds. I think as soon as man begins to utilize the wind and air currents, there will be another great step made in the development of flying."[140]

The Chicago press discussed airplanes and flying activities from around the country regularly, but most Chicagoans had not seen an airplane fly over their city, because it was costly to bring famous aviators to town. After flying his new Hudson-Fulton machine in St. Louis, Glenn Curtiss and Tom Baldwin decided to visit Chicago and stage a privately sponsored air show without the inducement of cash prizes.[141] On October 16, 1909, after the motorcycle race at the Hawthorne Race Track in Stickney, a western suburb of Chicago, Curtiss flew his airplane and Baldwin his airship; Chanute sat with one of his daughters in the grandstand. "After the engine was started, Curtiss leaped into his seat, pulled the lever, and the machine slid forward. After traversing about 200 feet, it rose smoothly into the air, followed by a roar of cheers as the strange bird passed almost directly over the crowd. The first flight covered only about half a mile at a height of about 35 feet, but the spectators were cheering."[142] Chicago lived up to its reputation as the "Windy City," or as Chanute stated, "the usual Chicago half gale prevailed." During the three-day event, Curtiss made three flights in high winds, flying over the whole track and startling the crowd "with dips and tilts."[143] These airplane flights were the first recorded in Illinois and launched Chicago into the air age. The meet was a financial loss for the two aviators, but on the plus side, Curtiss and his wife Lena spent two evenings with Chanute, talking about aviation and the control mechanism he employed on his airplane.[144]

Unfortunately, Curtiss could not stay another day and attend the WSE meeting at which Chanute discussed "Recent Progress in Aviation." About 150 members and guests attended, in spite of inclement weather. WSE president Allen introduced the speaker: "It is a remarkable instance that just twelve years ago this evening, 20 October 1897, Mr. Chanute gave his first paper before this society on the subject of aviation, entitled 'Gliding Experiments.' . . . The opportunity comes to few men to appear before the same body twelve years after their predictions had been made, and be able to point to the fulfillment of those predictions, as can be done by Mr. Chanute tonight."[145] Chanute talked about his own gliding experiments with Avery and Herring flying in 1896 and 1897, followed by Maxim's work of 1894, Langley's activities between 1896 and 1903, the work of the Wrights, Dumont, Delagrange, Farman, Blériot, Bell, Curtiss, and others, bringing the subject completely up to date. In closing, Chanute showed a film of the events at Rheims that Curtiss had just given him. The WSE Awards Committee later awarded the "Chanute Medal" (see "The Western Society of Engineers in 1901" in chapter 6) to Chanute, the originator of the award, for the best-researched engineering paper in 1909.

Receiving books from authors, Chanute still enjoyed writing book reviews, including for Hildebrandt: *Airships, Past and Present*;[146] Maxim: *Artificial and Natural Flight*;[147] and Lougheed: *Vehicles of the Air*.[148] Using his extensive personal library, he also assisted Paul Brockett of the Smithsonian Institution in compiling the 940-page *Bibliography of Aeronautics*, published in 1910.[149]

For some time, James Means had contemplated a literary project to document how the glider development led to the invention of the airplane. This project partially materialized with the *Epitome*, appearing in June 1910, in which Means republished certain articles from his three *Annuals* and added new ones, including Chanute's "Soaring Flight." Continuous devotion to encouraging aeronautics through documentation secured Means solid respect among promoters of aeronautics.

Several people had urged Chanute to update his *Progress in Flying Machines*, but he was not inclined to spend effort on this project. Late in 1909, William Jackman, the former editor of the *Chicago Journal* who had visited Chanute's 1896 flying camp, assembled information for a how-to book on *Flying Machines, Construction & Operation*. Chanute gladly gave permission to include his article "Evolution of the Two-Surface Machine." Seeing the proof pages in late April 1910, the seventy-eight-year-old engineer walked to the publisher's office to offered suggestions; Chanute's article "Soaring Flight" became chapter 20.

The book, bound in red leather, sold for $1.50 and for $1.00 bound in cloth. Because it quickly sold out, an expanded edition was published two years later. In the second edition, Jackman acknowledged: "All this was 'a labor of love' on Mr. Chanute's part. He gave of his time and talents freely because he was enthusiastic in the cause of aviation, and because he knew the authors of this book and desired to give them material aid in the preparation of the work—a favor that was most sincerely appreciated."[150] This was Chanute's last literary contribution in the interest of aviation.

The Wright Controversy: Differing Philosophies

The mostly handwritten letters between Chanute and Wilbur or Orville Wright shed light on their colorful personalities, but also on how harmony waned. In the beginning, the Wrights generally confided in Chanute whenever they ran into problems; in later years, they used him to distribute select information to the press or requested his help to sell their "baby." Usually, Chanute did as asked, but he began to notice a growing reticence on the part of the Wrights to share information. At one time, Chanute told Spratt: "They are not the first young men I have helped into a fortune who have shown anxiety to forget me when they have seen it coming."[151]

With a curious public and an ever-hungry appetite for sensation by the media, news of the Wrights flying filtered through the press, and enterprising reporters used the readily available photos of Chanute's gliders or Langley's aerodrome to highlight their articles. The brothers did not appreciate seeing their name in the caption of another man's plane, but to the general public, all airplanes looked alike.

Chanute's trip to Europe in the spring of 1903 promoted the World's Fair in St. Louis, and he discussed the aeronautical activities in the United States in many lectures. At the same time, the Wrights had applied for their first patent in the United States and then submitted their claims to European agencies. The German Patent Office held up their application in 1905 and nullified the main claim in March 1912, stating that the information had been previously published which was sufficient to "teach anyone to fly."[152] The decision was later overridden, in spite of the fact that the German Patent Office in the years prior to World War I usually obstructed foreigners from taking out patents, often for nationalistic reasons.

Lonely after losing his wife, family members, and friends, Chanute enjoyed the company of a younger generation of flying machine experimenters. He liked talking about flying gliders of the past and appreciated occasional recognition for his accomplishments. Chanute, well into his 70s, was beginning to show his age. At that time the Wrights were at the top of their aeronautical careers and proud of their accomplishments, and they did not appreciate reading in the press that their knowledge came from their former mentor.

The aeronautic community stood firmly behind Chanute, the "Grand Old Man" of aeronautics. Graham Bell gave what appeared to be the public's opinion in a talk at the Washington Academy of Sciences in March 1907: "In all of the accounts which I have seen of the experiments of the Wright brothers, no mention has been made of the fact that the success of the Wrights has been built on the valuable work of Mr. Chanute, who for years carried on work in construction and testing of gliding machines, and who I understand furnished the Wright brothers with the design for their first gliding machine. There is perhaps no one who has made a closer study and has a more thorough understanding of the whole subject of aerodromics than Mr. Chanute, and I should like very much to see him given due credit for the very important work he has done."[153]

There was also a difference of opinion between Chanute and the Wrights about stability and the appropriate time to add power to obtain sustained flight. Chanute believed that the airplane should be eminently stable and self-righting, preferably using some automatic device, prior to adding power or elaborate directional controls. The former bicycle makers believed equally in safety, but felt that their newly invented control system overcame the need for a complex stabilizing mechanism; they applied power soon after their three-axis controls seemed to work. In the end, both parties probably agreed that a successful aeroplane needed both stability and control.

Besides being disturbed by what Chanute thought were plain signs that the Wrights were pursuing financial reward over aeronautical progress, he also worried about them losing their flying skills. He believed that he always tried to give the Wrights sound, objective advice and was deeply concerned that the brothers, complacent in their original success and involved in pointless (in Chanute's mind) litigation, were losing the race of aviation development, a recurring theme in their correspondence. "I still differ with you as to the possibility of your being caught up, if you rest on your oars. It is practice, practice, practice which tells and the other fellows are getting it," Chanute wrote in November 1906.[154] Wilbur responded by return mail: "Fear that others will produce a machine capable of practical service in less than several years does not worry us;"[155] he was confident that they would make their fortune in due time.

In January 1909, Katharine and the convalescing Orville joined Wilbur in Europe, demonstrating and selling their *Flyer*. Shortly after returning to America later in the spring, Wilbur shared their experiences with Chanute. Their wealthy former mentor replied that he "rejoiced over their triumphs" and wrote, probably with a twinkle in his eyes, that he "was particularly gratified with the sensible and modest way in which you accepted your honors. It encourages the hope that you will still speak to me when you become millionaires."[156]

In December 1908 the Smithsonian Institution established the Langley Medal, for "meritorious investigations" in the science of aeronautics and its application to aviation. Chanute accepted the chairmanship of the awards committee and asked Bell, Means, George Squier, and John Brashear to join him. But then Chanute learned that the award was to "recognize results obtained," so he replied to the secretary of the Smithsonian: "The results obtained by Wright Brothers are far superior to any others, but the resolution mentions 'meritorious investigations,' not achievements, and this seems to imply that those investigations should be given to the world, a thing which the Wrights have hitherto declined to do."[157] Committee members overrode Chanute's opinion and nominated the Wrights, who received the award on February 10, 1910; Graham Bell spoke for the committee[158] and highlighted Langley's contributions.

After fulfilling their contract with the Signal Corps in August 1909, Orville went to Europe to demonstrate their *Flyer* and vented his frustration in a letter to his brother Wilbur back in America. "Daily Mail of London and the Paris New York Herald are very much against us and give us a slap when ever possible. . . . I think it would be a good plan to give out an interview in which notice is made of suing all who have any connection with infringing machines."[159] The news of a patent fight went through the press like wildfire.

At the Aeronautic Society meeting in December 1909, Spratt discussed wing shapes found in nature and expressed his "gratitude for much encouragement and help received through many years from that kind-hearted, patient, and earnest worker and supporter of aeronautics, Mr. O. Chanute."[160] Spratt also wanted credit for sharing his aerodynamic findings and agreed to testify in the ongoing patent suits regarding the evolution of the Wright airplane. Hearing this from Spratt, Chanute replied that he hoped not to be asked.

As he did in the past, Chanute sent holiday greetings to his friends and wrote to the Wrights that he was sorry for not having seen them fly in the past year, but at almost seventy-eight years old, he felt less inclined to travel. Wilbur responded that he and Orville would attend the AAAS dinner in Chanute's honor in January and that they looked forward to seeing him.

About 130 Boston men attended the AAAS dinner in early January 1910, including the Wrights and Willis Moore of the Weather Bureau. Wilbur talked briefly, also mentioning their former mentor. The aging Chanute's unhappiness with the Wrights peaked at this event.

A week later, the *New York World* published an interview with Chanute, quoting him saying, "Wilbur and Orville Wright have brought to fruition the efforts of others in the field of aeronautics. They are justly entitled to a reward for the things that they themselves have actually performed." He recalled that the Wrights had stopped flying to keep their machine secret, but others had duplicated their efforts. "As to the wing-warping patent under which they demand damages from others for alleged infringement, this is not an original idea with them. On the contrary, many inventors have worked on it from the time of Leonardo da Vinci. Two or three have actually accomplished short glides with the basic warping idea embodied in their machines."[161]

Reading the report, Wilbur thought his mentor's advancing age had made him forget the facts. He wrote an angry letter and mailed it with the clipping to Chanute's old address, from where it was forwarded. Chanute read the letter and considered not replying; but he then pulled his letterpress books and read his early letters to Wilbur and then wrote probably the harshest letter he had ever written. To him, the write-up was as inaccurate as newspaper accounts usually were; instead of discussing its contents, he took up the principal issues. "I did tell you in 1901 that the mechanism by which your surfaces were warped was original with yourselves. This I adhere to, but it does not follow that it covers the general principle of warping or twisting wings; the proposal for doing this being ancient. . . . If the courts will decide that you were the first to conceive the twisting of the wings, so much the better for you, but my judgment is that you will be restricted to the particular method by which you do it."[162]

Being unhappy with Wilbur, who Chanute thought should show a little more gratitude for the help he had received in the past, Chanute continued:

> I am afraid, my friend, that your usual sound judgment has been warped by the desire for great wealth. If my opinions form a grievance in your mind, I am sorry, but this brings me to say that I also have a little grievance against you. In your speech at the Boston dinner on 12 January you began by saying that I "turned up" at your shop in Dayton in 1901 and that you then invited me to your camp. This conveyed the impression that I thrust myself upon you at that time and it omitted to state that you were the first to write to me in 1900, asking for information which was gladly furnished, that many letters passed between us, and that both in 1900 and 1901 you had written me to invite me to visit you, before I "turned up" in 1901. This has grated upon me ever since that dinner and I hope that in the future, you will not give out the impression that I was the first to seek your acquaintance, or pay me left handed compliments."[163]

Writing the letter made Chanute feel a little better.

Wilbur vented his frustration with another lengthy letter; he believed that the world owed them something as inventors. Neither brother appreciated hearing that they had obtained their first experiences on a Chanute machine and that they were his pupils. He saw nothing where their former mentor had contributed to their success and wanted a public correction of the widespread impression about their relationship.[164] This time the aging engineer did not respond, but he mentioned Wilbur's violent letter to Spratt and that he was reluctant to "engage in a row."[165]

Not having received a reply from Chanute, Wilbur composed a kinder letter late in April: "My brother and I do not form many intimate friendships and do not lightly give them up. I believe that unless we could understand exactly how you felt, and you could understand how we felt, our friendship would tend to grow weaker instead of stronger. . . . It is our wish that anything, which might cause bitterness, should be eradicated as soon as possible. If we discuss matters in this spirit I believe all misunderstandings can be removed."[166] As the increasingly frail Chanute prepared for his last trip to Europe, he was gratified to receive the letter, and wrote: "I hope, upon my return from Europe, that we will be able to resume our former relations." But his clock was running out; this was his last letter to Wilbur.

Historians, including Tom Crouch, and aerospace engineer John Anderson, have discussed the controversial history of wing warping, and the attitudes between the pioneers, at length. A good summary of the issue appears in *Introduction to Flight* by Anderson: "The history of ailerons is steeped in history and controversy. . . . Ideas of warping the wings or inserting vertical surfaces at the wing tips cropped up several times during the late 19th century and into the first decade of the 20th century, but always in the context of a braking surface that would slow one wing down and pivot the airplane about a vertical axis. . . . The Wright brothers' claim that they were the first to invent wing warping may not be historically precise, but clearly they were the first to demonstrate its function and to obtain a legally enforced patent on its use, combined with simultaneous rudder action for total control of banking."[167]

In their philosophical disagreements, both Chanute and the Wrights had vented their pent-up feelings and it was time to renew their former friendship. This so-called controversy had built up over time; it was a part of human relations, age difference, and egos, but in their final letters, both parties expressed their wishes to reestablish their former relationship.

A Life of Achievements Closes

Chanute arrived with his three daughters in Dover, England, on May 23, 1910, and heard that Jacques de Lesseps, whose father he had met decades earlier, had made the second crossing of the English Channel in a *Blériot* monoplane just two days earlier. How he wished he could have been at the landing site!

After spending a couple of weeks in Paris, the Chanutes continued their trip to Carlsbad in the Austro-Hungarian Empire. In late June Chanute came down with pneumonia and was transferred to the American Hospital in Paris; Lizzie canceled all her father's future appointments.

Three days later, the Aeronautical Society of Great Britain voted unanimously to award their Gold Medal to Chanute.[168] Not knowing how to reach him, the secretary contacted the American Embassy in Paris, who provided the hospital's address. Receiving the letter, Lizzie responded that her father hoped to acknowledge the touching inquiries of his friends soon. Because Chanute would not be able to receive the honor personally, the society shipped their medal to Paris.

By late September Chanute had regained enough strength to return home, and on October 1 the Chanutes boarded the French liner *La Savoie* at Le Havre. Even though Chanute was barely strong enough for the ocean voyage, it did not take him long to meet some fellow passengers, aeronauts who were en route to America and talk about aviation.[169] One of them was John Moisant, originally from Chicago, who talked about the biplane glider he had built years earlier and the *Blériot* monoplane he had just flown from Paris to London. Chanute knew that he had laid the groundwork for their aeronautical accomplishments and after ages of yearning, effort, and ridicule, "we have at last learned to fly" was his reported comment.

Back home in Chicago, Chanute started to feel a little better. Avery was a regular visitor, talking of old times, Jackman dropped off his book *Flying Machines*, and Means mailed the *Epitome*, all of which pleased the aging engineer. With a certain inner satisfaction he also admired the Gold Medal from the Aeronautical Society of Great Britain that finally reached him.

Chanute's shining light went out the day before Thanksgiving, on November 23, 1910. The long career of the seventy-eight-year-old civil engineer ended in the sleep that knows no awakening. Charley reported for the family that his father suffered no pain and was conscious a few minutes before dissolution took place. "His poor body was so worn out that when he fell asleep the machinery simply ran down, and he left us."[170] Funeral services were held two days later at the Chanute home on North Dearborn, and a number of prominent engineers and associates from around the country came to Chicago to pay their last respects. Bill Avery, William Glassford, and Wilbur Wright were three of his closest aeronautical friends in attendance.[171] Octave Chanute was buried beside his wife on December 2 in the James and Chanute family plot at Springdale Cemetery in Peoria, Illinois, with Avery acting as ceremonial pallbearer.

What Others Said

Chanute's long association with transportation had come to an end, leaving a lasting legacy.

James Means wrote in his eulogy: "Those who knew him will always remember his loveable character and will think of the oft-repeated saying: He was more willing to give credit to others than to claim any for himself. We may well believe that whenever in the future the history of aviation shall be reviewed, the name Chanute will stand forth as that of one of the great founders."[172]

After attending Chanute's memorial service in Chicago, Wilbur Wright wrote to the Signal Corps' Captain William Glassford that Chanute's influence would never be eliminated from airplane design.[173] "His books and correspondence have inspired and encouraged others to action, especially in France and America, to such an extent that I think no one of the older workers can justly claim to have influenced progress in the art more powerfully than he." Possibly thinking back to his pivotal talk to the Western Society of Engineers in Chicago, Wilbur stated in his eulogy that Chanute's labors "had vast influence in bringing about the era of human flight. His 'double-deck' modification of the old Wenham and Stringfellow machines will influence flying machine design so long as flying machines are made. . . . No one was too humble to receive a share of his time. In patience and goodness of heart he has rarely been surpassed. Few men were more universally respected and loved."[174]

Adding dimension to the personal qualities that guided Chanute, members of the Aero Club of America, the Aeronautic Society of New York, the American Societies of Civil Engineers, of Mechanical Engineers, of Electrical Engineers, and the New York Electrical Society, honored Chanute on January 6, 1911. Hudson Maxim, brother of Sir Hiram Maxim and president of the Aeronautic Society, stated eloquently:

> Octave Chanute was the veritable father of aviation, and he was always, in all things, a vedette of progress. He was one of those whom duty send far to the fore, where, unaided and lonesome, they make their landmarks, beckon their fellows to follow, and move further on; and, when the world comes up, then, and not until then, are their landmarks seen to be true and their labor found to be worth while. . . . Chanute was one of those rare intellectual giants big enough and generous enough to endow other inventors and workers with his knowledge and to lend a hand to help them utilize it, and all without jealousy or envy. Aviation was a thing dearer to his heart than any self-greatening. It is gratifying to know that before his lamp of life went out he had the satisfaction of seeing fairly accomplished that master achievement which he had so generously patronized and for which he had so long labored and prayed—the conquest of the air. . . . Always it has been a devoted few who have stood in the vanguard and fought the hard fight of progress. One of such few was Octave Chanute. He belonged to the true nobility of brains.[175]

THE END.

NOTES

Chapter 1. The Formative Years

1. Advertisement, "New York and Havre Packets," *The Albion* 6, no. 50 (Dec. 15, 1838): 400.

2. Leo Schelbert and H. Rappolt, *Alles ist ganz anders hier. Auswandererschicksale in Briefen aus zwei Jahrhunderten* (Olten, Switzerland: Walter-Verlag, 1977).

3. Etienne Gérin, letter to O. Chanute, dated July 17, 1862, Chanute Family Papers.

4. "Report & Manifest of passengers taking on board the Ship *Havre* at Havre in France & bound for New Orleans," 1838. New Orleans Ship Lists (1820–1859), arriving into New Orleans, Louisiana, microfilm #1231, Newberry Library, Chicago.

5. Edwind W. Fay, *History of Education in Louisiana* (Washington, D.C.: Government Printing Office, 1898).

6. Earl F. Niehaus, "Jefferson College, The Early Years," *Louisiana Historical Quarterly* 38, no. 4 (Oct. 1955): 63–89.

7. Fay, *History of Education in Louisiana*.

8. Henry Rightor, *Standard History of New Orleans, Louisiana* (Chicago: Lewis Publishing, 1900).

9. Elizabeth Chanute, "My Father," 1912, Chanute Family Papers.

10. "New Orleans Passenger and Immigration Lists (1820–1850)," http://www.ancestry.com, accessed Feb. 2004.

11. Joseph Chanut, letter to Emilie E. Fourchy, dated Feb. 1860, Chanute Family Papers.

12. Chanute Family Papers.

13. Joseph Chanut, *Histoire de France, depuis la mort de Louis XVI, jusqu'a la révolution de juillet 1830, inclusivement* (Paris: Suite et continuation d'Anquetil, 1830).

14. Chanute Family Papers.

15. "Traveling Scenes—Air Line Roads," *Chicago Daily Tribune*, June 3, 1853.

16. Wellington Williams, *The Traveler's and Tourist's Guide through the United States of America, Canada, etc.* (Philadelphia: Lippincott, Grambo, 1851).

17. "Wire Suspension Bridge over the Monongahela at Pittsburgh," *American Railroad Journal* 2, no. 24 (June 13, 1846): 376–79.

18. Solomon W. Roberts, *Account of the Portage Rail Road over the Allegheny Mountain in Pennsylvania* (Philadelphia: Nathan Kite, 1836). Abstracted and republished in *The Pennsylvania Magazine of History and Biography* as "Reminiscences of the First Railroad over the Allegheny Mountain," 2, no. 4 (Apr. 8, 1878), 370–93.

19. Charles F. Carter, *When railroads were new* (New York: Henry Holt, 1909).

20. Roberts, *Account of the Portage Rail Road*.

21. John C. Smith, *Illustrated hand-book: a new guide for travelers through the United States of America* (New York: Sherman & Smith, 1847).

22. Philip Nicklin, "Trip on the Allegheny Portage Railroad from Johnstown to Hollidaysburg via the Allegheny Portage Railroad," 1835, http://www.explorepahistory.com/odocument.php?docId=205, accessed Jan. 2005.

23. Charles Dickens, *American Notes and The Uncommercial Traveler* (Philadelphia: T. B. Peterson & Bros., 1842).

24. Theodore L. Condron, "Octave Chanute—Versatile Engineer and Aviation Pioneer," *Journal of the Western Society of Engineers* 50, no. 4 (Dec. 1945): 158–66.

25. Rossiter Johnson, "Coudert, Frederick Rene," in *The Twentieth Century Biographical Dictionary of Notable Americans* (Boston: Biographical Society, 1904).

26. "Instruction. Charles Coudert's Lyceum," *New York Times*, Sept. 5, 1851.

27. Editorial, "Railroad System of the United States," *Merchants' Magazine* 18, no. 1 (Jan. 1848): 98–99.

Chapter 2. The University of Experience

1. Alice Chanute Boyd, "Ancestry of Octave Chanute," 1912, Chanute Family Papers.

2. Board of Directors, Report, "Members who died in 1874–5: Henry A. Gardner." *American Society of Civil Engineers, Proceedings* 1, no. 12 (Dec. 1875): 335–37.

3. John B. Jervis, "Hudson River Railroad, a sketch of its history and prospective influence on the railway movement," *Merchants' Magazine* 22, no. 3 (Mar. 1850): 278–89.

4. Thomas C. Meyer, "The New Hamburg Tunnel on the Hudson River Railroad," *Hunt's Merchants' Magazine* 26, no. 1 (Jan. 1852): 111.

5. Richard P. Morgan, "Western Society of Engineers Annual Meeting," *Journal of the Association of Engineering Societies* 9, no. 3 (Jan. 8, 1890): 140–41.

6. Chanute, Good-Bye letter to Joseph Chanut, his father, Dec. 7, 1850, Chanute Family Papers.

7. "Hudson River Railroad. Opening of the road to Albany," *New York Times*, Oct. 9, 1851.

8. "What railroads are doing," *Brooklyn Eagle*, Jan. 15, 1850.

9. Octave Chanute, "President American Society of Civil Engineers," *Engineering News* 25, no. 23 (May 23, 1891): 496, full page portrait facing page 496.

10. State and Federal Governmental Records in the Illinois State Archives, http://www.sos.state.il.us/departments/archives/serv-sta.html, accessed 2005–2007.

11. Paul W. Gates, *The Illinois Central Railroad and its colonization work*, Series: *Harvard Economic Studies*, vol. 42 (Cambridge, Mass.: Harvard University Press, 1934).

12. Thomas C. Clarke, "The Building of a Railway," *Scribner's Magazine* 3, no. 6 (June 1888): 643–71.

13. Federal Census Records, Illinois, 1860, http://www.census-online.com/links/IL/1860.html, accessed May 2007.

14. Laws of the State of Illinois, "Chicago & Mississippi Railroad," in *Passed by the 17th General Assembly, Second Session* (Springfield, Ill.: Lanphier & Walker, 1852), 46–49.

15. Gene V. Glendinning, *Chicago & Alton Railroad, the only way* (De Kalb: Northern Illinois University Press, 2002).

16. *The History of Livingston County, Illinois* (Chicago: Wm. Le Baron Jr., 1878).

17. *Livingston County Deed Records*, 1853–67. Courthouse, Pontiac, Livingston County, Illinois.

18. Chanute Family Papers.

19. "Immigration Certificate, Nat. Rec.Pg.13," Apr. 17, 1854. Naturalization Service, McLean County Circuit Court, Illinois.

20. Alice Chanute Boyd, "Ancestry of Octave Chanute."

21. Ernest East, letter to Elizabeth Chanute, dated May 25, 1940, Chanute Papers, Peoria Public Library, Peoria, Illinois.

22. Newton Bateman and Paul Selby, ed., *Historical Encyclopedia of Illinois and History of Peoria County* (Chicago and Peoria: Munsell, 1900).

23. Abraham Lincoln, "Speech at Peoria, Illinois, in reply to Senator Douglas, Oct. 16, 1854," in John G. Nicolay and John Hay, ed., *Complete Works of Abraham Lincoln*, vol. 2 (New York: The Lamb Publishing Company, 1894), 190–262.

24. Chanute Family Papers.

25. Benjamin C. Neff, personal correspondence on Richard Fuller James, a first cousin of Annie Riddell James, 2008–2009, Lincoln, Nebraska.

26. Kettelle, Charles, "Rites of Marriage," 1857, Peoria, Illinois, County Court House, Record No. 3919.

27. Patricia Goitein, personal correspondence on the history of Peoria and its citizens, Civil War matters and slavery, 2000–2008, Peoria, Illinois.

28. Joseph Chanut, letter to Annie James Chanute, dated Feb. 18, 1858, Chanute Family Papers.

29. Chanute Family Papers.

30. "The 'Foam'," *Peoria Daily Transcript*, June 8, 1858.

31. "Grand Fancy Dress-Ball," *Peoria Daily Transcript*, Jan. 1, 1859.

32. Chanute Family Papers.

33. Paul Stringham, *Toledo, Peoria & Western. Tried, Proven & Willing* (Peoria, Ill.: Deller Archive, 1993).

34. Laws of the State of Illinois, "Peoria and Oquawka Railroad," in *Passed by the 17th General Assembly, Second Session* (Springfield, Ill.: Lanphier & Walker, 1852), 193–95.

35. "Another 'Cut-Off'," *Chicago Tribune*, Mar. 8, 1853.

36. Thomas C. Cochran, *Railroad Leaders, 1845–1890*, Harvard University. Research Center in Entrepreneurial History (New York: Russell & Russel, 1965).

37. Chicago, Burlington & Quincy Rail Road Company, "Report of the Board of Directors, presented to the Stockholders at the Annual Meeting, June 20th, 1862." (Chicago: Dunlop, Sewell & Spalding, 1862).

38. Gates, *Illinois Central Railroad*.

39. *The Past and Present of Woodford County, Illinois* (Chicago: Wm. Le Baron Jr., 1878).

40. "Emigration Westward," *Chicago Tribune*, Mar. 22, 1856.

41. "Partial destruction of the Rock Island Bridge," *Chicago Tribune*, May 9, 1856.

42. John B. Jervis, "Report of John B. Jervis, Civil Engineer, in relation to the Railroad Bridge over the Mississippi River at Rock Island," (New York: Wm. C. Bryant, 1857).

43. Abraham Lincoln "Argument in the Rock Island Bridge Case, Sept. 24, 1857," in Nicolay and Hay, ed., *Complete Works of Abraham Lincoln*, 340–54 .

44. "No Verdict," editorial, *Chicago Tribune*, Sept. 26, 1857.

45. "Over the river," *Peoria Daily Transcript*, Apr. 6, 1857.

46. Octave Chanute, "The longest railway draw-bridge in the world. Letter to the Editor," *American Railway Review* 2, nos. 4 and 6 (Feb. 9 and 16, 1860): 69, 74, 85. Abstract in *American Railway Times*, Feb. 25, 1860, 75. *New York Times*, Mar. 1, 1860.

47. "Financial Matters in New York," *Chicago Tribune*, Aug. 28, 1857.

48. *History of Livingston County.*

49. *Livingston County Deed Records*, 1853–67. Courthouse, Pontiac, Livingston County, Illinois.

50. *History of Livingston County.*

51. Chanute, letter to Cruger and Secor, dated Feb. 24, 1862, Chanute Papers, LoC.

52. "Accident at the Peoria Bridge," in *Chicago Tribune*, Mar. 8, 1859.

53. Robert G. Ingersoll, "The Illinois River Packet Co. vs. The Peoria Bridge Association," 1865. Chanute Papers, Peoria Public Library, Peoria, Ill. Collected by Ernest East.

54. *Journal of the Senate of the Twenty-Seventh General Assembly of the State of Illinois*, Mar. 28, 1872 (Springfield: Illinois Journal Printing Office, 1872): 770–75.

55. "Sale of the Peoria & Oquawka Railroad," *Chicago Tribune*, July 1, 1859.

56. Hiram W. Beckwith, *History of Iroquois County, together with Historic Notes on the Northwest* (Chicago: H. H. Hill, 1880).

57. "Opening of the Logansport, Peoria & Burlington Railroad," *Chicago Tribune*, Dec. 30, 1859.

58. Federal Census Records, Illinois, 1860, http://www.census-online.com/links/IL/1860.html, accessed May 2007.

59. Chanute Family Papers.

60. Chanute, letter to Samuel Gilman, dated Sept. 6, 1860, Chanute Papers, LoC.

61. Chanute, letter to William H. Cruger, dated Sept. 19, 1860, Chanute Papers, LoC.

62. Chanute, letter to E. Hiserodt, dated Sept. 18, 1860, Chanute Papers, LoC.

63. "Lord Renfrew in the field. The visit of the Prince of Wales and Suite to Dwight," *Chicago Tribune*, Sept. 29, 1860.

64. Alice Chanute Boyd, "Ancestry of Octave Chanute."

65. Chanute, letter to Charles A. Secor, dated Aug. 1, 1861, Chanute Papers, LoC.

66. Editorial, "What next? Flying," *Scientific American* 14, no. 3 (Sept. 25, 1858): 21.

67. Octave Chanute, "Experiments in Flying. An Account of the Author's own Inventions and Adventures," *McClure's Magazine* 15, no. 2 (June 1900): 127–33.

68. Editorial, "Flying," *Scientific American* 8, no. 25 (Mar. 5, 1853): 194.

69. Ainé N. Bescherelle, *Histoire des Ballons et des Locomotives Aériennes, depuis Dédale jusqu'a Petin.* (Paris: Marescq et Compagnie chez Gustave Havard, 1852).

70. David Young, *Chicago Aviation, an Illustrated History* (De Kalb: Northern Illinois University Press, 2003).

71. "Balloon Ascension," *Peoria Weekly Republican*, July 25 and Aug. 1, 1856.

72. "Balloon Ascension," *Illinois Gazette* (Lacon), Aug. 2, 1856.

73. Patricia Goitein, personal correspondence.

74. *The war of the rebellion: a compilation of the official records of the Union and Confederate Armies.* Prepared under the direction of the secretary of war, by BVT. Lieut. Col. Robert N. Scott, Third U.S. Artillery and published pursuant to an act of Congress approved June 16, 1880. Series 1, vol. 7 (Washington, D.C.: Government Printing Office, 1882).

75. Alice Chanute Boyd, "Some Memories of my Father," Chanute Family Papers, containing an account of his trip to New Orleans (1915, 1918).

76. *The war of the rebellion.*

77. "Various Southern Items," *Peoria Daily Transcript*, Dec. 15, 1861.

78. "Voyage a la Nouvelle-Orléans et Condition Actuelle du Sud," *L'Opinion Nationale*, Jan. 17 and 18, 1862, signed E. Fauchet. Albert Zahm had the French language newspaper article translated ("A voyage to New Orleans and the actual condition of the south"). Chanute Family Papers.

79. Cochran, *Railroad Leaders.*

80. John B. Jervis, *Railway property. A treatise on the construction and management of railways* (New York: Phinney, Blakeman & Mason, 1861).

81. Octave Chanute, "The Weight of Rails and the Breaking of Iron Rails," *American Society of Civil Engineers, Transactions* 3, no. 8 (1874): 111–17.

82. Chanute Family Papers.

83. Chanute Family Papers.

84. Chanute Family Papers.

85. Octave Chanute, "Best means of supplying the city of Peoria with water," Peoria City Clerk Report contained in 352.1, 1862–1871. Peoria, Illinois, Apr. 11, 1864, 5–17.

86. Glendinning, *Chicago & Alton Railroad.*

87. "Chicago & Alton Railroad," *Chicago Tribune.*

88. Octave Chanute, "Chief Engineer's Report," in *Second Annual Report of the President and Directors of Chicago & Alton Railroad Co. for the year ending Dec. 31, 1864* (Chicago: Evening Journal Book & Job Printing House, 1865).

89. *Chicago's First Half Century, the city as it was fifty years ago and as it is to-day* (Chicago: Inter Ocean Publishing, W. P. Dunn, 1883).

90. Elias Colbert and Everett Chamberlin, *Chicago and the Great Conflagration* (Chicago: J. S. Goodman, 1872).

91. Octave Chanute, "Chief Engineer's Report," in *Third Annual Report of the President and Directors of Chicago & Alton Railroad Co. for the year ending Dec. 31, 1865* (Chicago: Evening Journal Book and Job Printing House, 1866).

92. Chanute, letter to Alexander L. Holley, dated Nov. 10, 1866, Chanute Papers, LoC.

93. Chanute, "Chief Engineer's Report" (1866).

94. Octave Chanute, "Improvement in Repairing Railroad Rails" (No. 61,397), U.S. Patent Office, Jan. 22, 1867.

95. Chanute, letter to W. C. Dodge, dated Dec. 11, 1866, Chanute Papers, LoC.

96. Octave Chanute, "Steel and Steel-Topped Rails. Letter to the Editor," *The Manufacturer and Builder* 2, no. 2 (Feb. 1870): 37–38. Also published in *American Railway Times*, 22, no. 13 (Mar. 26, 1870): 99.

97. Chanute, "Chief Engineer's Report" (1865).

98. Chanute, letter to Thomas C. Meyer, dated Mar. 22, 1865, Chanute Papers, LoC.

99. Colbert and Chamberlin, *Chicago and the Great Conflagration.*

100. Chanute, letter to Henry A. Gardner, dated Jan. 6, 1865, Chanute Papers, LoC.

101. "The Presidents' Funeral. Reception of the Remains of Abraham Lincoln in Chicago," *Chicago Tribune*, May 2, 1865.

102. "The President's Obsequies," *New York Times*, May 4, 1865.

103. Octave Chanute, "Note on repairing a bridge during a flood," *Civil Engineers' Club of the Northwest* no. 47, presented on Nov. 9, 1874

104. Chanute, letter to Thomas C. Meyer, dated Mar. 19, 1866, Chanute Papers, LoC.

105. Jack Wing, *The Great Union Stock Yards of Chicago* (Chicago: Religio-Philosophical Publishing Association, 1865).

106. "Union Stock Yard and Transit Company," *Chicago Tribune*, Jan. 13, 1865.

107. Chanute, letter to Thomas C. Meyer, dated May 6, 1865, Chanute Papers, LoC.

108. Chanute, letter to John Burnham, dated Feb. 15, 1867, Chanute Papers, LoC.

109. Wing, *Great Union Stock Yards.*

110. Chanute, "Chief Engineer's Report" (1866).

111. Louise C. Wade, *Chicago's Pride: the Stockyards, Packingtown and Environs in the Nineteenth Century* (Urbana and Chicago: University of Illinois Press, 1987).

112. Francis E. Griggs, North Andover, Massachusetts, personal correspondence with the author, 2008–2009.

113. Octave Chanute, *Plans for Bridge across the Mo. River at St. Charles*, Chanute Papers, LoC.

114. Chanute, letter to Henry G. Prout, dated Nov. 15, 1898, Chanute Papers, LoC.

115. *A bill to authorize and establish certain post roads*, 39th Cong., 1st sess. (Mar. 27, Apr. 19, Apr. 25, Apr. 26, and May 1, 1866): S. 236.

116. "Proceedings and Report of the Board of Civil Engineers convened at St. Louis, in August, 1867, to consider the subject of the construction of a rail and highway bridge across the Mississippi River at St. Louis" (St. Louis: George Knapp, 1867). William McAlpine delivered the opening address (pages 3–8). Report of the committee on superstructure and approaches, with Chanute being one of the six committee members (pages 47–54).

117. Gouverneur K. Warren, "Report on bridging the Mississippi River between Saint Paul, Minn., and St. Louis, Mo." Report No.69, 45th Congress, 2d Session (Washington, D.C.: Government Printing Office, 1878.)

118. Robert W. Jackson, *Rails across the Mississippi* (Urbana and Chicago: University of Illinois Press, 2001).

119. Chanute, letter to James Eads, dated June 27, 1868, Chanute Papers, LoC.

120. Editorial, "Illinois & Saint Louis Bridge," *Journal of the Franklin Institute* 59, no. 3 (Mar. 1870): 145–48.

121. U.S. Senate, "A bill to authorize and establish certain post roads," S.236.

122. "The Mississippi Bridge at Quincy, Illinois," *New York Times*, Aug. 5, 1867.

123. Thomas C. Clarke, *Account of the Iron Railway Bridge across the Mississippi River at Quincy, Illinois* (New York: D. Van Nostrand, 1869).

124. "Quincy Railroad Bridge," *Chicago Tribune*, Nov. 9, 1868. Abstract in *American Railroad Journal*, Nov. 1868: 1144; *Journal of the Franklin Institute*, Dec. 1868: 364.

125. Chanute, letter to James F. Joy, dated Dec. 15, 1866, Chanute Papers, LoC.

126. Octave Chanute, "Chief Engineer's Report," in *Fourth Annual Report of the President and Directors of Chicago & Alton Railroad Co. for the year ending Dec. 31, 1866* (Chicago: Evening Journal Book and Job Printing House, 1867).

Chapter 3. Opening the West

1. Dionysius Lardner, *A treatise on the new Art of Transport, its Management, Prospects, and Relations, Commercial, Financial, and Social* (London: Taylor, Walton & Maberly, 1850).

2. George S. Morison and A. Bourn, "James Frederic Joy, F. Am. Soc. C. E.," *American Society of Civil Engineers, Transactions* 37, no. 6 (June 1897): 575–77.

3. "Kansas City—Completion of the Bridge Across the Missouri," *Chicago Tribune*, July 11, 1869.

4. "Le Kansas et la Guerre Civile aux États-Unis," *La Presse*, Oct. 2, 1856 Joseph Chanut regularly supplied the French press with information on American affairs.

5. Octave Chanute and George S. Morison, *The Kansas City Bridge with an account of the regimen of the Missouri River, and a description of methods used for founding in that river* (New York: D. Van Nostrand, 1870). Abstract in *Van Nostrand's Eclectic Engineering Magazine* (Sept. 1870): 225–31 and (Oct. 1870): 404–12. Note: H. P. Wright, an investment banker from Kansas City, received Chanute's personal copy in early 1911; he donated the book in the 1940s to the university library. Special Collections, University of Missouri, Kansas City, holds it today (available from http://wt.diglib.ku.edu/ titlepages/rhe635.htm).

6. Octave Chanute, "George Shattuck Morison—A memoir," *Journal of the Western Society of Engineers* 9, no. 1 (January–Feb. 1904): 83–88.

7. William P. Anderson and John A. L. Waddell, "Memoir of Joseph Tomlinson," *Canadian Society of Civil Engineers, Transactions* 19, no. 2 (October–Dec. 1905): 321–25.

8. Chanute and Morison, *Kansas City Bridge*.

9. Louis W. Potts and George F. W. Hauck, "Frontier Bridge Building: The Hannibal Bridge at Kansas City, 1867–1869," *Missouri Historical Review* 89, no. 2 (Jan. 1995): 139–61.

10. William H. Miller, *History of Kansas City, together with a sketch of the commercial resources of the country with which it is surrounded* (Kansas City: Birdsall & Miller, in conjunction with the Board of Trade, 1881).

11. Chanute and Morison, *Kansas City Bridge*.

12. Chanute, letter to James F. Joy, dated July 19, 1867, in Henry B. Joy Historical Research Records, Bentley Historical Library, University of Michigan, Ann Arbor.

13. Chanute and Morison, *Kansas City Bridge*.

14. Octave Chanute, "Kansas City and Cameron Railway Bridge," *American Railway Times* 19, no. 40 (Oct. 5, 1867): 319; "The Cameron Railway Bridge at Kansas City," *Chicago Tribune*, Aug. 22, 1867.

15. Octave Chanute, "Pneumatic Bridge Foundations," *Journal of the Franklin Institute of the State of Pennsylvania for the Promotion of the Mechanic Arts* 85 and 86, nos. 6, 1, and 2 (June, July, and Aug. 1868): 387–91, 17–28, and 89–99.

16. Chanute, letter to H. M. Merriweather, dated Feb. 15, 1901, Chanute Papers, LoC.

17. Octave Chanute and George S. Morison, "Improved Dredging-Machine" (No. 98,848), U.S. Patent Office, Jan. 18, 1870.

18. Chanute, "Pneumatic Bridge Foundations," 89–99.

19. Charles R. Suter, "Western River Improvements: Report on the bridge over the Missouri at Kansas City, dated Nov. 2, 1868." Washington, D.C.: Government Printing Office, 1869: 303–8.

20. Chanute, letter to Thomas C. Clarke, dated Mar. 1, 1869, Chanute Papers, LoC.

21. Chanute, letter to J. L. Williams, dated Apr. 7, 1868, Chanute Papers, LoC.

22. Chanute, letter to I. W. Vaughan, dated May 18, 1870, Chanute Papers, LoC.

23. "Common Council Meeting. An Invitation to participate in a Bridge Celebration," *Chicago Tribune*, June 22, 1869.

24. Octave Chanute, "Opening of Kansas City Bridge. Engineer's Test Report," *Journal of the Franklin Institute* 58, no. 3 (Sept. 1869): 177–78. Abstract in *The Daily Journal*, Kansas City, July 6, 1869.

25. "Kansas City Bridge Celebration," Kansas City *Daily Journal of Commerce*, July 6, 1869.

26. "Kansas City Bridge Opening," *Kansas City Weekly Journal of Commerce*, July 10, 1869.

27. "The Kansas City Bridge. Origin and History of the Enterprise," *New York Times*, July 11, 1869. From *St. Joseph (Missouri) Herald*, July 3, 1869.

28. "The Railway and Road Bridge at Echuca," *The Argus*, May 1, 1878, Melbourne, Victoria, Australia.

29. Octave Chanute, "The Kansas City Bridge in '69," *Journal of the Western Society of Engineers* 5, no. 1 (January–Feb. 1900): 58–59.

30. Charles N. Glaab, *Kansas City and the Railroads* (Madison: State Historical Society of Wisconsin, 1962).

31. Starting in 1852, Joseph Chanut's biographies of statesmen, authors, and politicians were published in the subsequent forty-six volumes of the *Nouvelle biographie universelle depuis les temps les plus reculés jusqu'à nos jours* (Paris: Firmin Didot Fréres).

32. Chanute, letter to Thomas C. Meyer, dated Dec. 2, 1870, Chanute Papers, LoC.

33. Craig Miner, "Border Frontier: The Missouri River, Fort Scott & Gulf Railroad in the Cherokee Neutral Lands, 1868–1870," *Kansas Historical Quarterly* 35, no. 2 (Summer 1969): 105–29.

34. Oklahoma Historical Society, "Chronicles of Oklahoma," 1936, http://digital.library.okstate.edu/chronicles/bookshelf.html, accessed July 2007.

35. Charles J. Kappler, *Indian Affairs. Laws and Treaties*, vol. 2, "Treaty with the Cherokee," July 19, 1866, 942–48, Apr. 27, 1868, 996–97. Clerk to the Senate Committee on Indian Affairs (Washington, D.C.: Government Printing Office, 1904).

36. Paul W. Gates, *Fifty Million Acres: Conflicts over Kansas Land Policy, 1854–1890* (Ithaca, N.Y.: Cornell University Press, 1954).

37. "Completion of the Kansas City & Cameron Railroad," *Chicago Tribune*, Nov. 23 and 25, 1867.

38. Donald D. Banwart, *Rails, Rivalry and Romance, 1864–1980* (Fort Scott, Kans.: Historic Preservation Association of Bourbon County, 1982).

39. Kansas & Neosho Railroad Company, "Minutes of the Board of Directors' and Stockholders' meetings," Olathe, Kansas, July 30, 1868.

40. Glaab, *Kansas City and the Railroads*.

41. Mary Firth, *Memoir of Frank Russell Firth* (Boston: Lee and Shepard, 1873). Review in *Railroad Gazette*, July 13, 1872, 293.

42. Gates, *Fifty Million Acres*.

43. William Nicks, personal correspondence in regard to the founding of Lenexa, Kansas, 2002–2006.

44. Chanute, letter to J. H. Jackson, dated Mar. 27, 1871, Chanute Papers, LoC.

45. "The great flood. Great destruction of property," *The Border Sentinel*, July 28, 1869.

46. Chanute, letter to J. I. Weld, dated Mar. 1, 1869, Chanute Papers, LoC.

47. Chanute, letter to Thomas C. Meyer, dated Nov. 15, 1869, Chanute Papers, LoC.

48. Banwart, *Rails, Rivalry and Romance*.

49. Chanute, letter to S. Bartlett, dated Dec. 13, 1869, Chanute Papers, LoC.

50. Chicago, Burlington & Quincy Rail Road Company, "Report of the Board of Directors, presented to the Stockholders at the Annual Meeting, June 30, 1869" (Chicago: Rounds & James, Book and Job Printers, 1869).

51. "Fort Scott and Her Men," *Girard Weekly Press*, Nov. 25, 1869.

52. Miner, "Border Frontier."

53. "A Letter from James Joy," *Labette County Paper*, July 6, 1868.

54. Gates, *Fifty Million Acres*.

55. "The Neutral Lands," *Fort Scott Press*, July 11, 1869.

56. Lula L. Brown, "Cherokee Neutral Lands Controversy," master's thesis, Kansas State Teachers College of Pittsburg, 1923. Published in Paul W. Gates, "The Fruits of Land Speculation," Arno Press, 1979.

57. "From the Neutral Lands. Progress of the Survey. Engineer's Camp," *Chicago Tribune*, Aug. 13, 1869.

58. J. V. Hanna, "Report of Committee No. II—On Ballasting," in *American Railway Engineering and Maintenance-of-Way Association Proceedings*, 1904. Chanute submitted his comment on ballasting, see 484–85.

59. Chanute, letter to H. W. S. Cleveland, dated Dec. 8, 1869, Chanute Papers, LoC.

60. Missouri River, Fort Scott & Gulf Railroad Company, "Report to the Bondholders and Stockholders for the year ending Dec. 31, 1877" (Boston: Alfred Mudge & Son, 1878).

61. "A Letter from James Joy," *Baxter Springs Herald*, Aug. 10, 1869.

62. "Excursion," *Fort Scott Monitor*, Feb. 16, 1870.

63. "The Railroad," *Cherokee Sentinel*, Mar. 19, 1870.

64. "Correspondence from James F. Joy," *Girard Weekly Press*, Mar. 31, 1870.

65. "Col. Chanute," *Cherokee Sentinel*, Apr. 2, 1870.

66. "Kansas," in *The American Annual Cyclopedia and Register of important events of the year 1870* (New York: D. Appleton, 1870), 418–21.

67. V. V. Masterson, *The Katy Railroad and the last frontier* (Norman: University of Oklahoma Press, 1952).

68. George W. Martin, "The boundary lines of Kansas," Preprint of Address, to be published in Volume 11, September–Oct. 1909.

69. "Baxter Springs and the Railroad Celebration," *Girard Weekly Press*, May 19, 1870.

70. Masterson, *Katy Railroad*.

71. "Railroad Rights in the Indian Territory," *New York Times*, May 29, 1870.

72. "Railroad through the Indian Territory," *New York Times*, July 22, 1870.

73. Chanute, letter to George B. Chase, dated Jan. 2, 1873, Chanute Papers, LoC.

74. Octave Chanute, "Chief Engineer's Report," in *First Annual Report to the Bondholders and Stockholders of the Missouri River, Fort Scott & Gulf Railroad Company for the year ending Dec. 31, 1870* (Boston: Alfred Mudge & Son, 1871).

75. James H. Lyles, *Official Railway Manual of the Railroads of North America* (New York: Lindsay Walton, 1870–71).

76. Miner, "Border Frontier," 106.

77. Octave Chanute, "Report of the Leavenworth, Lawrence & Galveston Railroad," *Railroad Gazette* 16, no. 30 (July 27, 1872): 324–25.

78. I. E. Quastler, *The railroads of Lawrence, Kansas, 1854–1900: a case study in the causes and consequences of an unsuccessful American urban railroad program* (Lawrence, Kans.: Coronado Press, 1979).

79. Deborah Barker, personal correspondence on the Leavenworth, Laurence & Galveston Railroad, Ottawa, Kansas, Sept. 2009.

80. William G. Cutler, *History of the State of Kansas* (Chicago: A. T. Andreas, 1883).

81. Chanute, letter to W. I. Weld, dated Jan. 16, 1871, Chanute Papers, LoC.

82. "Railroad Lands in Kansas." *Merchants' Magazine* 62, no. 3 (Mar. 1870): 231.

83. John Smeaton, "An experimental enquiry concerning the natural powers of wind and water to turn mills and other machines, depending on a circular motion" (London: I. and J. Taylor, 1759). Chanute saw this pamphlet referenced in J. Bennett, "Aerometry," *Journal of the Franklin Institute* (May 1860): 317–20.

84 "A Destructive Storm," *Lawrence Journal*, Sept. 8, 1872.

85. James M. Walker, "Outgoing letters in four volumes of letterpress books (1871–1881)," Chicago, Burlington & Quincy Railroad Archives, Newberry Library, Chicago.

86. Chanute, letter to James M. Walker, dated July 4, 1871, Chanute Papers, LoC.

87. Chanute, letter to O. H. Gray, dated Dec. 15, 1872, Chanute Papers, LoC.

88. Robert E. Hosack, "Chanute—The Birth of a Town," *Kansas Historical Quarterly* 41, no. 4 (Winter 1975): 468–87.

89. Chanute, letter to John W. Scott, dated Nov. 30, 1872, Chanute Papers, LoC.

90. "General News Items," *Chicago Tribune*, Dec. 17, 1872.

91. Octave Chanute, "Greetings," in *The Chanute Library and its Friends* (Chanute, Kans.: Chanute Tribune Printers, 1907), 11–12.

92. Chanute, letter to Thomas C. Meyer, dated Aug. 26, 1872, Chanute Papers, LoC.

93. "Leavenworth, Lawrence & Galveston Railway," *American Railway Times*, June 24, 1871, 195. Reprinted from *Kansas City Journal of Commerce*.

94. Chanute, letter to James M. Walker, dated May 19, 1871, Chanute Papers, LoC.

95. Chanute, letter to James M. Walker, dated July 28, 1871, Chanute Papers, LoC.

96. Chanute, letter to Joseph Tomlinson, dated Mar. 8, 1871, Chanute Papers, LoC.

97. Chanute, letter to Edwin S. Bowen, dated Oct. 25, 1871, Chanute Papers, LoC.

98. Octave Chanute, "Premiums to Locomotive Engineers," *American Railway Times* 24, no. 7 (Feb. 17, 1872): 54. Abstract in *Brotherhood of Locomotive Engineers' Monthly Journal*, Mar. 1872, 103; *Scientific American*, Mar. 2, 1872, 146.

99. "Railway Passes," *Brotherhood of Locomotive Engineers' Monthly Journal* 6, no. 1 (Jan. 1872): 24–25.

100. R. S. Elliott, "Forest Planting on the Great Plains," in *Landscape Architecture*. (Chicago: Janse, McClurg, 1873), 133–35. From a letter by Chanute to Professor Hayden, dated Nov. 8, 1871. Abstract in *Galveston News*, Jan. 7, 1872.

101. Chanute, "Report of the Leavenworth, Lawrence & Galveston Railroad" (1872).

102. Chanute, letter to Charles F. Adams, dated July 18, 1871, Chanute Papers, LoC.

103. Thomas C. Cochran, *Railroad Leaders, 1845–1890*, Harvard University, Research Center in Entrepreneurial History (New York: Russell & Russel, 1965).

104. Chanute, letter to W. I. Weld, dated Sept. 19, 1870, Chanute Papers, LoC.

105. Chanute, letter to James F. Joy, dated Mar. 24, 1871, Chanute Papers, LoC.

106. "Burlington & Missouri River Railroad in Nebraska's System," in *Second Annual Report of the Board of Transportation for the Year ending June 30, 1888* (Lincoln, Neb.: Journal Company, State Printers, 1888), 144–66.

107. Lewis C. Edwards, *History of Richardson County, Nebraska* (Indianapolis: B. F. Bowen, 1917).

108. Chanute, letter to Frank R. Firth, dated July 5, 1871, Chanute Papers, LoC.

109. "Fearful Railroad Accident," *Chicago Tribune*, June 15, 1872.

110. Chanute, letter to George S. Morison, dated July 16, 1872, Chanute Papers, LoC.

111. "Burlington & Missouri River Railroad in Nebraska's System."

112. Chanute, letter to James F. Joy, dated Apr. 19, 1871, Chanute Papers, LoC.

113. Cuthbert Powell, *Twenty years of Kansas City's live stock trade and traders* (Kansas City, Mo.: Pearl Printing, 1893).

114. Chanute, letter to James F. Joy, dated Feb. 20, 1871, Chanute Papers, LoC.

115. Chanute, letter to Daniel L. Halladay, dated June 19, 1871, Chanute Papers, LoC.

116. Octave Chanute, "Report of feasibility of establishing a barge line on the Missouri River, dated 6 May 1872," to the *Kansas City Board of Trade.* Republished in *Missouri Republican*, St. Louis, Mar. 13, 1877; Miller, *History of Kansas City*, 136–42 and *History of Jackson County, Missouri.*

117. Chanute, Report to the Mayor and Council, Kansas City, dated June 24, 1871, Chanute Papers, LoC.

118. Chanute, letter to Charles F. Adams, dated Mar. 15, 1873, Chanute Papers, LoC.

119. Chanute, letter to James M. Walker, dated June 7, 1872, Chanute Papers, LoC.

120. "News Items." *Railroad Gazette*, Mar. 15, 1873.

Chapter 4. At the Top

1. Chanute, letter to George B. Chase, dated Jan. 2, 1873, Chanute Papers, LoC.

2. "President Watson on Erie's Future," *Chicago Tribune*, May 9, 1873.

3. "Important improvements proposed by the Erie Company," *Chicago Tribune*, Jan. 30, 1873.

4. Letterpress books, Manuscript Division, Library of Congress. Chanute's time at the Erie had to be reconstructed by consulting his own publications, as well as contemporary newspapers and magazines. He probably continued copying his outgoing letters into letterpress books, but the existence of these books is not known, because they were not part of the donation by the Chanute estate to the Library of Congress in 1931.

5. Chanute, letter to Thomas C. Clarke, dated May 6, 1873, Chanute Papers, LoC.

6. Chanute, letter to Henry A. Gardner, dated July 20, 1873, Chanute Papers, LoC.

7. Octave Chanute, "Erie Railway Specifications for Iron Bridges," *Van Nostrand's Engineering Magazine*, 9 no. 58 (Oct. 1873) 371–74.

8. "Historical Sketch of the Development of American Bridge Specifications," in "Report of Committee No. 15—On Iron and Steel Structures." Proceedings of the *Sixth Annual Convention, American Railway Engineering and Maintenance-of-Way Association* 6, Chicago, Mar. 21–23, 1905, 199–217.

9. Henry Petroski, *Engineers of Dreams: Great Bridge Builders and the Spanning of America* (New York: Alfred A Knopf, 1995).

10. Paul L. Wolfel, R. Khuen, and O. E. Hovey, "Memoir of Charles Conrad Schneider," *American Society of Civil Engineers, Transactions* 81, no. 12 (Dec. 1917): 1665–70.

11. "The Erie Company Working for a Direct Connection to Chicago," *Chicago Tribune*, Oct. 26, 1873.

12. Octave Chanute, "The Elements of Cost of Railroad Freight Traffic," *American Society of Civil Engineers, Transactions* 2, no. 2 (Feb. 1874): 381–98. Abstract in *Chicago Tribune*, Mar. 26, 1874; *Journal of the Railway Association of America* (Oct. 1874): 25–45, 127–28; *New York Times*, Apr. 21, 1874; *Van Nostrand's Engineering Magazine* (May 1874): 471–83; *Railway World* (Oct. 18, 1879): 989. Abstract is also included in the *Sixth Annual Report of the Railroad Commissioners, Report No.29*, Boston, Mass., Jan. 1875, 35–39; State of Wisconsin Governor's Message, Jan. 13, 1876, 48–49.

13. James C. Spencer, "Report No. 463, to accompany bill S.485," Senate of the United States, 45th Congress 2d Session. June 4, 1878, 1–14, Government Printing Office, Washington, D.C.

14. "Mr. Gardner calls on the Baltimore & Ohio Officers," *Chicago Tribune*, Oct. 22, 1874.

15. Captain Henry Tyler, "Erie Railway. Capt. Tyler's Report to A. De Laski, Esq., Chairman, London Banking Association," *American Railroad Journal* 30, nos. 45–49 (Nov. 7, 14, 21, and 28, and Dec. 5, 1874): 1475–77, 1508–9, 1540–41, 1572–73, 1603–5, 1636–37.

16. Ashbel Welch, M. N. Forney, O. Chanute, and I. M. St. John, "On the Form, Weight, Manufacture and Life of Rails," (adopted June 10, 1874) *American Society of Civil Engineers, Transactions* 3 (1875): 87–106. Abstract in *Railroad Gazette* July 11, 1874, 269–70. A second report was presented on May 5, 1875, published in *Transactions* 4 (1875): 136–41; the third and final report was presented on June 15, 1876, published in *Transactions* 5 (1876): 327–29.

17. Octave Chanute, "Rail Section and Fish Plate for the Erie Railway," *Railroad Gazette* 19, no. 6 (Feb. 6, 1875): 51. Drawings of the Chanute head, as adopted by the Erie, are shown as "doodles No.64" at http://www.sil.si.edu/exhibitions/doodles/, accessed Jan. 2008.

18. Octave Chanute, "Foot Prints of Locomotives," *Railroad Gazette* 20, no. 16 (Apr. 21, 1876): 171–72. Abstract in *Appleton's Cyclopedia of Applied Mechanics* (1891): 638–39.

19. Octave Chanute, "The Weight of Rails and the Breaking of Iron Rails," *American Society of Civil Engineers, Transactions* 3, presented July 15, 1874 (1875): 111–17.

20. Henry G. Prout, "The Development of the Steel Rail in the United States," *Engineering Magazine* 13, nos. 4 and 5 (July and Aug. 1897): 567–79, 704–16.

21. Octave Chanute, "On the Theoretical Resistance of Railroad Curves, Discussion," *American Society of Civil Engineers, Transactions* 7, no. 4 (Apr. 1878): 97–101. Abstract in De Volson Wood (1894) "Key and Supplement to Elementary Mechanics" (Philadelphia: John Wiley & Sons, 1894), 62–63.

22. Octave Chanute, "The Erie Joint-Fastening," *Railroad Gazette* 19, no. 9 (Feb. 27, 1875): 83.

23. Octave Chanute, "On Steel Rails, Discussion," *American Institute of Mining Engineers, Transactions* 9, no. 2 (Feb. 1881): 578–92.

24. Chanute, letter to J. E. Watkins, curator U.S. National Museum, dated June 24, 1891, Chanute Papers, LoC.

25. "The Ice Gorges. Port Jervis threatened with destruction," *New York Times*, Mar. 3, 1875.

26. "The Delaware Ice Gorge," *Georgia Weekly Telegraph and Journal & Messenger*, Mar. 16, 1875.

27. "Breaking of the great ice gorge at Port Jervis, NY," *Chicago Tribune*, Mar. 18, 1875; "The flooded valleys. Review of last week's freshets," *New York Times*, Mar. 22, 1875.

28. "Capacity and ability of the Erie Railway," *Pomeroy's Democrat*, Apr. 3, 1876.

29. "The effort to cut a channel through the Ice," *New York Times*, Mar. 26, 1875.

30. "Portage Bridge on the Erie Railway totally destroyed by fire," *New York Times*, May 7, 1875.

31. George S. Morison, "The New Portage Bridge," *American Society of Civil Engineers, Transactions* 5 (Jan. 1876): 1–8. *Discussion*, 235–39.

32. "The new Iron Bridge at Portage," *New York Times*, Aug. 1, 1875.

33. Octave Chanute, "Travel Diaries," 1875. Transcribed by Pearl Young. Chanute Family Papers.

34. Ibid.

35. J. Dutton Steele, O. Chanute, and C. Fisher, "Railway Signals," *American Society of Civil Engineers, Transactions* 4, no. 5 (May 1875): 147–61.

36. Octave Chanute, "Factors of Safety," *Engineering News* 7, no. 4 (Jan. 31, 1880): 41–42. Abstract in *Stoddart's Encyclopedia Americana* (1883), 636.

37. Octave Chanute, J. G. Barnard, and Q. A. Gillmore, *Report of Board of Consulting Engineers, appointed to recommend a plan for the New York and Long Island Bridge across the East River, at Blackwell's Island* (New York: New York and Long Island Bridge Company, 1877).

38. Octave Chanute, "Adhesion and tractive power of locomotives and the resistance of trains," *Railroad Gazette* 26, no. 12 (Mar. 24, 1882): 180. Reprinted in Charles H. Haswell, *Mechanics' and Engineers' Pocket-Book of Tables, Rules, and Formulas pertaining to Mechanics, Mathematics, and Physics* (New York: Harper & Brothers, 1890): 681–84.

39. Octave Chanute, "Repairs of Masonry," *American Society of Civil Engineers, Transactions* 10, no. 9 (Sept. 1881): 291–308. Abstract in *Railroad Gazette*, June 24, 1881, 348–49; *American Architect & Building News*, Nov. 26, 1881, 253–54.

40. "Railroad Interests," *New York Times*, Feb. 26, 1876.

41. "Financial Affairs," *New York Times*, Mar. 24, 1876.

42. Edward H. Mott, *Between the Ocean and the Lakes: the Story of Erie* (New York: John S. Collins, 1899).

43. "Payment of Erie Railway Employees," *New York Times*, Nov. 17, 1876.

44. Emory R. Johnson, *American Railway Transportation*, Appleton's business series (New York: A. Appleton, 1904).

45. Octave Chanute, "Memoir of Albert Fink, Past President, Am. Soc. C. E.," *American Society of Civil Engineers, Transactions* 41, no. 6 (1899): 626–38.

46. "Visit of dissatisfied employees of the Erie Railway to Receiver Jewitt," *New York Times*, June 29, 1877.

47. Joseph A. Dacus, *Annals of the Great Strikes in the United States* (Chicago: L. T. Palmer, 1877), 192–204, 258–71.

48. Elizabeth Chanute, "My Father."

49. "The Striker Donahue in Court," *New York Times*, July 29, 1877.

50. Elizabeth Chanute, "My Father."

51. "Trains on the Erie Abandoned. Complete blockade at and west of Hornellsville," *New York Times*, July 21, 1877.

52. "Affairs along the Erie Road. All trains stopped at Hornellsville," *New York Times*, July 22, 1877.

53. "The Great Railroad Riots. The Erie Road in Possession of the Strikers," *New York Times*, July 23, 1877.

54. "The Trouble on the Erie Road. Conference between the Strikers and Railroad Officials," *New York Times*, July 24, 1877.

55. Octave Chanute, "Report," in *New York, Lake Erie & Western Railroad Company. Report of the Board of Directors to the Stockholders for the four months ending Sept. 30, 1878* (New York: Martin B. Brown, 1878).

56. Octave Chanute, "Report," in *New York, Lake Erie & Western Railroad Company. Report of the Board of Directors to the Stockholders for the fiscal year ending Sept. 30, 1879* (New York: Martin B. Brown, 1879).

57. Octave Chanute, "General Specifications for Iron Bridges, issued in 1879 by the

New York, Lake Erie, and Western Railroad Company," in *Mechanics of Girders: A treatise on bridges and roofs*, ed. John Davenport Crehore (New York: John Wiley & Sons, 1886): 413–22.

58. Octave Chanute, "The Use of Steel for Bridges, Discussion," *American Society of Civil Engineers, Transactions* 8, no. 10 (Oct. 1879): 279–94.

59. "An iron truss gives way," *New York Times*, July 20, 1879.

60. Octave Chanute, "Report," in *New York, Lake Erie & Western Railroad Company. Report of the Board of Directors to the Stockholders for the fiscal year ending Sept. 30 1880* (New York: Martin B. Brown, 1880).

61. Octave Chanute, "Stability of Stone Structures, Discussion," *American Society of Civil Engineers, Transactions* 8, no. 9 (Sept. 1879): 251–62.

62. "Erie Railway Reorganized," *New York Times*, Apr. 28, 1878.

63. "Report of James C. Spencer, Referee, in the Supreme Court, County of New York. The People of the State of New York, Plaintiff, against the Erie Railway Company and others, Defendants," New York, Oct. 31, 1879.

64. "The Erie Railway Taxation Case," *Trenton Daily State Gazette*, Aug. 5, 1879.

65. "Railroad Freight Rates," *New York Times*, Oct. 14, 1879. Abstract in *Railway World*, Oct. 18, 1879, 989.

66. Octave Chanute, "Notes of a Visit to the Keely Workshop," *Engineering News* 40, no. 26 (Dec. 20, 1898): 418.

67. "Erie's Monster Locomotives," *New York Times*, Oct. 13, 1879.

68. Octave Chanute, "Uniformity in Railway Rolling Stock," *American Society of Civil Engineers, Transactions* 11, no. 9 (Sept. 1882): 291–313. Abstract in *School of Mines Quarterly*, Jan. 1883, 145–46; *Mémoires et Compte Rendu des Travaux de la Société des Ingénieurs Civils*, 1883, 580–81.

69. George M. Bond, *Standards of Length and their Practical Application* (Hartford, Conn: Pratt & Whitney, 1887).

70. Matthias N. Forney, "Standard Nuts and Bolts. Discussion," *25th Annual Convention of the American Railway Master Mechanics' Association* (June 20–22, 1892): 144–49.

71. O. Dockstader, "Engine Coal Pockets on the Erie," *Railroad Gazette* 26, no. 37 (Oct. 5, 1882): 647–48. Abstract in Walter G. Berg, *Buildings and Structures of American Railroads* (New York: Wiley & Sons, 1893).

72. Francis M. Wilder, "Train Resistance and the Dynamometer, with discussions," *14th Annual Convention of the American Railway Master Mechanics' Association. Proceedings* 14 (June 14–16, 1881): 34–66.

73. Octave Chanute, "Cantilever Bridge at Niagara Falls, Discussion," *American Society of Civil Engineers, Transactions* 14, no. 12 (Dec. 1885): 594–95.

74. Chanute, "Uniformity in Railway Rolling Stock."

75. Chanute, "Report," *New York, Lake Erie & Western Railroad Company*, 1880.

76. "The laying of the third rail," *Cincinnati Commercial*, Jan. 4, 1879.

77. "Railroad War," *New York Times*, July 20, 1881.

78. Octave Chanute, "Wind Pressure upon Bridges, Discussion," *American Society of Civil Engineers, Transactions* 10, no. 5 (May 1881): 169–71.

79. Octave Chanute, "Leading Engineering Works," *American Society of Civil Engineers, Transactions* 24, no. 5 (June 1891): 397–429. Abstract in *Railroad Gazette*, May 29, 1891, 368–69; *Scientific American*, June 6, 1891, 352–53; *Scientific American Supplement*, Aug. 8 and 15, 1891, 13002–4, 13016–18. Also published as a pamphlet.

80. Thomas C. Clarke, "The Kinzua Viaduct, 1882," *American Society of Mechanical Engineers* 21st Meeting, vol. 11 (May 1890): 961–73.

81. "The Kinzua Viaduct," *Scientific American*, 46, no. 19 (May 13, 1882): 287–88.

82. C. R. Grimm, "The Kinzua Viaduct of the Erie Railroad Company," *American Society of Civil Engineers, Transactions* 46, no. 1 (Jan. 1901): 21–77.

83. Pennsylvania State Parks, "Kinzua Bridge State Park," 2003, http://www.dcnr.state .pa.us/STATEPARKS/PARKS/kinzuabridge.aspx, accessed Apr. 2005.

84. "The Erie's New Coal Branch," *New York Times*, Feb. 26, 1883.

85. Octave Chanute, "Report," in *New York, Lake Erie & Western Railroad Company. Report of the Board of Directors to the Stockholders for the fiscal year ending Sept. 30, 1883* (New York: Martin B. Brown, 1883).

86. Chanute, letter to the De La Vergne Refrigerating Co., dated July 14, 1883, Chanute Papers, LoC.

87. Octave Chanute, "Refrigerating Machines," 1884, Chanute Papers, LoC.

88. Octave Chanute, "Report," in *New York, Lake Erie & Western Railroad Company. Report of the Board of Directors to the Stockholders for the fiscal year ending Sept. 30 1884* (New York: Martin B. Brown, 1884).

Chapter 5. Self Realization

1. Chanute, Good-Bye letter to Joseph Chanut, dated Dec. 7, 1850, Chanute Family Papers.

2. Octave Chanute, "Engineering Progress in the United States," *American Society of Civil Engineers, Transactions* 9, no. 6 (June 1880): 217–58. Abstract in *Californian*, Sept. 1880, 278–79; *Manufacturer & Builder*, Sept. 1880, 204; *Mémoires et Compte Rendu des Traveux de la Société des Ingénieurs Civil*, Aug. 1880, 217–24; *Railroad Gazette*, June 4, 1880, 297–99 and June 11, 1880, 313; *Scientific American*, Aug. 7, 1880, 86, and Aug. 14, 1880, 105.

3. Thomas Telford, 1828. "Charter of the Institution of Civil Engineers," ICE Archives, London.

4. Raymond H. Merritt, *Engineering in American Society, 1850–1875* (Lexington: University Press of Kentucky, 1969).

5. Board of Direction Report, "Minutes of Meetings," *American Society of Civil Engineers. Proceedings* 4, no. 7 (July 1878): 89–98. The various amendments to the constitution were presented at the Sept. 4 meeting and accepted on Nov. 6, 1878.

6. "Annual Convention of the American Society of Civil Engineers," *Chicago Tribune*, June 6, 1872.

7. "Biographical Sketch of Robert H. Thurston," *The Manufacturer and Builder* 26, no. 6 (Sept. 25, 1894): 121.

8. Octave Chanute, "Reports of Committees. On Founding a Testing Laboratory," *American Society of Civil Engineers. Proceedings* 1 (June 3, 1874): 97–98. Final report was presented on June 10, 1875, 294.

9. Octave Chanute, "Iron and Steel considered as Structural Material—A Discussion," *American Institute of Mining Engineers, Transactions* 10 (Feb. 1882): 375–80. Abstract in *Engineering Magazine*, Sept. 1882, 375–78.

10. Octave Chanute, "On the necessity of Government aid in organizing a system of tests of materials used for structural purposes, Discussion," *American Institute of Mining*

Engineers, Transactions 10 (Feb. 1882): 377–79. Abstract in *Engineering Magazine*, Sept. 1882, 182–85.

11. Chamber of Commerce of the State of New York, *Rapid transit in New York City and in the other great cities* (New York: Blumenberg Press, 1905).

12. Octave Chanute, Matthias N. Forney, Ashbel Welch, Charles K. Graham, and Francis Colingwood, "Rapid Transit and Terminal Freight Facilities," *American Society of Civil Engineers, Transactions* 4, (Apr. 1875): 1–80. Abstract in *Railroad Gazette*, Feb. 20, 1875, 76–77; *New York Times*, Feb. 3 and 19, 1875; *Manufacturer & Builder*, Oct. 1875, 234–35; *Frank Leslie's Illustrated Newspaper*, May 25, 1878, 201. Also published as an eighty-one-page pamphlet.

13. Octave Chanute, "Water Front Warehouses and Railroads for New York," *Railroad Gazette* 34, no. 34 (Aug. 29, 1902): 605.

14. Octave Chanute, "The Rapid Transit Question in New York City," *Railroad Gazette* 35, nos. 41, 42, and 43 (Oct. 9, 16, and 23, 1891): 700–703, 720–22, 745–46. Abstract in *Engineering News*, Oct. 10, 17, and 24, 1891; *New York Times*, Sept. 18, 1891

15. Jerome B. Crabtree, *The marvels of modern mechanism and their relations to social betterment* (Springfield, Mass.: King-Richardson, 1901).

16. Octave Chanute, "Exhibition of American Bridge Construction at the Centennial," *Railroad Gazette* 20, no. 6 (Feb. 11, 1876): 66.

17. William P. Shinn, O. Chanute, and F. De Funiak, "Report of Committee on Uniform Accounts and Returns of Railroad Corporations," *American Society of Civil Engineers, Proceedings* 3, no. 11 (Nov. 7, 1877): 123–24. A follow-up report was read on May 31, 1879, published in *Proceedings*, 5, 33–34.

18. Trevelyan Ridout LLB, "Agreement between the Northern Railway Company of Canada and the Credit Valley Railway Company, made 25th November, 1880," in *Statutes Special and General, relating to the Northern Railway Company of Canada* (Toronto: Hunter, Rose, 1883), 135–38, 453–58.

19. "Report of Committees: On Founding an Engineering Library and Museum," *American Society of Civil Engineers. Proceedings* 1 (Apr. 25, 1875): 224–29.

20. "Arrival of M. De Lesseps. A talk with him on his great project," *New York Times*, Feb. 25, 1880.

21. Octave Chanute, "Inter-Oceanic Canal Projects. Introduction of Ferdinand de Lesseps, followed by a discussion," *American Society of Civil Engineers, Transactions* 9, no. 3 (Mar. 1880): 87–88, 96–98. Abstract in *Professional Papers on Indian Engineering*, 1880, 1–35 and others.

22. "M. De Lesseps sight-seeing," *New York Times*, Feb. 26, 1880.

23. Octave Chanute, "The radical enlargement of the artificial water-way between the Lakes and the Hudson River, Discussion," *American Society of Civil Engineers, Transactions* 14, no. 3 (Mar. 1885): 115–17.

24. Matthias N. Forney, "American Society of Civil Engineers, Report of the 12th Annual Convention," *Railroad Gazette* 24, nos. 22 and 23 (June 4 and 11, 1880): 297–99, 313.

25. Chanute, "Engineering Progress in the United States," 1880.

26. Octave Chanute, "On the Relation of Science to the Industrial Arts," in AAAS, *Executive Proceedings* at the St. Louis, Missouri, meeting, Aug. 1878 (Salem, Mass.: Salem Press, 1879), 357–58.

27. Octave Chanute, "The Passenger Steamers of the Thames, the Mersey and the

Clyde. Discussion," Institute of Civil Engineers, Minutes of the Proceedings, Dec. 2, 1879, 152–55.

28. Robert H. Thurston, "President's Inaugural Address," *American Society of Mechanical Engineers, Transactions* 1, no. 1 (Nov. 1880): 13–29.

29. *"Memorial of Alexander Lyman Holley, C.E.* (New York: American Institute of Mining Engineers, 1884); Chanute was one of forty-nine contributors, see pages 39–40.

30. Chanute, letter to William Metcalf, dated Mar. 7, 1888, Chanute Papers, LoC.

31. Octave Chanute, "Scientific Invention, the Progress of Mechanical Science," in *AAAS, Section D, Mechanical Science and Engineering. Proceedings*, Buffalo, New York (Aug. 19, 1886), 174–82.

32. "Letter from Mark Twain," *Sacramento Daily Alta*, Aug. 1, 1869.

33. *Tenth Annual Report of the Aeronautical Society of Great Britain for the Year 1875* (Greenwich: Henry H. Richardson).

34. Theodore G. Ellis, "Rise and Progress in American Engineering," *American Society of Civil Engineers, Proceedings* 2, no. 6 (June 13, 1876): 73–79.

35. Octave Chanute, "Mechanical Flight," unpublished manuscript, dated Sept. 15, 1878, Chanute Papers, LoC.

35. Chanute, "Engineering Progress in the United States," 1880.

36. "Octave Chanute discusses Aviation in 2009," *St. Louis Post-Dispatch*, Oct. 10, 1909.

37. Octave Chanute, "Wind Pressure upon Bridges, Discussion," *American Society of Civil Engineers, Transactions* 10, no. 5 (May 1881): 169–71.

38. Robert H. Thurston, "Our Progress in Mechanical Engineering," *American Society of Mechanical Engineers, Transactions* 2, no. 2 (Nov. 1881): 425–53.

39. Ashbel Welch, "Annual Address," *American Society of Civil Engineers, Transactions* 11, no. 11 (Nov. 1882): 153–80.

40. J. W. Leonard, *The Industries of Kansas City. Historical, descriptive and statistical.* (Kansas City: J. M. Elstner, 1888).

41. Chanute, letter to C. B. Warring, dated Sept. 17, 1883, Chanute Papers, LoC.

42. Pearl I. Young, *Octave Chanute, 1832–1910* (San Francisco: E. L. Sterne, aeronautical bookseller, 1963).

43. Alice Chanute Boyd, "Ancestry of Octave Chanute," 1912, Chanute Family Papers.

44. Ibid.

45. Octave Chanute, "On the Increased Efficiency of Railways for the Transport of Freight. Discussion," *American Society of Civil Engineers, Transactions* 12, nos. 4, 5, and 6 (April, May, June 1883): 138–44, 180–88, 195–201.

46. Octave Chanute, "The Cost of Railroad Freight Traffic," *Railway Review* 25, nos. 18 and 19 (May 2 and 9, 1885): 205–6, 209, 217–18, 222. Abstract in *New York Times*, June 1, 1885.

47. Octave Chanute, "The Sewerage of Kansas City," *Kansas City Review of Science and Industry* 7, no. 9 (Jan. 1884): 519–27. Abstract in *Engineering News*, Feb. 16, 1884, 81–82.

48. Robert Moore, "On the sewerage of Kansas City, being a review of a paper on the same subject by O. Chanute," *Journal of the Association of Engineering Societies* 3, nos. 5 and 6 (Mar. 12, 1884): 67–75. Octave Chanute, Discussion and Reply, 75–87.

49. R. T. Horn, "The Street Paving Question," *Kansas City Review of Science and Industry* 7, no. 10 (Feb. 1884): 631–33.

50. "The Civil Engineer who was called 'The Father of Aviation,'" *Kansas City Star,* July 10, 1927.

51. Chanute, letter to John C. Goodridge, dated Jan. 3, 1889, Chanute Papers, LoC.

52. Octave Chanute, "The South Pass Jetties—Ten years' practical teachings in river and harbor hydraulics, Discussion," *American Society of Civil Engineers, Transactions* 15, no. 4 (Apr. 1886): 250–55.

53. Bureau of Harbors and Water-Ways, "Council of Engineering Societies on National Public Works. Introduced as H R 4923," Mar. 31, 1888.

54. Octave Chanute, "Clippings and pamphlets on Galveston Harbor (1872–1886), 1888." CrMS-290 (2v), Crerar Manuscript Collection. University of Chicago Library, Chicago.

55. James B. Eads, C. S. Smith, I. M. St. John, T. C. Clarke, J. Owen, A. J. Boller, O. Chanute, C. Macdonald, and J. W. Adams, "On the means of averting bridge accidents," *American Society of Civil Engineers, Transactions* 4, no. 1 (Apr. 1875): 124–35.

56. Octave Chanute, John A. L. Waddell, and William H. Breithaupt, "Report on Committee on Bridge Reform," *Journal of the Association of Engineering Societies* 8, no. 1 (Jan. 1889): 52–53. Manuscript copy, "An Act to Promote the Safety of Bridges," Chanute Papers, LoC, letterpress book, microfilm reel No 8, 517–25. Abstract in *Railroad Gazette,* Jan. 6, 1888, 9–10.

57. "Abstract of Minutes of the Society," *Journal of the Western Society of Engineers* 6, no. 2 (Feb. 1901): 84–85.

58. Chicago, Burlington & Quincy Railroad Company, "Thirty-Third Annual Report of the Board of Directors to the Stockholders for the year ending Dec. 31, 1885" (Cambridge, Mass.: John Wilson and Son, University Press, 1886.)

59. Chanute, letter to George B. Harris, dated Feb. 12, 1890, with a report on status of bridges along the line, Chanute Papers, LoC.

60. Chanute, letter to Albert E. Tonzalin, dated Oct. 22, 1885, Chanute Papers, LoC.

61. "Warring Railroads. The fight over the crossing on First Street North. Expert Testimony," *The St. Paul Globe,* Jan. 10, 1886. The legal fight of the right-of-way suit was first discussed in the same paper on Oct. 4, 1885.

62. "A Magnificent New Bridge," *Chicago Tribune,* Feb. 16, 1886. Abstract in *Railroad Gazette,* Feb. 26, 1886, 154.

63. Chanute, letter to Albert E. Tonzalin, dated Feb. 17, 1886, Chanute Papers, LoC.

64. Chanute, letter to George B. Harris, dated Mar. 15, 1890, Chanute Papers, LoC.

65. Atchison, Topeka & Santa Fe Railroad Company, "Sixteenth Annual Report of the Board of Directors to the Stockholders for the year ending Dec. 31, 1887" (Boston: Press of Geo. H. Ellis, 1888).

66. Keith L. Bryant, *History of the Atchison, Topeka & Santa Fe Railway* (New York: Macmillan, 1974).

67. Chanute, letter to William H. Breithaupt, dated Jan. 4, 1889, Chanute Papers, LoC.

68. William H. Breithaupt, "Lattice Girder Overhead Crossing, Chicago, Santa Fe & California Railway," *Journal of the Association of Engineering Societies* 8, no. 1 (Jan. 1889): 1–6.

69. Walter W. Curtis, "The Fort Madison Bridge across the Mississippi River," *Engineering News* 15, nos. 22 and 23 (June 2 and 9, 1888): 440–41, 466–67.

70. Park Morrill, *Floods of the Mississippi River,* Bulletin E, prepared under direction

of Willis L. Moore, chief of the Weather Bureau. (Washington, D.C.: U.S. Department of Agriculture, Weather Bureau, 1897).

71. Octave Chanute, J. F. Wallace and W. H. Breithaupt, "The Sibley Bridge," *American Society of Civil Engineers, Transactions* 21, no. 9 (Sept. 1889): 97–132. Abstract in *Minutes of Proceedings of the Institution of Civil Engineers*, London, 1889–1890, 414–15.

72. W. M. Camp, "Memorial Meeting for John Findley Wallace," *23rd Annual Convention, American Railway Engineering and Maintenance-of-Way Association. Proceedings* 23, no. 1 (Mar. 14–16, 1922): 55–57.

73. Glenn D. Bradley, *The Story of the Santa Fe* (Boston: Richard G. Badger, The Gorham Press, 1920).

74. Chanute, Wallace, and Breithaupt, "Sibley Bridge."

75. Chanute, letter to Albert A. Robinson, dated Dec. 18, 1888, Chanute Papers, LoC.

76. Ibid.

77. Chanute, letter to Albert A. Robinson, dated Apr. 5, 1888, Chanute Papers, LoC.

78. Chanute, letter to Albert A. Robinson, dated Jan. 15, 1889, Chanute Papers, LoC.

79. Octave Chanute, "The Ethics of Consulting Practice." *Engineering News* 28, no. 46 (Nov. 10, 1892): 444–46.

Chapter 6. A New Industry

1. Octave Chanute, "The flow of the Sudbury River, Massachusetts, for the years 1875 to 1879, and rainfall and the flow of streams. Discussion," *American Society of Civil Engineers, Transactions* 10, no. 7 (July 1881): 245, 247.

2. De Volson Wood, *A treatise on the resistance of materials and an appendix on the preservation of timber* (New York: John Wiley & Son, 1875).

3. John Bogart, "American Society of Civil Engineers Circular," May 17, 1879, in Pamphlets on Preserving Wood, Octave Chanute Holding, University of Chicago Library, Chicago.

4. Octave Chanute, "Preservation of Timber. Preliminary Report," *American Society of Civil Engineers, Transactions* 11, no. 10 (Oct. 1882): 325–44.

5. Octave Chanute, "Report of the Committee on the Preservation of Timber," Transaction No 309, *American Society of Civil Engineers, Transactions* 14, no. 7 (July 1885): 247–96.

6. Octave Chanute, "Report of the Committee on the Preservation of Timber," *American Society of Civil Engineers, Proceedings* 9, no. 1 (Jan. 17, 1883): 108. Abstract in *Railroad Gazette*, Jan. 12, 1883, 26; *American Architect & Building News*, Mar. 31, 1883, 155.

7. Octave Chanute, "National Exposition of Railway Appliances—Exhibit of Preserved Timber," *American Society of Civil Engineers*, Circular Communication 18, no. 391 (June 23, 1883): 296–97. Abstract in *American Architect & Building News*, June 23, 1883, 296–97.

8. Octave Chanute, "Report of the Committee on the Preservation of Timber, *Transaction* No 309; Appendix to the Committee Report, *Transaction* No 310; Discussions, *Transaction* No 312," *American Society of Civil Engineers Transactions* 14, nos. 7, 8, and 9 (July, Aug., and Sept. 1885): 247–96, 350–60, 372–98. Abstract in *Railroad Gazette*, June 12, 1885, 371; *Engineering Magazine*, Aug. 1885, 172; *Minutes of the Proceedings*, ICE, 1886, 447–50.

9. Henry G. Prout, "The Preservation of Timber," *Railroad Gazette* 30, no. 7 (Feb. 19, 1886): 121.

10. Chanute, letter to W. Stevenson, dated Oct. 24, 1884, Chanute Papers, LoC.

11. Joseph P. Card, "Preserving Wood" (No. 254, 274), U.S. Patent Office, Washington, D.C., Feb. 28, 1882.

12. Chanute, letter to Joseph P. Card, dated July 22, 1883, Chanute Papers, LoC.

13. Chanute, letter to W. Stevenson, Oct. 24, 1884.

14. Chanute, letter to S. Meade, dated Nov. 21, 1883, Chanute Papers, LoC.

15. Octave Chanute, "Tie Preserving," *Railroad Gazette* 35, no. 31 (July 31, 1891): 536.

16. Octave Chanute, "Report of Committee No. III—On Ties. Discussion," *American Railway Engineering and Maintenance-of-Way Association. Proceedings 6th Annual Convention* (Mar. 21–23, 1905): 775–80.

17. F. Meredith Jones, "Tie and Timber Preserving Works at Las Vegas, N. M., A. T. & S. F. RY," *Engineering News* 32, no. 11 (Sept. 13, 1894): 204.

18. Committee, "In Memoriam—Samuel M. Rowe. Died May 22, 1910," *Journal of the Western Society of Engineers* 15, no. 6 (November–Dec. 1910): 831–33.

19. Octave Chanute, "Report of Committee No. III—On Ties," *American Railway Engineering and Maintenance-of-Way Association, Proceedings 3rd Annual Convention*, Mar. 18–20, 1902.

20. Octave Chanute, "Report of the Committee on the Preservation of Timber," Transaction No 309, *American Society of Civil Engineers, Transactions* 14, no. 7 (July 1885): 247–96.

21. Chanute, letter to S. R. Callaway Esq., dated Oct. 21, 1885, Chanute Papers, LoC.

22. Octave Chanute, "Tie-Preserving Works at Laramie," *Railroad Gazette* 30, no. 43 (Oct. 29, 1886): 736–37.

23. Chanute, letter to Charles F. Adams, dated June 4, 1888, Chanute Papers, LoC.

24. Octave Chanute, "Report of Committee No. III—On Ties," *American Railway Engineering and Maintenance-of-Way Association. Proceedings 2nd Annual Convention*, Mar. 12–14, 1901, 103–33. Abstract in *Railway Age*, Mar. 15, 1901, 193, 341–348.

25. Chanute, letter to George R. Lockwood, dated June 12, 1887, Chanute Papers, LoC.

26. Chanute, letter to W. Snyder, dated Feb. 18, 1886, Chanute Papers, LoC.

27. U.S. Forest Service, "Report on the relation of railroads to forest supplies," Bulletin No.1 (1887); "Preliminary report on the use of metal track on railways as a substitute for wooden ties," Bulletin No.3 (1889); "Timber Physics," Preliminary report No. 6 (1892); "Report on the use of metal railroad ties and on preservative processes and metal tie-plates for wooden ties," Bulletin No.9 (1894)." Washington, D.C.: Government Printing Office.

28. Chanute, letter to Joseph B. Card, dated Apr. 14, 1906, Chanute Papers, LoC.

29. "Fighting the Smoking Chimneys," *Chicago Tribune*, Aug. 22, 1890.

30. J. W. Kendrick, "Ties and Tie Preservation," *Railway Age* 39, no. 4 (Feb. 3, 1905): 150–53. Also see letter report by R. Angst, dated Feb. 14, 1906, in *Proceedings of the Sixteenth Annual Convention of the Association of Railway Superintendents of Bridges and Buildings*, Oct. 16–18, 1906, 235.

31. Chanute, letter to Hermann von Schrenk, dated May 16, 1904, Chanute Papers, LoC.

32. Octave Chanute, "Memoirs of Deceased Members. Joseph P. Card," *American Society of Civil Engineers, Proceedings* 21, no. 1 (Jan. 1895): 68.

33. Octave Chanute, "The Artificial Preservation of Railroad Ties by the Use of Zinc Chloride, Discussion," *American Society of Civil Engineers, Transactions* 42 (Dec. 1899): 366–74. Abstract in *Railway Age*, Sept. 1, 1899, 640.

34. Octave Chanute, "Precautions to be observed in burnettizing ties." In *Wood Preservers' Association, Proceedings 5th Annual Meeting*, Jan. 17–19, 1909, 637 (appendix B).

35. Chanute, letter to H. A. Parker, dated Nov. 7, 1898, Chanute Papers, LoC.

36. Octave Chanute, "Preservation of Wood," *Cassier's Magazine* 7, no. 4 (Feb. 1895): 301–7.

37. Chanute, letter to James Means, dated July 14, 1898, Chanute Papers, LoC.

38. Samuel M. Rowe, *Hand Book of Timber Preservation* (Chicago: Pettibone, Sawtell, 1904).

39. Chanute, letter to Hermann von Schrenk, dated May 12, 1904, Chanute Papers, LoC.

40. Chanute, letter to Wilbur Wright, dated Jan. 1, 1902, Chanute Papers, LoC.

41. Kendrick, "Ties and Tie Preservation."

42. Chanute, letter to P. A. Bonebrake, dated May 2, 1900, Chanute Papers, LoC; Octave Chanute, "Preservative Treatment of Timber," *Journal of the Western Society of Engineers* 5, no. 4 (Apr. 1900): 100–126, 198–207 (discussion).

43. Octave Chanute, "Preserving Timbered Structures" (No. 430,068), U.S. Patent Office, June 10, 1890. Canadian Patent issued on June 10, 1890, as "Process for Preserving Wood Artificially against Decay" (No. 34,507).

44. Octave Chanute, "Travel Diaries," 1889. Chanute Family Papers.

45. "Current Foreign Topics," *New York Times*, June 28, 1889.

46. "Greeting the big ships. The fast trips of the Teutonic and the City of New York," *New York Times*, Aug. 16, 1889.

47. Octave Chanute, "Preserving Wood against Decay," in AAAS, *Section D, Mechanical Science and Engineering. Proceedings 38th Meeting*, Toronto (Aug. 1889): 199; abstract in *American Architect & Building News*, Sept. 21, 1889.

48. Octave Chanute, "The Paris Exposition of 1889," *Journal of the Association of Engineering Societies* 9, no. 7 (July 1890): 341–51. Abstract in *Chicago Tribune*, Sept. 27, 1890.

49. Chanute, letter to Edward T. Jeffery, dated Sept. 30, 1889, Chanute Papers, LoC.

50. Edward T. Jeffery, *Paris Universal Exposition* (London: On Behalf of the Citizens' Executive Committee of Chicago, 1890). Self-published.

51. Hubert H. Bancroft, *The Book of the Fair; an historical and descriptive presentation of the world's science, art, and industry, as viewed through the Columbian exposition at Chicago* (Chicago and San Francisco: Bancroft, 1893).

52. Octave Chanute, "Committee Report on the International Engineering Congress." Proceedings, Western Society of Engineers, *Journal of the Association of Engineering Societies* 11, no. 1 (Jan. 1892): 104–7.

53. "The World's Fair Tower," *Railroad Gazette* 35, no. 43 (Oct. 23, 1891): 747.

54. "It will Investigate. Society of Western Engineers to Examine the Tunnel," *Chicago Tribune*, Feb. 4, 1892. See also "Proceedings, Western Society of Engineers," *Journal of the Association of Engineering Societies* 11, no. 3 (Mar. 1892): 170–71.

55. Octave Chanute, "Foundations and Floors for the Buildings of the World Columbian Exposition, Discussion," *Journal of the Association of Engineering Societies* 10, no. 12 (Dec. 1891): 579–84.

56. Octave Chanute, "The Railway Problem of Chicago, in relation to Terminals, Rapid Transit, Marine Commerce and related Interests, Discussion," *Journal of the Association of Engineering Societies* 11, no. 9 (Sept. 1892): 470–83.

57. Octave Chanute, "The Multiple Dispatch Railway, Discussion," *Journal of the Association of Engineering Societies* 10, no. 1 (Jan. 1891): 68–79.

58. Chanute, letter to Max E. Schmidt, dated Apr. 28, 1894, Chanute Papers, LoC.

59. Octave Chanute, "Leading Engineering Works," *American Society of Civil Engineers, Transactions* 24, no. 5 (June 1891): 397–429. Abstract in *Railroad Gazette*, May 29, 1891, 368–69; *Scientific American*, June 6, 1891, 352–53; *Scientific American Supplement*, Aug. 8 and 15, 1891, 13002–4, 13016–18. Also published as a pamphlet.

60. Octave Chanute, "Engineering Congress and Engineering Headquarters, Columbian Exposition, 1893," *Journal of the Association of Engineering Societies* 12, no. 2 (Feb. 1893): 95–98.

61. Jeffery, *Paris Universal Exposition*.

62. "The International Engineering Congress," *American Society of Civil Engineers, Proceedings* 19, no. 8 (July 31–Aug. 5, 1893): 134–47. Abstract in *Engineering Education*, Proceedings of Section E (Columbia, Mo.: E. W. Stephens, 1894), 1–4 (opening session), 331–38 (closing session).

63. "The International Engineering Congress," *American Society of Civil Engineers, Proceedings* 19, no. 7 (July): 134–48.

64. "Oil Jumps to $2.54. Boom in petroleum appears to gain ground steadily," *Chicago Tribune*, Apr. 17, 1895.

65. Chanute, "Artificial Preservation of Railroad Ties."

66. Chanute, letter to F. Holbein, dated Oct. 27, 1898, Chanute Papers, LoC.

67. Chanute, letter to John D. Isaacs, dated Oct. 29, 1898, Chanute Papers, LoC.

68. Octave Chanute, "A New Portable Tie Treating Plant," *Engineering News* 42, no. 7 (Aug. 17, 1899): 108–9. A similar write-up was published in *Railroad Gazette*, Aug. 18, 1899, 581–82.

69. Octave Chanute, "The Zinc-Creosoting Process in the United States," *Railroad Gazette* 34, no. 25 (June 20, 1902): 455.

70. Octave Chanute, "Preservation of Wood, Appendix F," in *The materials of construction, a treatise for engineers on the strength on engineering materials*, ed. John B. Johnson (New York: John Wiley & Sons, 1900): 776–80.

71. Octave Chanute, "Report of Committee No. III—On Ties," *American Railway Engineering and Maintenance-of-Way Association. Proceedings*, Mar. 14–15, 1900, 11–13, 64–65, 71–87, 182–85. Abstract in *Railway Age*, Mar. 16, 1900, 302–3.

72. Octave Chanute, "Preservation of Timber," *Railroad Gazette* 32, no. 30 (Aug. 27, 1900): 507, 509.

73. Chanute, letter to Ernest Pontzen, dated Sept. 15, 1886, Chanute Papers, LoC.

74. Octave Chanute, "Preservative Treatment of Timber. Lecture delivered before the students of the Rensselaer Polytechnic Institute," *The Polytechnic* 7, no. 3 (Nov. 29, 1890): 43–49. Abstracts in *Railroad Gazette*, Dec. 15, 1890, 853–56; *Engineering News*, Dec. 13, 1890, 528; *Science*, Dec. 12, 1890, 326–27; *Scientific American*, Jan. 3, 1891, 5; *Scientific American Supplement*, Feb. 14, 1891, 12600–12601.

75. Chanute, letter to American Steel & Wire Company, dated Feb. 2, 1899, Chanute Papers, LoC.

76. Chanute, letter to F. R. McIntyre, dated Apr. 9, 1899, Chanute Papers, LoC.

77. Chanute, "New Portable Tie Treating Plant."

78. Great Northern Railway Company, *Twelfth Annual Report for the fiscal year ended June 30, 1901* (Boston: Press of Geo. H. Ellis, 1901): 18.

79. "A big wire trust planned," *New York Times*, Oct. 20, 1889.

80. Chanute, letter to Charles D. Chanute, dated Feb. 2, 1901, Chanute Papers, LoC.

81. Octave Chanute, "Proposed stamping of ties to be treated by Chicago Tie Preserving Company," July 25, 1901, Chanute Papers, LoC.

82. Jeff Oaks, *Date Nails and Railroad Tie Preservation*. Special Report #3, 3 vols. (Indianapolis: Archeology and Forensic Laboratory, 2007).

83. Chanute, "Preservative Treatment of Timber"; Octave Chanute, "The Preservation of Railway Ties in Europe," *American Society of Civil Engineers, Transactions* 45, no. 6 (June 1900): 498–549, abstract in *Railroad Gazette*, Oct. 19, 1900, Jan. 4, and Feb. 1, 1901, as "German specifications for the preservation of railroad cross-ties," in J. B. Johnson, *"Engineering Contracts and Specifications,"* (New York: Engineering News Publishing, 1902), 485–91.

84. Chanute, "Report of Committee No. III — On Ties," 1901.

85. Hermann von Schrenk, "Report on the condition of treated timber laid in Texas, Feb. 1902," U.S. Department of Agriculture, Bureau of Forestry, Bulletin No. 51, 1904.

86. Octave Chanute, "Discussion of Paper on 'Comparative Value of Cross-Ties of different materials,'" *American Railway Engineering and Maintenance-of-Way Association. Bulletin No. 78* (Aug. 1906): 22–24.

87. Chanute, letter to Charles D. Chanute, dated Oct. 17, 1900, Chanute Papers, LoC.

88. Octave Chanute, "Process of Preserving Wood (No. 688,932)," U.S. Patent Office, Dec. 17, 1901.

89. Chanute, letter to George P. Whittlesey, dated June 13, 1901, Chanute Papers, LoC.

90. Octave Chanute, "Report of Committee No. III — On Ties," *American Railway Engineering and Maintenance-of-Way* Association, Proceedings 4th Annual Convention (Mar. 17–19, 1903). Also published in Bulletin no. 39, Mar. 1903. Abstract in *Railroad Gazette*, Apr. 24, 1903.

91. Octave Chanute, "Report of Committee No. III — On Ties," American Railway Engineering and Maintenance-of-Way Association. Proceedings 5th Annual Convention (Mar. 15–17, 1904), 66–70.

92. Chanute, letter to Howard F. Chappell, dated Apr. 16, 1904, Chanute Papers, LoC.

93. Warren R. Roberts, "A Personal Reminiscence of Octave Chanute. Adopted from a letter, dated Dec. 18, 1942," *Midwest Engineer* 45, no. 1 (1992–93): 1, 29–31.

94. James E. Cronin, *Hermann von Schrenk, a biography* (Chicago: Kuehn Publisher, 1959).

95. Hermann von Schrenk, "Factors which cause the decay of wood," *Journal of the Western Society of Engineers* 6, no. 2 (Apr. 1901): 89–103.

96. Octave Chanute, "Address of Mr. Chanute, the Retiring President," *Journal of the Western Society of Engineers* 7, no. 1 (Feb. 1902): 1–9.

97. "Kin of Octave Chanute discovered in Chicago," *Chicago School's Skylines. Army Air Forces Technical Training Command*, Feb. 4, 1943, 1, 8.

98. Chanute, letter to Hermann von Schrenk, dated Nov. 19, 1904, Chanute Papers, LoC.

99. Octave Chanute, "Travel Diaries," 1902–3, Chanute Family Papers.

100. Octave Chanute, "George Shattuck Morison," *Journal of the Western Society of Engineers* 9, no. 1 (Feb. 1904): 83–88.

101. Chanute, letter to Thomas Rodd, dated June 12, 1902, Chanute Papers, LoC.

102. Chanute, letter to Thomas Rodd, dated May 12, 1903, Chanute Papers, LoC.

103. Chanute, letter to E. O. Falkner, dated Feb. 18, 1905, Chanute Papers, LoC.

104. Chanute, letter to J. H. Fristor, dated Jan. 27, 1907, Chanute Papers, LoC.

105. "Absentees easy for Reviewers," *Chicago Tribune*, Sept. 4, 1901.

106. Henry E. Riggs, "The Valuation of Public Service Corporation Property," *American Society of Civil Engineers. Proceedings* 36, no. 9 (Nov. 1910): 1369–1538.

107. W. M. Camp, "Memorial Meeting for John Findley Wallace."

108. Octave Chanute, "International Engineering Congress. Report of Section H—Miscellaneous," *American Society of Civil Engineers, Proceedings* 30, no. 9 (Nov. 1904): 443–44.

109. Octave Chanute, "The International Railway Congress of 1905, Washington, DC," *Journal of the Western Society of Engineers* 10, no. 5 (Oct. 1905): 601–14.

110. Chanute, letter to W. G. Raymond, dated May 10, 1904, Chanute Papers, LoC. Letter was published in *Proceedings of the 12th Annual Meeting of the Society for the Promotion of Engineering Education*, St. Louis (Sept. 1–3, 1904): 118.

111. Chanute, letter to George W. Saathoff, dated Nov. 16, 1905, Chanute Papers, LoC.

112. "Urbana Installs New President," *Chicago Tribune*, Oct. 19, 1905.

113. Chanute Family Papers.

114. Octave Chanute, "The Steaming of Timber," *Wood Preservers' Association. Proceedings 3rd Annual Meeting* (Jan. 15–17, 1907): 2–64. Abstract in *Engineering News*, Jan. 31, 1907, 138–40; *Railroad & Engineering Review*, Mar. 2, 1907; *Scientific American Supplement*, Nov. 23, 1907; Samuel Rowe, *Hand Book of Timber Preservation*, 302–7.

115. Octave Chanute, "Minutes of the Annual Meeting," *Journal of the Western Society of Engineers* 14, no. 1 (Jan. 1909): 114–19.

116. Octave Chanute, "History of Wood Preservation in America," *Wood Preservers' Association. Proceedings 5th Annual Meeting* (Jan. 17–19, 1909): 11–19.

117. Octave Chanute, "Wood Preservation from an Engineering Standpoint, Discussion," *Journal of the Western Society of Engineers* 15, no. 3 (June 1910): 346–66.

118. Octave Chanute, "Report of Committees," *Wood Preservers' Association. Proceedings 6th Annual Meeting* (Jan. 18–20, 1910): 114–20.

119. Chanute, letter to J. H. Fristor, dated Jan. 21, 1910, Chanute Papers, LoC.

120. Octave Chanute, Travel Diaries, 1910. Chanute Family Papers. Transcribed by Pearl Young.

121. Octave Chanute, "Report of Committee XVII—On Wood Preservation. Sub-Committee 'A,' Revision of Manual." American Railway Engineering and Maintenance-of-Way Association, *Bulletin No. 131* (Jan. 1911): 49–60.

Chapter 7. From the Locomotive to the Aeromotive

1. Octave Chanute, "Scientific Invention," AAAS, Section D, *Mechanical Science and Engineering. Proceedings*, Buffalo, New York, Aug. 19, 1886, (1887): 174–82. Abstract in *American Architect & Building News*, Aug. 13, 1887; *Science*, Aug. 27, 1886; *Science Supplement*, Sept. 3, 1886.

2. Octave Chanute, "Gliding Experiments," *Journal of the Western Society of Engineers*

2, no. 5 (Dec. 1897): 595–628. Republished in *Scientific American Supplement*, Jan. 8, 1898, 18368–70; Jan. 15, 1898, 18380–82; Jan. 22, 1898, title page, 18390–91. Abstract in *Chicago Tribune*, Nov. 14, 1897.

3. Baden F. S. Baden-Powell, "The Late Mr. F. W. Brearey," *Aeronautical Journal* 1, no. 1 (Jan. 1899): 9–11.

4. Chanute, letter to Israel Lancaster, dated Apr. 26, 1886, Chanute Papers, LoC.

5. Chanute, "Scientific Invention."

6. Israel Lancaster, "The Problem of the Soaring Bird," *American Naturalist* 19, nos. 11 and 12 (Nov. and Dec. 1885): 1055–58, 1162–71.

7. Octave Chanute, "Langley's Contribution to Aerial Navigation," in *Langley, Samuel Pierpont. Secretary of the Smithsonian Institution, 1887–1906* (Washington D.C.: Smithsonian Institution, 1907): 30–35.

8. "Proceedings, Section of Mechanical Science and Engineering," *Science* 8, no. 187 (Sept. 3, 1886): 215–17.

9. "Papers read before the American Association for the Advancement of Science, Buffalo Meeting, Aug. 1886," *Scientific American* 55, no. 10 (Sept. 4, 1886): 154.

10. Gary Bradshaw, "Invention and Artificial Intelligence," in *Discovery Science. Lecture Notes in Computer Science* (Singapore: Springer Verlag, 2005): 1–13.

11. Chanute, letter to Charles Latimer, dated Jan. 14, 1888, Chanute Papers, LoC.

12. Octave Chanute, "Experiments with glider model." In William J. Jackman and Thomas H. Russell, *Flying Machines, Construction & Operation* (Chicago: Charles C. Thompson, 1910): 13–14.

13. Chanute, letter to Charles W. Hastings, dated Dec. 2, 1888, Chanute Papers, LoC.

14. Chanute, letter to De Volson Wood, dated Dec. 8, 1888, Chanute Papers, LoC.

15. Jules Verne, *The Clipper of the Clouds* (London: S. Low, Marston, Searle & Rivington, 1887).

16. Octave Chanute, "The Problem of Aerial Navigation. Campbell's Airship a Step Towards its Solution," *Chicago Tribune*, Dec. 28, 1888. Letter to the editor in response to "A ship for the skies," dated Dec. 24, 1888.

17. Chanute, letter to Robert Thurston, dated Feb. 4, 1889, Chanute Papers, LoC.

18. Samuel P. Langley, "Story of Experiments in Mechanical Flight," in *The Aeronautical Annual*, ed. James Means (Boston: W. B. Clarke, 1897): 11–25.

19. Chanute, letter to Matthias N. Forney, dated Nov. 10, 1886, Chanute Papers, LoC.

20. Octave Chanute, "The Latest Rapid Transit Scheme," *Railroad and Engineering Journal* 63 (vol. 3, new series), no. 4 (Apr. 1889): 199.

21. Octave Chanute, "Note sur la Résistance de l'Air aux Plans Obliques," *L'Aéronaute* 22, no. 9 (Aug. 1, 1889): 197–214. Abstract in *Zeitschrift für Luftschiffahrt*, Jan. 1890, 144–46.

22. Octave Chanute, *Proceedings of the International Conference on Aerial Navigation, held in Chicago, Aug. 1, 2, 3 and 4, 1893* (New York: American Engineer and Railroad Journal, 1894).

23. Octave Chanute, "Resistance of air to inclined planes in motion," in *AAAS Proceedings, Section D, Mechanical Science and Engineering*, Toronto (Aug. 1889): 198–99.

24. Octave Chanute, "Aerial Navigation," *The Crank* 4, no. 8 (May 1890): 1, 2–6.

25. Albert F. Zahm, "Aerial Navigation," *Journal of the Franklin Institute* 138, nos. 4 and 5 (Oct. and Nov. 1894): 265–87, 347–56. Note: Chanute was asked to lecture; he in turn asked Zahm to present it. Abstract in *Engineering World*, Mar. 1894.

26. Octave Chanute, "Aerial Navigation," *Railroad and Engineering Journal* 64, nos.

7–11 (July–Nov. 1890): 316–18, 365–67, 395–97, 442–44, 498–501. Reprinted as thirty-six-page pamphlet. Reviews in *Chicago Tribune*, Feb. 9, 1891; *Engineering News*, Jan. 31, 1891, 108–9; *Manufacturer & Builder*, Mar. 1891; *Military Service Institution of the United States*, Nov. 1892, 1193–1207, Jan. 1893, 154–65; *Popular Science Monthly*, June 1891, 564; *Science*, Apr. 10, 1891.

27. Octave Chanute, "Motors for aerial machines," *Scientific American Supplement* 35, no. 897 (1893): 14281–82.

28. Octave Chanute, "Progress in Aerial Navigation," *Engineering Magazine* 2, no. 2 (Oct. 1891): 1–13.

29. "Progress in Flying Machines. Inventors of such structures no longer thought to be insane," *Chicago Tribune*, Oct. 11, 1891.

30. Octave Chanute, "Aerial Navigation," in *Modern Mechanism* (New York: Appletons' Cyclopaedia of Applied Mechanics, 1892): 1–9

31. Robert P. Woods, "Early Presidents of the Society, XXI. Octave Chanute, 1832–1910," *Civil Engineering* 7, no. 12 (Dec. 1937): 871–73.

32. Octave Chanute, "Progress in Flying Machines," *Railroad and Engineering Journal* 65, no. 10 (Oct. 1891): 461–65.

33. Octave Chanute, "Air resistance at very high velocities," *Railroad Gazette* 34, no. 52 (Dec. 26, 1890): 887. Letter to the editor in reply to a correspondent, dated Dec. 12, 1889.

34. Octave Chanute, "On the Soaring of Birds," *Railroad and Engineering Journal* 65, no. 3 (Mar. 1891): 117–19. Abstracted from *Nature* 42, no. 1086 (Aug. 21, 1890): 397–98.

35. Octave Chanute, "A new flying machine." *Railroad and Engineering Journal* 65, no. 9 (Sept. 1891): 405–6. Translated and abstracted from *L'Illustration*, June 20, 1891.

36. Charles H. Gibbs-Smith, *Aviation. An Historical Survey from its Origins to the end of World War II* (London: Her Majesty's Stationery Office, 1970).

37. Octave Chanute, "Proceedings of the Annual Meeting," *Journal of the Western Society of Engineers* (Jan. 7, 1891): 1–11. Abstract in *Railroad and Engineering Journal*, May 1891, 240.

38. Octave Chanute, "The Effect of Invention upon the Railroad and other Means of Intercommunication," in *American Patent Centennial Celebration. Proceedings and Addresses* (Washington, D.C.: Press of Gedney & Roberts, 1891): 161–73. Abstract in *Engineering News*, May 2, 1891, 419–20.

39. Chanute, letter to J. Brown Goode, dated Apr. 10, 1896, Chanute Papers, LoC. This letter is reprinted in the *Aeronautical Annual* (1897) in "Samuel Pierpont Langley," 8.

40. Matthias N. Forney, "Editorial," *Railroad and Engineering Journal* 65, no. 10 (Oct. 1891): 433.

41. Octave Chanute, "Progress in Flying Machines," in *Railroad and Engineering Journal*. Twenty-seven articles on the "Progress in Flying Machines" were published between Oct. 1891 and Jan. 1894.

42. John D. Anderson Jr., "Infancy of Aerodynamics. To Lilienthal and Langley" (chapter 4), in *A History of Aerodynamics and Its Impact on Flying Machines* (Cambridge, U.K.: Cambridge University Press, 1997).

43. Chanute, letter to Francis H. Wenham, dated Sept. 13, 1892, Chanute Papers, LoC.

44. Chanute, letter to Otto Lilienthal, dated Jan. 12, 1894, Chanute Papers, LoC. This and the other letters were translated and are part of *Otto Lilienthals flugtechnische Kor-*

respondenz, assembled and annotated by Werner Schwipps, published by the Lilienthal-Museum in Anklam, Germany (1993).

45. Otto Lilienthal, "Why is artificial flight so difficult an invention," *American Engineer and Railroad Journal* 68, no. 12 (Dec. 1894): 575–76. Translation of "Weshalb ist es so schwierig das Fliegen zu erfinden?" published in *Prometheus* 6, no. 261 (Jan. 1894): 7–10.

46. Chanute, letter to Lawrence Hargrave, dated Sept. 26, 1893, Chanute Papers, LoC.

47. Chanute, "Experiments with glider model," in *Flying Machines*.

48. Chanute-Mouillard correspondence on the subject of flight, Apr. 16, 1890 to May 20, 1897. A. F. Zahm had many letters translated in the 1930s and the remainder was translated under Pearl I. Young's guidance by Eugene Moritz and M. Louis Kraus, edited by Juliette Bevo-Higgins. The digitized correspondence is available at http://invention. psychology.msstate.edu/inventors/i/Chanute/library/Chanute_Mouillard/Chanute-Mouillard.html.

49. Louis-Pierre Mouillard, letter to Chanute, dated Oct. 22, 1890, Chanute-Mouillard correspondence.

50. Chanute, letter to Louis-Pierre Mouillard, dated Nov. 20, 1890, Chanute-Mouillard correspondence.

51. Octave Chanute, "Progress in Flying Machines," *American Engineer and Railroad Journal* 67, no. 1 (Jan. 1893): 38–40.

52. Mouillard, letter to O. Chanute, dated June 14, 1892, Chanute-Mouillard correspondence.

53. Chanute, letter to George P. Whittlesey, dated Dec. 12 1892, Chanute Papers, LoC.

54. Louis-Pierre Mouillard, assignor of one-half to Octave Chanute, of Chicago, "Means for Aerial Flight" (No. 582,757), U.S. Patent Office, May 18, 1897.

55. Chanute, letter to Louis-Pierre Mouillard, dated Sept. 16, 1895, Chanute-Mouillard correspondence.

56. Chanute, letter to Louis-Pierre Mouillard, dated Dec. 31, 1895, Chanute-Mouillard correspondence.

57. Chanute, letter to Louis-Pierre Mouillard, dated Oct. 16, 1896, Chanute-Mouillard correspondence.

58. Wilbur Wright, "What Mouillard Did," *Aero Club of America Bulletin* 1, no. 3 (Apr. 1912): 3–4. Abstract in *New York Times*, Mar. 3, 1912; *New York Tribune*, Apr. 14, 1912.

59. Octave Chanute, "Not Things, But Men. The World's Congress Auxiliary of the World's Columbian Exposition of 1893." Circular published and mailed in Dec. 1892. (Chicago: Department of Engineering. General Division of Aerial Navigation).

60. Chanute, letter to Carl E. Myers, dated Nov. 30, 1892, Chanute Papers, LoC.

61. Octave Chanute, "Aerial Navigation Conference," *American Engineer and Railroad Journal* 67, no. 9 (Sept. 1893): 416–17.

62. Samuel P. Langley, "The Internal Work of the Wind," in *Proceedings of the International Conference on Aerial Navigation. Aug. 1–4, 1893*, (New York: *American Engineer and Railroad Journal*, 1894): 66–104. Typescript copy, CrMS-177. Republished in *Smithsonian Contributions of Knowledge*, No. 884.

63. Albert Zahm, "Diary of conference on aerial navigation." Albert F. Zahm Papers. Archives of the University of Notre Dame, Notre Dame, Indiana.

64. Chanute, letter to Louis-Pierre Mouillard, dated Feb. 17, 1893, Chanute-Mouillard correspondence.

65. Matthias N. Forney, *Aeronautics*, (Oct. 1893 through Sept. 1894). Republished in book format as *Proceedings of the International Conference on Aerial Navigation, held in Chicago, Aug. 1, 2, 3 and 4, 1893*.

66. Senator Brice "Report to secure aerial navigation to accompany bill S. 1344." *Report of Committees of the Senate of the United States for the 3rd Session, 53rd Congress*, No.992 (Washington, D.C.: Government Printing Office, 1895). Abstract in James Means' *Aeronautical Annual* (1896): 80–85, 154.

67. David Young, *Chicago Aviation, an Illustrated History* (De Kalb: Northern Illinois University Press, 2003).

68. Chanute, *Progress in Flying Machines* (New York: American Engineer and Railroad Journal, 1894). Review in *Electrical Engineer* (May 9, 1894): 411.

69. Chanute, "Conclusions," *Progress in Flying Machines*.

70. Chanute, letter to Matthias N. Forney, dated Mar. 22, 1894, Chanute Papers, LoC.

71. John J. Montgomery, "Correspondence and other papers of and relating to Montgomery," University of North Carolina, Chapel Hill, Southern Historical Collection, 1958.

72. Chanute, letter to John J. Montgomery, dated Mar. 30, 1894, Chanute Papers, LoC.

73. Octave Chanute, "Aeronautical Notes — Snap-shot at a Gull," *American Engineer and Railroad Journal* 69, no. 2 (Feb. 1895): 97.

74. Chanute, letter to Wilhelm Kress, dated Mar. 14, 1895, in *The Papers of Wilhelm Kress* (Vienna, Austria: Technisches Museum Wien).

75. Chanute, letter to Finje Van Salverda, dated Sept. 20, 1894, Chanute Papers, LoC.

76. Chanute, letter to John C. Trautwine, dated Jan. 10, 1894, Chanute Papers, LoC.

77. Engineer, "Dr. Lilienthal," *American Architect & Building News* 49, no. 1026 (Aug. 24, 1895): 74.

78. Octave Chanute, "Mr. Maxim's Flying Machine," *American Engineer and Railroad Journal* 68, no. 9 (Sept. 1894): 405.

79. Matthias Forney, "Editorial," *American Engineer and Railroad Journal* 68, no. 10 (Oct. 1894): 433.

80. Hermann W. L. Moedebeck, *Taschenbuch zum praktischen Gebrauch für Flugtechniker und Luftschiffer* (Berlin: W. H. Kühl, 1895). Chanute reviewed this book in "Recent Aeronautical Publications," *American Engineer*, July 1895, 340.

81. Charles F. Marvin, "Monograph on the mechanics and equilibrium of kites" (Washington, D.C.: Secretary of Agriculture, 1897). Prepared with the approval of Willis L. Moore, chief of the Weather Bureau.

82. James Means, *The Aeronautical Annual* (Boston: W. B. Clarke, 1895). Chanute reviewed the book in *American Engineer*, Mar. 1895, 147–48.

83. Herring, letter to Frank I. French, business manager, *American Engineer and Railroad Journal*, dated Sept. 19, 1894, in *Research Notes*, collected by Sherwin Murphy, CrMs-171 Crerar Manuscript Collection, University of Chicago Library, Chicago

84. Herring, letter to O. Chanute, dated Jan. 1, 1895. Transcribed letters are in *Research Notes*, collected by Sherwin Murphy, CrMs-171 Crerar Manuscript Collection, University of Chicago Library, Chicago.

85. Chanute, letter to Augustus M. Herring, dated Dec. 31, 1894, Chanute Papers, LoC.

86. Chanute, letter to Samuel P. Langley, dated May 10, 1895, Chanute Papers, LoC.

87. Herring, letter to O. Chanute, dated May 25, 1895, Chanute Papers, LoC.

88. Chanute, letter to Augustus M. Herring, dated May 28 1895, Chanute Papers, LoC.

89. Langley, letter to Augustus M. Herring, dated Aug. 8, 1895, in *Scrapbook of clippings, magazine articles by S. P. Langley*, and typed material collected by Octave Chanute, CrMs-178, Crerar Manuscript Collection, University of Chicago Library, Chicago.

90. Octave Chanute, "Sailing Flight," in *The Aeronautical Annual* (Boston: W. B. Clarke): 60–76 (part 1, 1896), 98–127 (part 2, 1897).

91. Chanute, "Gliding Experiments," 598.

92. John D. Anderson Jr., *The Airplane: A history of its technology* (Reston, Va.: American Institute of Aeronautics and Astronautics, 2002).

93. Octave Chanute, "The present status of aerial navigation," *Engineering Magazine* 11, no. 1 (Apr. 1896.): 47–58.

94. Octave Chanute, "Recent Experiments in Gliding Flight," in *Aeronautical Annual*, ed. James Means (Boston: W. B. Clarke, 1897), 30–53.

95. Octave Chanute, "Soaring Machine" (No. 582,718), U.S. Patent Office, May 18, 1897.

96. Chanute, letter to William A. Glassford, dated Sept. 3, 1895, Chanute Papers, LoC.

97. William Avery, "Some little success of the aeroplane in aerial navigation," *The Cherry Circle* 14, no. 1 (Jan. 1908): 36–43.

98. "Men Fly in Midair," *Chicago Tribune*, June 24, 1896.

99. "O. Chanute and his Air Ship," *Kansas City Journal*, June 25, 1896.

100. "Trying to Fly," *Westchester Tribune*, June 27, 1896.

101. William A. Glassford, "Military Aeronautics," *Journal of the Military Service Institution of the United States* 19, no. 7 (May 1896): 561–76.

102. "Steal the Birds' Art," *Chicago Record*, June 29, 1896.

103. Octave Chanute, "Diary of the glides in 1896," Chanute Papers, LoC. Transcribed by Marvin McFarland, in *The Papers of Wilbur and Orville Wright, including the Chanute-Wright letters*, Appendices 4, Chanute Documents (1953).

104. Chanute, letter to Charles H. Lamson, dated Oct. 13, 1896, gives recipe for pyroxelene varnish, adapted from British Patent No.2249, dated Sept. 15, 1860, Chanute Papers, LoC.

105. Chanute, "Recent Experiments in Gliding Flight."

106. Chanute, letter to James Means, dated July 7, 1896, Chanute Papers, LoC.

107. Chanute, letter to Francis H. Wenham, dated July 7, 1896, Chanute Papers, LoC.

108. Chanute, letter to Thomas Moy, dated Aug. 13, 1897, Chanute Papers, LoC.

109. Octave Chanute, "Improvements in and relating to Flying Machines (No. 13,372)," British Patent Office, Apr. 2, 1898.

110. "His invention cost him his life," *Chicago Tribune*, Aug. 12, 1896.

111. Chanute, "How Lilienthal was killed," in *Flying Machines*, 14–15.

112. Robert H. Thurston, "Lilienthal, the Aviator," *Science* 4, no. 88 (Sept. 4, 1896): 303.

113. Chanute, letter to Samuel Cabot, dated Jan. 5, 1898, Chanute Papers, LoC.

114. Walter H. Allport, "Octave Chanute and H. T. Ricketts—A Sidelight," *Scientific American*, Sept. 23, 1911, 275. Edgar J. Goodspeed gives additional information in "Howard Taylor Ricketts," published in *The University Record* 8, no. 2 (Apr. 1922): 93–120.

115. Chanute, "Diary of the glides in 1896."

116. Octave Chanute, "Evolution of the 'two-surface' flying machine," *Aeronautics* 3, nos. 3 and 4 (Sept. and Oct. 1908): 9–10, 28–29. Reprinted as chapter 1 in *Flying Machines*, 7–18.

117. "Use Wings in Flight," *Daily Inter Ocean* (Chicago), Aug. 25, 1896.

118. "Go coasting in the air," *Chicago Tribune*, Sept. 8, 1896. While staying in the Chanute camp, reporter Henry Bunting wired his reports to the office: "Flying machines tested on the Indiana sand dunes," Sept. 9; "Hope to try flying machine today," Sept. 11; "Air ship works well," with drawing of Butusov and the Albatross ready to be launched from the ramp, Sept. 12; "Wings chilled by dense lake fog," Sept. 13; "Excursionists ashore in a gale," Sept. 13; "Albatross takes to lake breeze," Sept. 16; "Ship fails to fly," Sept. 27; "Hopes for the airship," Sept. 28. Abstracts were printed not only in American papers, but also worldwide, including in *Nature*, Sept. 24, 1896, 518; *Nature*, Oct. 15, 1896, 577; *La Nature*, Nov. 7, 1896, 562; *Evening Post* (Wellington, New Zealand), Nov. 21, 1896.

119. Henry S. Bunting, "Primitive Birdmen," Naples, Florida, 1946. Written for a writer's contest in the *Atlantic Monthly*, called "America's First." The twenty-page typed manuscript was published as "Eyewitness to Octave Chanute's Secretive Flights" in *Aviation History*, Sept. 1997, 22–28.

120. Chanute, "Recent Experiments in Gliding Flight."

121. "Airship works well," *Chicago Tribune*, Sept. 12, 1896.

122. Chanute, letter to Edward C. Huffaker, dated Sept. 30, 1896, Chanute Papers, LoC.

123. "Airship's Final Test," *Chicago Record*, Sept. 28, 1896.

124. Octave Chanute, "Experiments in Flying. An Account of the Author's own Inventions and Adventures," *McClure's Magazine* 15, no. 2 (June 1900): 127–33. Review in *Biloxi Daily Herald*, Aug. 26, 1900.

125. Chanute, letter to James Means, dated Jan. 28, 1897, Chanute Papers, LoC.

126. Frank F. Fowle, "Octave Chanute: Pioneer Glider and Father of the Science of Aviation," *Indiana Magazine of History* 32, no. 3 (Sept. 1936): 226–30.

127. David Young "The Wright Stuff. 100 years ago Octave Chanute paved the way for the first airplane flight in modern aviation," *Sunday Chicago Tribune*, June 23, 1996.

128. Tom D. Crouch, "Octave Chanute, Pioneer of Flight," *NSM Historical Journal* 18, no. 2 (1996): title page, 4–14, 16.

129. "The Albatross, Queen of the Air," *Chicago Times-Herald*, Sept. 27, 1896.

130. "Chicagoan comes back from the dead," *Chicago Tribune*, Apr. 16, 1911.

131. William P. Butusov, "Soaring Machine" (No. 606,187), U.S. Patent Office, June 28, 1898. Assignor of one-half to Octave Chanute, of Chicago, Illinois. The assignment from Butusov is recorded in Liber Z 53, page 346 of transfer of patents and provides that the patent shall be issued jointly (see letter from Chanute to Munday, Evarts & Adcock, dated June 20, 1998).

132. Steve Spicer, "The Octave Chanute Pages and Carr's Beach and the Carr Family of Miller," http://www.spicerweb.org/Chanute/Cha_index.aspx, accessed 1995–2010.

133. Bunting, "Hope to try flying machine today," Sept. 11, 1896.

134. Chanute, "Diary of the glides in 1896."

135. "The Gull," Editorial, *Chicago Tribune*, Oct. 1, 1896.

136. "O. Chanute and his Air Ship," *Kansas City Journal*, June 25, 1896.

137. William P. Butusov, "Memorandum of Agreement," 1896. Chanute Papers, LoC.

138. Chanute, letter to Carl E. Myers, dated Nov. 10, 1898, Chanute Papers, LoC.

139. "To soar at 1,000 feet," *Chicago Times-Herald*, Sept. 29, 1897.

140. Augustus M. Herring, "Recent advances toward a solution of the problem of the century," in Means, ed., *Aeronautical Annual (1897)*: 54–75.

141. Paul Dees, "The 100-Year Chanute Glider Replica, an adventure in education," in *Society of Automotive Engineers*, World Aviation Congress, Anaheim, Calif., AIAA and SAE, Oct. 13–16, 1997.

142. Simine Short, "Birth of American Soaring Flight: A New Technology," *American Institute of Aeronautics and Astronautics* 43, no. 1 (Jan. 2005): 17–28.

143. Eugene Husting, "Augustus M. Herring," in WW1 *Aero, the Journal of the Early Aeroplane* (Nov. 1990): 3–20.

144. Augustus M. Herring, scrapbook (1916), Kroch Library, Cornell University, Ithaca, New York.

145. Octave Chanute and Augustus M. Herring, "Improvements in or relating to means and appliances for effecting aerial navigation (No. 15,221)," British Patent Office, July 25, 1897.

146. Chanute, letter to James Means, dated June 20, 1897, Chanute Papers, LoC.

147. "That Flying Machine. Story probably another prodigious fake," *Salt Lake Herald*, Nov. 24, 1896.

148. Octave Chanute, "Aviation," *Sibley Journal of Engineering* 11, no. 7 (Apr. 1897): 266–68. Frontispiece of the June 1898 issue shows "Two modern flying machines," Chanute's *Katydid* and the biplane glider.

149. Chanute, letter to Milton B. Punnett, dated May 25, 1898, Chanute Papers, LoC.

150. Chanute, letter to Samuel Cabot, dated Sept. 19, 1897, Chanute Papers, LoC.

151. Chanute, "Experiments in Flying."

152. Octave Chanute, "Notes," Chanute Papers, LoC.

153. Herring, "Recent advances toward a solution."

154. Chanute, letter to James Means, dated Jan. 3, 1907, Chanute Papers, LoC.

155. "How it Feels to Fly," *Chicago Times-Herald*, Sept. 8, 1897.

156. Robert Kronfeld, *On Gliding and Soaring* (London: John Hamilton, 1934).

157. Octave Chanute, "Some American Experiments," *Aeronautical Journal* 2, no. 5 (Jan. 1898): 9–11, frontispiece; Octave Chanute, "Gleitflugversuche in Nordamerika," *Prometheus* 9, no. 458 (Sept. 9, 1898): 662–64.

158. Herring, letter to Chanute, dated Mar. 17, 1901, in Exhibit 319, *Herring-Curtiss Company vs. Glenn Curtiss et al.*, Steuben County, New York. Olin Library, Cornell University, Ithaca, New York.

159. Herring scrapbook. Olin Library, Cornell University, Ithaca, New York.

160. Octave Chanute, "Aeronautics," in *The New Volumes of the Encyclopaedia Britannica*, ed. T. S. Spencer (London, Edinburgh, and New York: Adam & Charles Black, 1902): 100–104.

161. Herring, letter to O. Chanute, dated Mar. 17, 1901. Chanute Papers, LoC.

162. Chanute, letter to Augustus M. Herring, dated Mar. 24, 1901, Chanute Papers, LoC.

163. Albert F. Zahm, "Invention of the 'Chanute Glider," *American Aeronaut and Aerostatist* 2, no. 6 (June 1908): 250–51.

164. Chanute, "Evolution of the 'two-surface' flying machine."

165. Chanute, letter to Ernest L. Jones, dated Oct. 22, 1909, Chanute Papers, LoC.

166. "First American to Fly—Tells of 12-Winged Plane," *Chicago Daily News*, Oct. 4, 1940.

167. Chanute, letter to James Means, dated Sept. 29, 1897, Chanute Papers, LoC.

168. "Airship Exhibit at Museum," *Washington Post*, Oct. 7, 1906.

169. Chanute, letter to Samuel P. Langley, dated June 11, 1897, Chanute Papers, LoC.

Chapter 8. Encouraging Progress in Flying Machines

1. Octave Chanute, "Experiments in Flying. An Account of the Author's own Inventions and Adventures," *McClure's Magazine* 15, no. 2 (June 1900): 127–33.

2. Octave Chanute, "American Gliding Experiments. Amerikanische Gleitflug-Versuche (translated by Rittmeister Warder)," *Illustrirte Aeronautische Mittheilungen*. 2, no. 1 (Jan. 1898): 4–8, 9–12.

3. Octave Chanute, "Conditions of Success in the Design of Flying Machines. Die Bedingungen des Erfolges im Entwurf von Flugapparaten" (translated by Rittmeister Warder), *Illustrirte Aeronautische Mittheilungen* 3, no. 4 (Apr. 1899): 37–41, 41–46.

4. Chanute, letter to Hermann W. L. Moedebeck, dated July 22, 1901, Chanute Papers, LoC.

5. Chanute, letter to Francis H. Wenham, dated Nov. 25, 1900, Chanute Papers, LoC.

6. Chanute, letter to Thomas Moy, dated Apr. 29, 1901, Chanute Papers, LoC.

7. Gary Bradshaw, "To Fly is Everything." Simine's U.S. Aviation Patent Database, 1998, http://invention.psychology.msstate.edu/patents/index.html, accessed 1995–2010.

8. Octave Chanute, "Soaring Flight, how to perform it," *Aeronautics* 6, no. 4 (Apr. 1909): 134–37. Reprinted in *Flight*, June 26, 1909, 384–85, July 3, 1909, 395–96; William Jackman, *Flying Machines, Construction & Operation* (1910): 179–90; James Means, *Epitome* (Boston: W. B. Clarke, 1910), 76–82.

9. Charles H. Lamson, 1896–1897, Scrapbook of clippings. Collected by O. Chanute. CrMS-176, Crerar Manuscript Collection, University of Chicago Library, Chicago.

10. Chanute, letter to James Means, dated July 23, 1898, Chanute Papers, LoC.

11. Chanute, letter to Samuel P. Langley, dated Dec. 6, 1898, Chanute Papers, LoC.

12. Tom D. Crouch, *A Dream of Wings. Americans and the Airplane, 1875–1905* (New York and London: W. W. Norton, 1981).

13. Chanute, letter to Samuel P. Langley, dated Dec. 28, 1897, Chanute Papers, LoC.

14. Langley, letter to O. Chanute, dated Dec. 1, 1897, Chanute Papers, LoC.

15. Chanute, letter to Samuel P. Langley, dated Dec. 4, 1897, Chanute Papers, LoC.

16. Ansel E. Talbert, "America's Air Navy of the Future," *Interavia* 4, no. 9 (Sept. 1949): 544–47.

17. Chanute, letter to George A. Spratt, dated Oct. 15, 1899, Chanute Papers, LoC.

18. Chanute, letter to George A. Spratt, dated Oct. 6, 1899, Chanute Papers, LoC.

19. "Talks of Zeppelin's Airship," *Chicago Tribune*, July 4, 1900.

20. Octave Chanute, "Aerial Navigation," *The Independent* 52, no. 2682 and 2683 (Apr. 26 and May 3, 1900): 1006–7, 1058–60. Abstract in *Boston Globe*, May 13, 1900, 47; *Zeitschrift für Luftschiffahrt*, July 1900, 164–65.

21. Werner Schipps, *Die Schule Lilienthal's*, Gesammelte Vorträge des Internationalen Symposiums im Sept. 1991 (Berlin: Museum für Verkehr und Technik, 1992).

22. Octave Chanute, "How to learn to fly," *American Aeronaut* 1, no. 6 (June 1908): 199–203. Article was reprinted in Oct. 1909, 119–22.

23. W. Wright, letter to O. Chanute, dated May 13, 1900, Wright Papers, Manuscript Division, Library of Congress, Washington, D.C.

24. Chanute, letter to Wilbur Wright, dated May 17, 1900, Chanute Papers, LoC.

25. Chanute, "First Proposed by Wenham," in *Flying Machines.*

26. Chanute, letter to James Means, dated Dec. 24, 1900, Chanute Papers, LoC.

27. Chanute, letter to Edward C. Huffaker, dated Feb. 7, 1901, Chanute Papers, LoC.

28. W. Wright, letter to O. Chanute, dated May 12, 1901, Wright Papers, LoC.

29. W. Wright, letter to Reuchlin Wright, dated July 3, 1901, in Wright Papers: Family Correspondence, Manuscript Division, Library of Congress, Washington, D.C.

30. Chanute, letter to Wilbur Wright, dated June 29, 1901, Chanute Papers, LoC.

31. Chanute, letter to Edward C. Huffaker, dated June 1, 1901, Chanute Papers, LoC.

32. Chanute, letter to George A. Spratt, dated July 4, 1901, Chanute Papers, LoC.

33. W. Wright, letter to Milton Wright, dated July 26, 1901, in Wright Papers: Family Correspondence, LoC.

34. Octave Chanute, "Gliding Machines, the latest aeronautical experiments," *Illustrated Scientific News*, Feb. 1903, 73.

35. Tom D. Crouch, "The Recovered Legacy. Octave Chanute and his photos of the Wright Brothers Experiments at Kill Devil Hills. 1901–1902," http://international.loc.gov/master/ipo/qcdata/qcdata/wrightold/wb005.html, accessed Jan. 2001.

36. Edward C. Huffaker, "Experiments with gliding models conducted by E. C. Huffaker for O. Chanute," 1899. Collected by O. Chanute. CrMS-173, Crerar Manuscript Collection. University of Chicago Library, Chicago.

37. Edward C. Huffaker, "Diary kept for Mr. O. Chanute at Kill Devil Hills near Kitty Hawk, NC, of experiments with gliding models," 1901. Wilbur and Orville Wright Papers, LoC. Available from http://hdl.loc.gov/loc.mss/mwright.01002.

38. Kevin Kochersberger, Robert Ash, Colin Britcher, Drew Landman, and Ken Hyde, "An evaluation of the Wright 1901 glider using full scale wind tunnel data," *Journal of Aircraft* 40, no. 3 (May–June 2003): 417–24.

39. Chanute, letter to George A. Spratt, dated Nov. 28, 1901, Chanute Papers, LoC.

40. Chanute, letter to Hermann W. L. Moedebeck, dated Aug. 17, 1901, Chanute Papers, LoC.

41. Katharine Wright, letter to Milton Wright, dated Sept. 25, 1901, in Wright Papers, Family Correspondence, LoC.

42. Wilbur Wright, "Some Aeronautical Experiments," *Journal of the Western Society of Engineers* 6, no. 6 (Dec. 1901): 490–510. Reprinted in Smithsonian Report for 1902, 133–48. Abstract in *Science*, Apr. 18, 1902, 632–33; *Scientific American*, Feb. 22, 1902, 125 and others.

43. W. Wright, letter to O. Chanute, dated Nov. 2, 1901, Wright Papers, LoC.

44. Chanute, letter to Hermann W. L. Moedebeck, dated Feb. 15, 1902, Chanute Papers, LoC.

45. Chanute, letter to Charles F. Marvin, dated Dec. 18, 1901, Chanute Papers, LoC.

46. Chanute, letter to Wilbur Wright, dated Nov. 18, 1901, Chanute Papers, LoC.

47. Chanute, letter to George A. Spratt, dated Jan. 21, 1902, Chanute Papers, LoC.

48. Chanute, letter to Charles H. Lamson, dated Feb. 8, 1900, Chanute Papers, LoC.

49. Chanute, letter to Wilbur Wright, dated Mar. 4, 1902, Chanute Papers, LoC.

50. Chanute, letter to Augustus M. Herring, dated May 30, 1902, Chanute Papers, LoC.

51. Chanute, letter to George A. Spratt, dated Sept. 17, 1902, Chanute Papers, LoC.

52. Kevin Kochersberger, "An evaluation of the Wright 1902 glider using full scale wind tunnel data," *AIAA*-2003–0096, Jan. 6–9, 2003.

53. Orville Wright, Diary, 1902. Wright Papers, LoC.

54. Chanute, letter to Charles H. Lamson, dated Oct. 14, 1902, Chanute Papers, LoC.

55. Tom D. Crouch, "The Recovered Legacy."

56. Chanute, letter to Samuel P. Langley, dated Oct. 21, 1902, Chanute Papers, LoC.

57. Chanute, letter to Edward C. Huffaker, dated Oct. 20, 1902, Chanute Papers, LoC.

58. Chanute, letter to Hermann W. L. Moedebeck, dated Oct. 21, 1902, Chanute Papers, LoC.

59. Chanute, letter to B. F. S. Baden-Powell, dated Dec. 9, 1902, Chanute Papers, LoC. See Presidential Address, General Meeting of the Aeronautical Society of Great Britain, *Aeronautical Journal* 7, no. 1 (Jan. 1903): 2–8.

60. "Airships as Pleasure Craft," *St. Louis Post Herald*, Oct. 29, 1902.

61. Chanute, letter to Wilbur Wright, dated Mar. 7, 1903, Wright Papers, LoC.

62. Chanute, letter to Wilbur Wright, dated Mar. 13, 1903, Wright Papers, LoC.

63. Victor Silberer, "Der Stand der Luftschiffahrt zu Anfang 1904" (Wien: Verlag der Allgemeinen Sport-Zeitung, 1904).

64. Octave Chanute, "Der Kunstflug. Fortschritte und neuere Erfahrungen im Kunstflug," in *Taschenbuch zum praktischen Gebrauch für Flugtechniker und Luftschiffer*, ed. Hermann W. L. Moedebeck (Berlin: W. H. Kühl, 1904).

65. "Weltausstellung in St. Louis," *Illustrierte Aeronautische Mitteilungen* 7, no. 3 (Mar. 1903): 192–93.

66. Octave Chanute, "Gliding Machines"; "Die neusten Fortschritte in der praktischen Fliegekunst," in *Illustrierte Zeitung*, Mar. 8, 1903, Leipzig, Germany, information submitted by Raimund Nimführ; F. Le Beschu, "Locomotion Aérienne en Amérique," *Le Monde Illustré*, Mar. 28, 1903, 299.

67. Chanute, letter to Wilbur Wright, dated Apr. 4, 1903, Wright Papers, LoC.

68. Ernest Archdeacon, "M. Chanute á Paris," *La Locomotion* 30, no. 80 (Apr. 11, 1903): 225–27. Translation in McFarland, *Papers of Wilbur and Orville Wright*, 654–57.

69. Georges Besançon, "Diner-Conférence du 2 Avril 1903, M. Chanute á l'Aéro-Club," *L'Aérophile* 11, no. 4 (Apr. 1903): 81–86.

70. Gertrude Bacon, *Memories of Land and Sky* (London: Methuen, 1928).

71. Gertrude Bacon, "Pigs That Fly. How 'Blindfold Pigs' indicate Character," *The Strand Magazine* 44, no. 264 (Dec. 1912): 733–38. Abstract in *Kansas City Star*, Jan. 16, 1913.

72. Octave Chanute, "La Navigation Aérienne aux Etats-Unis," *L'Aérophile* 11, no. 8 (Aug. 1903): 171–83. Translated text in McFarland, *Papers of Wilbur and Orville Wright*, as "Mr. Chanute in Paris," 654–73.

73. Chanute, letter to Wilbur Wright, dated June 30, 1903, Chanute Papers, LoC.

74. W. Wright, letter to O. Chanute, dated July 2, 1903, Wright Papers, LoC.

75. W. Wright, letter to O. Chanute, dated July 14, 1903, Wright Papers, LoC.

76. W. Wright, letter to O. Chanute, dated July 24, 1903, Wright Papers, LoC.

77. Chanute, letter to Wilbur Wright, dated July 27, 1903, Chanute Papers, LoC.

78. Octave Chanute, "L'Aviation en Amérique," *La Revue Générale des Sciences* 14, no. 22 (Nov. 30, 1903): 1133–42.

79. O. Wright, letter to Milton and Katharine Wright, dated Nov. 19, 1903, Wright Papers, LoC.

80. Chanute, cablegram to Katharine Wright, dated Dec. 17, 1903, and telegram to Wright Brothers, dated Dec. 18, 1903, Chanute Papers, LoC.

81. W. Wright, telegram to O. Chanute, dated Dec. 28, 1903, Wright Papers, LoC.

82. William T. Magruder, "Aeronautics. Section D—Mechanical Science and Engineering," *Science* 19, no. 479 (Mar. 4, 1904): 367.

83. Octave Chanute, "Aerial Navigation," *Popular Science Monthly* 64, no. 25 (Mar. 1904): 385–93. Reprinted in Smithsonian Report for 1903, 173–181. Abstract in *Scientific American Supplement*, Mar. 26, 1904; *Engineering Magazine*, May 1904, 267–69; *Aeronautical Journal*, July 1904, 61–62; *Engineering World*, Aug. 1906, 222; *American Catholic Quarterly Review*, Apr. 1904, 397–99.

84. Wilbur Wright, letter to the family, dated Nov. 23, 1903, Wright Papers, LoC.

85. Chanute, letter to Albert A. Merrill, dated July 21, 1902, Chanute Papers, LoC.

86. William Avery, letter to Percy Hudson, dated Jan. 24, 1905, Chanute Papers, LoC.

87. Chanute, letter to William Avery, dated Sept. 23, 1904, Chanute Papers, LoC.

88. Chanute, letter to William Avery, dated Oct. 16, 1904, Chanute Papers, LoC.

89. Chanute, letter to William Avery, dated Oct. 23, 1904, Chanute Papers, LoC.

90. Octave Chanute, "Means for Aerial Flight (No. 834,658)," U.S. Patent Office, Oct. 30, 1906.

91. Chanute, letter to Wilbur Wright, dated Dec. 16, 1905, Chanute Papers, LoC.

92. W. Wright, letter to O. Chanute, dated Jan. 31, 1906, Wright Papers, LoC.

93. Pepilla Gueydan, letter to Pearl I. Young, dated Mar. 9, 1964, Chanute Family Papers.

94. Chanute, letter to F. Crowninshield, dated Mar. 12, 1906, Chanute Papers, LoC.

95. Octave Chanute, "Wiener Flugtechnischer Verein." Letter from O. Chanute to the Directors of the Vienna Aero Club," *Illustrierte Aeronautische Mitteilungen* 10, no. 3 (Mar. 1906): 142–43.

96. W. Wright, letter to O. Chanute, dated Oct. 10, 1906, Wright Papers, LoC.

97. Chanute, letter to Wilbur Wright, dated Oct. 15, 1906, Chanute Papers, LoC.

98. Octave Chanute, "An Opening for Sportsmen," *Automotor Journal* 11, no. 2 (Jan. 13, 1906): 42–43.

99. Handy Man, "How to build a Chanute-type glider," *Scientific American* 100, no. 17 (Apr. 24, 1909): 319. See also, "How to Make a Gliding Machine," *Scientific American Supplement*, Apr. 28, 1906, 25353; "How to make a Glider," *Popular Mechanics*, Apr. 1909, 386–88; "Some Experiments in Gliding Flight, the type to avoid and practical conclusions," *Flight*, Dec. 4, 1909, 775–77; and others.

100. Laurence J. Lesh, "Flying, as it was. Memories of Octave Chanute," *Sportsman Pilot* (May 15, 1938): 18, 36–37.

101. Ibid.

102. Octave Chanute, Maxwell-McClure Correspondence, 1905–10. Collection of original letters not included in the letterpress books. Owned by Bill Nicks, Lenexa, Kansas.

103. "He Tells of a Real Airship," *Kansas City Star* and *Kansas City Times*, Oct. 11, 1906.

104. "Said Mr. Chanute," Chanute, *Daily Tribune*, Dec. 21, 1906.

105. Octave Chanute, "Langley's Contribution to Aerial Navigation," in *Langley, Samuel Pierpont. Secretary of the Smithsonian Institution, 1887–1906* (Washington, D.C.: Smithsonian Institution, 1907): 30–35.

106. Octave Chanute, "Artificial Flight. Part 3," in *Pocket-Book of Aeronautics*, ed. Hermann W. L. Moedebeck (London: Whittaker, 1907): 295–316.

107. Chanute, letter to Hermann W. L. Moedebeck, dated Apr. 15, 1907, Chanute Papers, LoC.

108. Octave Chanute, "Recent Aeronautical Progress in the United States," *Aeronautical Journal* 12, no. 47 (July 1908): 52–55.

109. Octave Chanute, "The Wright Brothers' Motor Flyer," in *Navigating the Air*, eds. Israel Ludlow, William J. Hammer, and Augustus Post (New York: Aero Club of America, Doubleday, Page, & Company (1907): 3–5.

110. "Problems of the Air. Aerial Flight Sure to Come—Interview with Alexander Graham Bell," *New York Herald*, Apr. 29, 1907. Abstract in *Chicago Tribune*, Apr. 30, 1907; *Los Angeles Times*, Apr. 30, 1907.

111. "Five Years' Study of Flying Machines—Interview with Octave Chanute," *New York Herald*, May 19, 1907. Submitted as a letter, dated Feb. 5, 1907. Abstract in *Idaho Statesman*, May 19, 1907; *Scientific American Supplement*, June 1, 1907, 26262.

112. Chanute, letter to William B. Stout, dated June 7, 1907, Chanute Papers, LoC.

113. William B. Stout, *So away I went!* (Indianapolis: Bobbs-Merrill, 1951).

114. Octave Chanute, "Conditions of Success with Flying Machines," *American Magazine of Aeronautics* 1, no. 1 (July 1907): 7–9. Abstract in *General Aviation Magazine* (Russia), 84–86; in *Wiener Luftschiffer Zeitung* (Austria), Dec. 1907, 263–64.

115. Octave Chanute, "Pending European Experiments in Flying," *American Aeronaut and Aerostatist* 1, no. 1 (Oct. 1907): 13–16.

116. Octave Chanute, "Aerial Navigation," *Engineering World* 4, no. 8 (Aug. 1906): 222.

117. Willis L. Moore, "International Aeronautical Congress, President's Address," *Aeronautics* 1, no. 11 (Nov. 1907): 19–23. Prepared by Chanute (see letters to Willis L. Moore, dated Oct. 5 and 15, 1907, Chanute Papers, LoC).

118. Octave Chanute, "International Aeronautical Congress, Discussions," *Aeronautics* 2, nos. 1–4 (January–Apr. 1908): 24. Chanute submitted discussions on papers by A. V. Roe, January, 24; by E. W. Smith, February, 229; by G. A. Spratt, March, 17.

119. Octave Chanute, *"Future Uses of Aerial Navigation,"* *Aeronautics* 2, no. 6 (June 1908): 15–16. Abstract in *Fly*, Nov. 1909, 18.

120. Octave Chanute, "Development and Future of Flying Machines," *The City Club Bulletin* 2, no. 15 (Nov. 18, 1908): 191–94.

121. "Circumstances that are destined to make Chicago the Aeronautical Center of America," *Chicago Tribune*, July 18, 1909.

122. Chanute, letter to Laurence J. Lesh, dated Jan. 6, 1908, Chanute Papers, LoC.

123. E. Husting, letter to Paul A. Schweizer, dated Apr. 14, 1985, in Schweizer Papers, National Soaring Museum, Elmira, New York.

124. "Chanute's 1/4 scale models of his gliding machines." The 1897 biplane, the 1896 *Katydid*, and the 1902 oscillating wing glider were shipped to the National Museum in Washington and are (as of spring 2009) on display in the "Early Flight Gallery" in the downtown Mall Museum. The family donated the second set after Chanute's death to the Chicago Academy of Science. They were given in the late 1920s to the Museum of Natural History, now the Museum of Science and Industry. The museum refurbished and displayed these models in conjunction with the one-hundredth anniversary of glider flying in 1996. The four models went back into storage ten years later.

125. Chanute, letter to John A. Schnaane, dated Oct. 21, 1909, Chanute Papers, LoC.

126. Alice Chanute Boyd, "Some Memories of my Father," Chanute Family Papers.

127. Chanute, "Soaring Flight" (1909).

128. Chanute, "Future Uses of Aerial Navigation."

129. Chanute, "Recent Aeronautical Progress in the United States."

130. Ibid.

131. Octave Chanute, "The Wright Brothers' Flights," *The Independent* 64, no. 3105 (June 4, 1908): 1287–88; Octave Chanute, "Bevorstehende Flugversuche in Amerika," *Illustrierte Aeronautische Mitteilungen* 12, no. 13 (July 1, 1908): 345–49.

132. Octave Chanute, A. Graham, J. Means, and A. Lawrence Rotch, "International Sport with Flying Machine," *Aeronautics* 2, no. 5 (May 1908): 4–5.

133. C. R. Roseberry, *Glenn Curtiss, Pioneer of Flight* (Garden City, New York: Doubleday, 1972.)

134. Ibid.

135. P. Lahm, "The First United States Army Aircraft Report (Sept. 1908)," *War Department Office of the Chief Signal Officer, Aeronautical Division*, from Defense Technical Information Center, Washington, D.C., Feb. 19, 1909, http://www.dtic.mil/dtic/, accessed Oct. 2009.

136. Octave Chanute, "First Steps in Aviation and Memorable Flights," *Aeronautics* 4, no. 1 (Jan. 1909): 24. Abstract in *Philadelphia Inquirer*, Jan. 10, 1909; republished in *The World Almanac & Encyclopedia*, 1910, 432–38.

137. "Curtiss Flies Nearly an Hour," *Chicago Tribune*, July 18, 1909.

138. "Das erste Heim für Flugmaschinen und Gleitflieger in Deutschland," *Deutsche Zeitschrift für Luftschiffahrt* 13, no. 14 (July 1909): 374–77.

139. Octave Chanute, "Captain Ferber Killed in a Fall," *Aeronautics* 5, no. 5 (Nov. 1909): 187.

140. "Octave Chanute discusses Aviation in 2009," *St. Louis Post-Dispatch*, Oct. 10, 1909.

141. Roseberry, *Glenn Curtiss, Pioneer of Flight*.

142. Octave Chanute, "Curtiss Flies in Chicago," *Aeronautics* 5, no. 6 (Dec. 1909): 216.

143. "Chicago Sees Curtiss Fly," *Chicago Tribune*, Oct. 17, 1909.

144. Curtiss, letter to Alexander G. Bell, dated Oct. 27, 1909, Papers of Alexander Graham Bell, Manuscript Division, Library of Congress, Washington, D.C.

145. Octave Chanute, "Recent Progress in Aviation and Chronology of Aviation," *Journal of the Western Society of Engineers* 15, no. 2 (Apr. 1910): 111–47. Reprinted in *Scientific American Supplement*, July 23, 1910, 56–58; July 30, title page and 72–74; Aug. 6, 88–90. *Chronology*, Aug. 13, 106–8; in Smithsonian Institution Report for 1910, 145–67; Abstract in *Encyclopedia Britannica*, 1910, 260–70.

146. Octave Chanute, "Scientific Books: Airships, Past and Present," *Science* 28, no. 705 (July 3, 1908): 20–21.

147. Octave Chanute, "Scientific Books: Artificial and Natural Flight," *Science* 30, no. 765, (Aug. 27, 1909): 282–83.

148. "Book Review: Board-D, Vehicles of the Air," *Chicago Tribune*, June 24, 1910.

149. Paul Brockett, *Bibliography of Aeronautics*, Smithsonian Miscellaneous Collection (Smithsonian Institution, Washington, D.C., 1910). The second edition covers the time span between 1909 and 1916, (Washington, D.C.: National Advisory Committee for Aeronautics, 1921).

150. Jackman, "In Memoriam," in *Flying Machines*, 2nd edition (1912), 4.

151. Spratt, letter to Orville Wright, dated Nov. 27, 1922, Wright Papers, LoC.

152. Hart Berg, letter to R. Von Kehler, dated Mar. 10, 1912, Wright Papers, LoC.

153. Alexander Graham Bell, "Aerial Locomotion," in *Washington Academy of Sciences* (Washington, D.C.: Washington Academy of Sciences, 1907).

154. Chanute, letter to Wilbur Wright, dated Nov. 28, 1906, Chanute Papers, LoC.

155. W. Wright, letter to O. Chanute, dated Dec. 1, 1906, Wright Papers, LoC.

156. Chanute, letter to Wilbur Wright, dated May 19, 1909, Chanute Papers, LoC.

157. Chanute, letter to Chas. Walcott, dated Jan. 27, 1909, Chanute Papers, LoC.

158. "Presentation of the Langley Medal to the Wright Brothers," *Science* 31, no. 792 (Mar. 4, 1910): 334–37.

159. O. Wright, letter to Wilbur Wright, dated Aug. 24, 1909, Wright Papers, Family Correspondence, LoC.

160. George A. Spratt, "A Report of Gliding Machine Tests," *Aeronautics* 4, no. 3 (Mar. 1909): 132.

161. "Dr. Chanute denies Wright Flying Claim," *New York World*, Jan. 17, 1910. Abstract in *Philadelphia Ledger*, Jan. 17, 1910.

162. Chanute, letter to Wilbur Wright, dated Jan. 23, 1910, Chanute Papers, LoC.

163. Ibid.

164. W. Wright, letter to O. Chanute, dated Jan. 29, 1910, Wright Papers, LoC.

165. Chanute, letter to George A. Spratt, dated Feb. 2, 1910, Chanute Papers, LoC.

166. W. Wright, letter to O. Chanute, dated Apr. 28, 1910, Wright Papers, LoC.

167. John D. Anderson Jr., "Historical Note: The Development of Flight Controls" (chapter 7.21), in *Introduction to Flight* (New York: McGraw-Hill, 2004): 567–69

168. T. O'B. Hubbard, letter to O. Chanute, dated Aug. 24, 1910. Chanute Family Papers.

169. "Aviators Gather for Belmont Flight," *New York Times*, Oct. 9, 1910.

170. Charles D. Chanute, "Letter to Lee S. Burridge, The Aeronautical Society, New York, NY, dated 5 Dec. 1910," *Aeronautics* 8, no. 1 (Jan. 1911): 43.

171. W. Wright, letter to George Spratt, dated Dec. 19, 1910, Wright Papers, LoC.

172. James Means, "Octave Chanute," *Science* 33, no. 846 (Mar. 17, 1911): 416–18.

173. W. Wright, letter to Colonel William A. Glassford, dated Nov. 30, 1910, in McFarland, *Papers of Wilbur and Orville Wright*.

174. Wilbur Wright, "The Life and Work of Octave Chanute," *Aeronautics* 8, no. 1 (Jan. 1911): 4.

175. Hudson Maxim, "In Honor of Octave Chanute," *Aircraft* 1, no. 12 (Feb. 1911): 432–33.

REFERENCES

INFORMATION FOR ANY SCHOLARLY BOOK comes from many different sources and this book is no different. Possibly the foremost source of information is the University of Chicago library system, which owns the Octave Chanute Collection, donated by the Chanute estate to the original John Crerar Library in 1911. These books, pamphlets, and scrapbooks are now available for research, at either the Crerar or Regenstein Libraries.

Chanute's letters, more scrapbooks, and other personal papers were donated to the Library of Congress in Washington, D.C., in late 1932; they are part of the Papers of Octave Chanute (http://lcweb2.loc.gov/cgi-bin/query/D?faid:1:./temp/~faid_DNpG::) at the Manuscript Division, while the books were incorporated into the general holdings. Dr. Leonard Bruno from the Manuscript Division mentioned to me in 1999 that his division functions as a library rather than a museum, serving any member of the general public doing serious research. With me in the process of gathering background information on Chanute and living more than seven hundred miles away, Len was always available to take a quick look at something specific in the collection and supply me with a copy if the requested page contained pertinent information. His knowledge of the material available at the Manuscript Division was a great help. Chanute's fragile letterpress books were copied in the 1960s to twenty-four rolls of microfilm and are available through interlibrary loan from the Library of Congress (http://www.loc.gov/rr/mss/f-aids/mssfa.html#c). These letters were frequently consulted because they reveal much insight into the man and are referenced as the Chanute Papers, LoC.

In 2003, the Chanute correspondence with the Wrights became available as part of the Wilbur and Orville Wright Papers (http://lcweb2.loc.gov/ammem/wrighthtml/wrighthome.html). They are referenced as Wright Papers, LoC. Len Bruno was responsible for selecting material for this extensive digital collection, providing not only information about the Wright brothers but also Chanute. Thanks are due to Len for helping to make this wealth of information readily available to all of us.

In 2006, the Library of Congress created a new presentation, called "The Dream of Flight" (http://www.loc.gov/exhibits/treasures/wb-dream.html). Because Chanute's album, showing photos of his gliding experiments in 1896 and 1897 and his bird flight studies, is one of the "American Treasures" at the Library of Congress, the photos were scanned and the album is available on their Web site. Chanute's photos taken at the Wright brothers' camp in 1901 and 1902 were digitized in 1999 as "The Recovered Legacy" and are now part of the Wilbur and Orville Wright Papers.

In the 1940s, Pearl I. Young, an employee of the National Advisory Committee for Aeronautics (now NASA), began researching Chanute and other aviation pioneers. She interviewed Chanute's two younger daughters and worked closely with Elaine Chanute Hodges; in 1963 she published a booklet with a three-page bibliography and a listing of some of Chanute's publications. Under Pearl's leadership, selected letters from Chanute's aeronautical correspondence were transcribed in the 1960s; these volumes are at the National Air and Space Museum archives. After her death in 1969, Pearl's research material was shipped to Elaine Hodges and subsequently donated to the Denver Public Library (http://eadsrv.denverlibrary.org/sdx/pl/search-s.xsp?q=octave).

The National Air and Space Museum owns much information on Octave Chanute, including material from the William Avery family, donated in the 1940s. Finding aids are available and many of the Chanute-related photos have been digitized and are available for sale.

In my five years of writing the manuscript, the Internet, with its variety of search engines, became a tremendous help. Information could be retrieved from digitized local newspapers, such as the *Brooklyn Eagle* (http://www.brooklynpublic library.org/digital/), *Chicago Tribune* (http://pqasb.pqarchiver.com/chicago tribune/), *New York Times* (http://www.nytimes.com/) and from the Library of Congress' Chronicling America: Historic American Newspapers (http://chroniclingamerica.loc.gov/). In the nineteenth century the media focused their interest on "movers and shakers," and especially those who had built impressive structures and machines. Thus, Chanute was frequently quoted and mentioned in magazines, and I retrieved pertinent information from periodicals, scanned by Proquest, and from Cornell University's Making of America Web site (http://cdl.library.cornell.edu/moa/) and from the University of Michigan's Making of America Web site (http://quod.lib.umich.edu/m/moa/). The digitized book site (http://books.google.com/) and American Patents (http://www.google.com/patents) were also helpful.

On a more personal note, there were two books that gave me a start on this biography, Marvin McFarland's *The Papers of Wilbur and Orville Wright, including the Chanute-Wright Letters and other Papers of Octave Chanute* and Tom Crouch's *The Dream of Wings*. As Tom stated, his book is the story that culminated "in the Wright brothers taking the great leap that brought men into the skies, but only because they were launched from the shoulders of giants."

Chanute was one of these giants, and Tom's book kindled my interest in this civil engineering giant.

As mentioned previously, personal information on Octave Chanute and his immediate family came largely from his descendents, and is noted as Chanute Family Papers. The family's willingness to collaborate in this project is truly appreciated.

INDEX

Page numbers in *italics* refer to illustrations.

SIMINE SHORT is an aviation historian who has researched
and written extensively on the history of motorless flight. Her
first book, *Glider Mail: An Aerophilatelic Handbook*, received
numerous research awards worldwide and is considered a standard
reference by aerophilatelists and aviation researchers. She lives
with her husband outside Chicago, Illinois.

The University of Illinois Press
is a founding member of the
Association of American University Presses.

Designed by Jim Proefrock
Composed in 10.5/13 Electra LH Std
with Bodoni Std display
by Jim Proefrock
at the University of Illinois Press
Manufactured by Sheridan Books, Inc.

University of Illinois Press
1325 South Oak Street
Champaign, IL 61820-6903
www.press.uillinois.edu